WEED AND CROP RESISTANCE TO HERBICIDES

WEED AND CROP RESISTANCE TO HERBICIDES

Edited by

R. De Prado

Pesticide Action Research Group,
Department of Agricultural Chemistry,
ETSIAM, University of Cordoba, Spain

J. Jorrín

Agricultural and Plant Biochemistry Research Group,
Department of Biochemistry and Molecular Biology,
ETSIAM, University of Cordoba, Spain

and

L. García-Torres

Weed Science Research Group,
Institute of Sustainable Agriculture,
Spanish Research Council, CSIC, Cordoba, Spain

KLUWER ACADEMIC PUBLISHERS
DORDRECHT / BOSTON / LONDON

A C.I.P. Catalogue record for this book is available from the Library of Congress

ISBN 0-7923-4581-9

Published by Kluwer Academic Publishers,
P.O. Box 17, 3300 AA Dordrecht, The Netherlands.

Sold and distributed in the U.S.A. and Canada
by Kluwer Academic Publishers,
101 Philip Drive, Norwell, MA 02061, U.S.A.

In all other countries, sold and distributed
by Kluwer Academic Publishers,
P.O. Box 322, 3300 AH Dordrecht, The Netherlands.

Printed on acid-free paper

Acknowledgements

This book has been written and edited on the bases of the lectures presented at the International Symposium on Weed and Crop Resistance to Herbicides, held in Córdoba in April, 1995. Despite numerous dificulties it has become a reality. Thank you very much to all those who in very different ways have contributed to the success of the above-mentioned symposium and the publication of the book.

The costs of the publication have been in partly covered by the EWRS, SEMh and the HRAC.

We dedicate this book to our families.

R. De Prado
J. Jorrín
L. García-Torres

PREFACE

Due to both foresight and concern about problems from resistance to herbicides, Rafael de Prado and his colleagues convened a «small» meeting to deal with these issues, in beautiful Córdoba, Spain in the spring of 1995. More than a hundred equally worried scientists registered and attended, bringing with them data on this burgeoning problem. This group presented more than one hundred posters demonstrating the breadth of the issues. This is twice the number of posters presented at a similar symposium hosted by John Caseley and his colleagues at Long Ashton in 1989. The content of the posters are published in a sister volume to this (Proceedings of the International Symposium on Weed and Crop Resistance to Herbicides- R. De Prado, J. Jorrín, L. García-Torres and G. Marshall eds., University of Córdoba, 276 pp., 1996).

Information from the plenary lectures of the Cordoba meeting are provided in the 36 chapters of the present volume. Comparison of this volume with the 12-chapter book edited by S.B. Powles and J.A.M. Holtum (Herbicide Resistance in Plants- Biology and Biochemistry, Lewis, Boca Raton, 353 pp., 1994) indicates how much happened during only a year or so. New cases of resistance have been found to previously resistant-herbicides, mechanisms of other resistances were elucidated, and additional insights into resistance management have been gained.

The growing rate of evolution of herbicide resistance does not bode well for agriculture, which has grown so dependent on herbicides for cost-effective, labor-efficient, and soil-conserving weed management. Herbicides are the backbone of weed control in developed world agriculture. In this arena, herbicides have grown to 70% or more of the pesticide market. The large number of new incidences of resistance in developing world agriculture suggest how underestimated the role of herbicide has been in these countries and what the effects of resistance will be.

Our colleagues dealing with insecticide and fungicide resistance dominate international symposia on pesticide resistance. This book shows that the field situation is far worse with herbicide than other pesticide resistances. Furthermore, far more is known about the evolutionary and biochemical causes of herbicide resistance than of the related pesticide resistances.

The organizers have clearly justified holding a meeting dealing with herbicide resistance alone, even though each of the pest disciplines have much to learn from the others. This written record shows how much herbicide resistance needs the depth and breadth of coverage given at this symposium. We thank and compliment Rafael De Prado and his colleagues for bringing this symposium to fruition. While only the two of us have been requested to write this preface, we are certain that we reflect the views and gratitude of the weed science community. The weed science community needs a new group of volunteers to follow in their footsteps, and organize the next herbicide resistance workshop.

Jonathan Gressel,
President International Weed Science Society

Stephen Duke,
President Weed Science Society of America

CONTENTS

Part III
Genetics and Biology of Herbicide Resistant Weeds

Part IV
Biotechnological Approaches to Develop Herbicide Resistance in Crops. Problems and Possibilities

Part V
Managing or Avoiding Herbicide Resistance: Integrated Mechanical, Chemical and Biological Methods for Weed Control

Conclusion

CONTRIBUTORS

Z. AMSELLEM. Plant Genetics, Weizmann Institute of Science, Rehovot, Israel.

R. BAERG. Department of Agronomy, University of Kentucky, Lexington, Kentucky, USA 40546-0091.

M. BARRETT. Department of Agronomy, University of Kentucky, Lexington, Kentucky, USA 40546-0091.

J.M. BECERRIL. Universidad del Pais Vasco, Facultad Ciencias, Departamento Biologia Vegetal y Ecologia, Apdo. 644, E-48080 Bilbao, Spain.

I.B. BRYAN. Zeneca Agrochemicals, Jealott's Hill Research Station, Bracknell, Berkshire RG12 6EY, U.K.

V.F. CAREY III. Valent USA Corp., Seymour, IL, USA.

A. CONA. IBEV-Area della Ricerca di Roma, CNR,. Via Salaria km. 29.3, 00016 Monterotondo Scalo, Italy.

J. COSTA. Monsanto España, S.A., Avda de Burgos 17, 2ª, 2803 Madrid,. Spain.

H. DARMENCY. Laboratoire de Malherbologie, INRA, BV1540, 21034 Dijon, France.

R. DE PRADO. Departamento Química Agrícola y Edafología, E. T. S. I. A.M., Universidad de Córdoba, Apdo. 3048, 14080 Córdoba, Spain.

M.D. DEVINE. Department of Crop Science and Plant Ecology, University of Saskatchewan, Saskatoon, Saskatchewan, Canada S7N OWO.

G. DONN. Hoechst Schering AgrEvo GmbH, Biochemical Research, 65926 Frankfurt am Main, Germany.

S.O. DUKE. USDA-ARS, Southern Weed Science Laboratory, Stoneville, MS, USA.

M.V. DUKE. USDA-ARS, Southern Weed Science Laboratory, Stoneville, MS, USA.

F. DURST. Laboratoire d' Enzymologie Cellulaire et Moléculaire, Institute de Biologie Moléculaire des Plantes-CNRS, 28 rue Goethe, F-67083 Strasbourg, Cedex.

W.E. DYER. Department of Plant, Soil and Environmental Sciences, Montana State University, Bozeman, Montana.

C.V. EBERLEIN. University of Idaho, PO Box AA, Aberdeen, ID 83210.

R. EDWARDS. University of Durham, Department of Biological Sciences, Durham, DH1 3LE, UK.

R.P. FINCH. Scottish Agricultural College, Department of Plant Science, Auchincruive, Ayr, KA6 5HW, U.K.

J. GASQUEZ. I.N.R.A., Laboratoire de Malherbologie, B.V. 1540, 21034 Dijon, Cedex, France.

B. GEIKEN. IBEV-Area della Ricerca di Roma, CNR,. Via Salaria km. 29.3, 00016 Monterotondo Scalo, Italy.

M. T. GIARDI. IBEV-Area della Ricerca di Roma, CNR,. Via Salaria km. 29.3, 00016 Monterotondo Scalo, Italy.

G.S. GILL. University of Adelaide, Roseworthy S.A. 5371, Australia.

R. GIMENEZ-ESPINOSA. Departamento Química Agrícola y Edafología, E. T. S. I. A.M., Universidad de Córdoba, Apdo. 3048, 14080 Córdoba, Spain.

J.C. GRAHAM. Herbicide Resistance Action Committee, Monsanto Europe S.A., Avenue de Tervuren 270-272, 1150 Brussels, Belgium.

J. GRESSEL. Plant Genetics, Weizmann Institute of Science, Rehovot, Israel.

J. W. GRONWALD. Plant Science Research Unit, USDA-ARS and Department of Agronomy and Plant Genetics, University of Minnesota, St. Paul, Minnesota 55108 USA.

M.J. GUTTIERI. University of Idaho, PO Box AA, Aberdeen, ID 83210.

K.K. HATZIOS. Department of Plant Pathology, Physiology and Weed Science, Virginia Polytechnic Institute and State University, Blacksburg, Virginia 24061-0330, USA.

T.R. HAWKES. Zeneca Agrochemicals, Jealott's Hill Research Station, Bracknell, Berkshire RG12 6EY, U.K.

R.E. HOAGLAND. USDA-ARS, Southern Weed Science Laboratory, Stoneville, MS, USA.

B.L. HOFFER. Biosciences Research Laboratory, USDA, Agricultural Research Service, State University Station, Fargo, North Dakota 58105-5674, USA.

N.J. JACOBS. Department Microbiology, Dartmouth College, Hanover, NH 03755, USA.

J.M. JACOBS. Department Microbiology, Dartmouth College, Hanover, NH 03755, USA.

D.M. JOEL. Department of Weed Research, ARO, Newe-Ya'ar Research Center, Haifa 31900, Israel.

J. JORRIN. Departamento de Bioquímca y Biología Molecular, ETSIAM, Universidad de Córdoba, Apdo 3048, 14080 Córdoba, Spain.

Y. KLEIFELD. Department of Weed Research, ARO, Newe-Ya'ar Research Center, Haifa 31900, Israel.

H.J. LEE. USDA-ARS, Southern Weed Science Laboratory, Stoneville, MS, USA.

N. LOPEZ-MARTINEZ. Departamento Química Agrícola y Edafología, E. T. S. I. A.M., Universidad de Córdoba, Apdo. 3048, 14080 Córdoba, Spain.

R. MACKENZIE. University of Liverpool, Department of Environmental and Evolutionary Biology, PO Box 147, Liverpool, L69 3BX. U.K.

G.MARSHALL. Scottish Agricultural College, Department of Plant Science, Auchincruive, Ayr, KA6 5HW, U.K.

H. MATSUMOTO. Institute Applied Biochemistry, University Tsukuba, Ibaraki 305, Japan.

C. A. MAXWELL. DuPont Agricultural Products, Stine-Haskell Research Center, P.O. Box 30, Newark Delaware U.S.A.

C.A. MALLORY-SMITH. University of Idaho, PO Box AA, Aberdeen, ID 83210.

J. MENENDEZ. Departamento Química Agrícola y Edafología, E. T. S. I. A.M., Universidad de Córdoba, Apdo. 3048, 14080 Córdoba, Spain.

D. MICHAELI. Plant Genetics, Weizmann Institute of Science, Rehovot, Israel.

A.M. MORTIMER. University of Liverpool, Department of Environmental and Evolutionary Biology, PO Box 147, Liverpool, L69 3BX. U.K.

S. MOSS. IACR-Rothamsted, Harpenden, Herts AL5 2JQ, UK.

U.B. NANDIHALLI. Hazleton Laboratories, P.O. Box 7545, Madison, Wl, 53707, USA.

M.D.K. OWEN. Iowa State University, Ames, IA 50011 USA.

G.M. PASTORI. Laboratorio de Fisiología Vegetal, Facultad de Ciencias Exactas, Físicas y Naturales, Universidad de Córdoba, 5000 Córdoba, Argentina.

N. POLGE. Department of Agronomy, University of Kentucky, Lexington, Kentucky, USA 40546-0091.

C. PONELEIT. Department of Agronomy, University of Kentucky, Lexington, Kentucky, USA 40546-0091.

S.B. POWLES. Department of Crop Protection, The University of Adelaide, Waite Campus, PMB 1, Glen Osmond, 5064 Australia.

C. PRESTON. Department of Crop Protection, The University of Adelaide, Waite Campus, PMB 1, Glen Osmond, 5064 Australia.

P.D. PUTWAIN. University of Liverpool, Department of Environmental and Evolutionary Biology, PO Box 147, Liverpool, L69 3BX. U.K.

K.N. REDDY. USDA-ARS, Southern Weed Science Laboratory, Stoneville, MS, USA. B. RUBIN. Department of Field Crops, Vegetables and Genetics, Faculty of Agriculture, the Hebrew University of Jerusalem, Rehovot 76100, Israel.

L.L. SAARI. DuPont Agricultural Products, Stine-Haskell Research Center, P.O. Box 30, Newark Delaware U.S.A.

J.P. SALAÜN. Laboratoire d' Enzymologie Cellulaire et Moléculaire, Institute de Biologie Moléculaire des Plantes-CNRS, 28 rue Goethe, F-67083 Strasbourg, Cedex.

D.L. SHANER. American Cyanamid Co., P. O. Box 400, Princeton, NJ, USA 08543-0400.

A. SHARON. Botany Department, Tel-Aviv University, Tel-Aviv, Israel.

T.D. SHERMAN. USDA-ARS, Southern Weed Science Laboratory, Stoneville, MS, USA.

R.H. SHIMABUKURO. Biosciences Research Laboratory, USDA, Agricultural Research Service, State University Station, Fargo, North Dakota 58105-5674, USA.

R.J. SMEDA. USDA-ARS, Southern Weed Science Laboratory, Stoneville, MS, USA.

R.E. TALBERT. University. of Arkansas, Fayetteville, AR, USA.

F. J. TARDIF. Department of Crop Protection, The University of Adelaide, Waite Campus, PMB 1, Glen Osmond, 5064 Australia.

D.C. THILL. University of Idaho, PO Box AA, Aberdeen, ID 83210.

K.C. VAUGHN. USDA-ARS, Southern Weed Science Laboratory, Stoneville, MS, USA.

D. WERCK-REICHHART. Laboratoire d' Enzymologie Cellulaire et Moléculaire, Institute de Biologie Moléculaire des Plantes-CNRS, 28 rue Goethe, F-67083 Strasbourg, Cedex.

F. ZIMMERLIN. Laboratoire d' Enzymologie Cellulaire et Moléculaire, Institute de Biologie Moléculaire des Plantes-CNRS, 28 rue Goethe, F-67083 Strasbourg, Cedex.

INTRODUCTION

CHAPTER 1
BURGEONING RESISTANCE REQUIRES NEW STRATEGIES

J. GRESSEL
Plant Genetics, Weizmann Institute of Science, Rehovot, Israel

Occam's Philosophy of KISS and Weed Control

In years gone by mixed farming was practiced, and weeds were kept at bay with a modicum of success by crop rotation with lots of cultivation. Crop varieties then available were far more competitive with the weeds than those presently grown. Yields were low because the crop competitiveness has a high cost in yield (cf. Herms and Mattson, 1992). Energy costs either as fuel for tractors, fodder for animals, or food for laborers were high. The advent of high yield-index crops and their greater relative responsiveness to fertilizer and water than their grandparents brought weeds, with a vengeance. A lower, non-competitive stature is a direct outcome of high yield indices. Herbicides were and are a necessity to keep the crop yields up and the weeds down in the new, wimpy crops. Such herbicide use is wholly justified when one considers the alternative: famine. We don't eat the straw, just the grain, so why grow straw? The herbicides made it too easy. The fields were clean, so why not grow just the high value cash crops; no need to rotate, and with no-till and herbicides, why ever cultivate and cause compaction and erosion? All this followed a scientific "razor" (principle), widely found in text books, by the 14th century English cleric Richard of Occam. As often stated Occam's Razor is: "The simplest solution to a problem is most likely to be the best one". Monoculture of major cash crops using a single herbicide over wide areas was the simplest solution to weed problems in crop food production. Industry, extension, and academia used Occam's Razor and exhorted farmers to KISS, where KISS stood for "keep it simple, stupid". My high school agronomy instructor 40 years ago ranted against monoculture, always stating that in the long run, rotation always pays. To him KISS meant: "keep it sophisticated, smarty" and time and again this has been proven right.

The poor agricultural practice of monoculture has led to soil and environmental problems, uncontrolled insects and diseases, and herbicide resistance. Almost all problems of resistance have evolved as a result of monoculture. If we have to plan theoretical strategies for preemptive resistance management we can theoretically pack up and go home. We just need to say: stop monoculture and all will be well. This is impossible; the world lives in constraints and the biggest is nutrition. Dislike it as we may, monoculture is with us. Much of the worlds wheatlands will grow only small grains; we no longer need oats to feed farm animals. Wheat and rice must be grown to prevent starvation. We may disapprove of monoculture just as the physician disapproves of unsafe sex, but the diseased patient must be treated, and hopefully educated. Our task is also to treat and educate. The above analogy is not far fetched in another respect: many cases of resistance are preventable, some are curable, but often when resistance is ensconced in a field, it is high impossible to get rid of the scourge vis a vis the selecting herbicide, and in many cases, even with other herbicides. Thus, prevention is a key, as not many curatives are available. Beware of the pesticide salesman who says: "Don't worry about resistance; if your weeds become resistant to one herbicide, we have lots of replacements". Exile such salesmen to Australia wheatlands for their education.

A Confluence of Problems: Resistances to Old and New. High Risk and Low Risk Herbicides

It will not be attempted to catalog the numbers of cases of resistance or estimate areas affected, this has been done recently (cf. Powles and Holtum, 1994). The current situation can be compared with what was known 12 years earlier (LeBaron and Gressel, 1982). Such cataloging usually provides gross guestimates, at best. Aerial monitoring for resistance has yet to be used, though it clearly could and should. Past history is an indicator of trends. We must learn from what has happened elsewhere, as evolution has clearly shown that it can come up with mechanisms that we researchers could not imagine and, in many cases, can still not explain.

This author has often been called a prophet of doom. It would be great for agriculture if our past prophecies had not turned out as they had. To some extent the trends are worse than envisioned. We have become almost blasé about new cases of resistance, and worse yet do little about prevention or spread. No longer does every instance of resistance become published whenever a previously known resistance is discovered to have evolved in a new country, state or province. The only way to measure the increases in triazine resistant weeds in many maize growing areas is to measure the increases in use of dicamba, pyridate or, more recently, other herbicides. The triazines are still widely used, though abandoned by fiat in parts of Europe, and they are under regulatory pressure elsewhere. The triazines were classified as a high risk group of herbicides vis a vis resistance, but we are in for an unwanted surprise, their major modern replacements, the inhibitors of acetolactate synthase (ALS) and acetylCoA carboxylase (ACCase) are at far greater risk than the triazines to have resistant populations evolve.

Many herbicides had been thought to pose a lower risk vis a vis evolution of resistant weed populations. Such lower risks were based on one or more of the following factors: are non-persistent; lack a single or known target; not widely used; resistance did not evolve despite many years of use. Nature has ways of dealing with man's desire for monotony. The utility of having Bordeaux mixture (copper sulfate/lime) as a fungicide for more than a hundred years without a single pathogen evolving resistance has been dampened by finding that the levels of residual copper in the soil are becoming phytotoxic. Farmer persistence has overcome the lack of paraquat persistence; repeated use for every flush of germinating weeds brought resistance. Single-flush grass weeds showed that persistence refers to biology and not chemistry; persistence refers to the longevity of active chemical residue in relation to the pattern of appearance of the weeds. A persistence of 14 days is long, when the weeds all appear in 7 days. The evolution of metabolic and/or other yet unknown mechanisms in weeds has allowed the evolution of resistance to multi-site herbicides, and to some single site ones as well. Additionally, nature came up with "problem soils" that quickly degrade some repeatedly used herbicides allowing weeds to grow, even if they were not inherently resistant. Smugness about the powers of nature to abhor continuous use of the same weed management system, whether the lawnmower, hoe, hands, disc, or herbicide, guarantees failure.

We now see a confluence of problems; resistance to new herbicides and to their less resistance prone older relatives. Herbicides inhibiting single targets: e.g. photosystem II, ALS, tubulin and ACCase have fallen prey to resistance on a grand scale of millions of hectares. Meanwhile, the "golden oldie", off patent herbicides, used heavily especially in the less developed parts of the world (including in the first world), as well as the low persistence PSII inhibitors (e.g. isoproturon, propanil), phenoxys (2,4-D, MCPA, dicamba), chloroacetamides (butachlor), thiocarbamates (triallate, molinate) have resistant weeds. This is very frightening; we are losing the compounds we thought we could depend upon should target site resistance evolve. This like the Australian liana with thorns pointing in two directions. It is called "lawyer vine" because "it gets you whether you are coming or going". Indeed, this describes the patterns of evolution of resistance, and the utility of the present pre-emptive management practices.

Conditions Leading to Herbicide Failure

Monoherbicide culture is not the sole factor leading to the appearance of resistant populations of weeds, although this is a common denominator. All the early models describing the parameters governing evolution were described around single gene inherited target site resistances, where herbicides exerted high selection pressure, gave excellent weed control, until they seemingly suddenly crashed.

Resistances now gaining ground are to the older herbicides that gave just adequate control. Resistance appeared as a slow, gradual loss of control, seen as a function of the number of times herbicide was used (Figure 1). Such slow incremental increases are typical of polygenically inherited traits (where the "poly" can also be the amplification of a gene). These cases were more typical of situations where the selection pressure was low and where the herbicide was used at a just lethal dose rate. Indeed, some of the cases of this new type of resistance are even to herbicides where target site resistance is common. Simazine resistance appeared in *Senecio* (Figure 1C) germinating at a time when most of the simazine had dissipated. The diclofop-methyl rate used in Australia to control *Lolium* (Figure 1 A, B) is a third that used to control *Lolium* in the U.S.A. and Canada.

To a certain extent the geographic distribution of target site vs. polygenic resistances is expected. The first world has a philosophy of near-perfect weed control, with manufacturer liability for only full labelled dose rates. There, high selection pressure will only allow for resistance by a single major gene. The likelihood of a large enough compliment of resistant polygenes to cope with such high doses is very unlikely (Gressel, 1995). Conversely, in an extensive agricultural situation, dose cutting to give just adequate control is more typical, especially where there is no manufacturers' liability. Thus, we get documented cases of the creep in resistance level, as in Figure 1 or hear of them in farmers' reports. Too often, the polygenically inherited resistances lead to cross-resistances to herbicides that are unrelated, both in chemical grouping and in target site of action, even when only one herbicide performed all the selection pressure. This had better turn on warning lights where herbicide rates are being dropped to accede to requirements to use less herbicides. If farmers stick to monocultures with reduced rates of the same herbicide, new resistant problems are likely. Where crop rotation is not possible, chemical abstinence using other techniques, might be an advisable part of a long term system.

A model has been developed showing that rotating low rates with moderate rates will delay polygenic resistance (Figure 2) and they should be checked in the field to ascertain if it holds up in reality. This could easily be done using seed from where resistance has just been verified at different levels.

Mixtures as Resistance-Preventing Strategies

The mixing of herbicides should at least in theory considerably delay the simultaneous appearance of two target resistances. To be effective, the components of such mixtures should: (a) Control the same weeds. Weeds controlled by only one component, need evolve resistance to only it; they are naturally resistant to the other; (b) Have the same persistence. (Otherwise some flushes of weeds will see only one herbicide); (c) Have different sites of action and can also be degraded by different enzymes. Indeed, a compound that prevents herbicide degradation (i.e. a "synergist") is an excellent mixing partner.

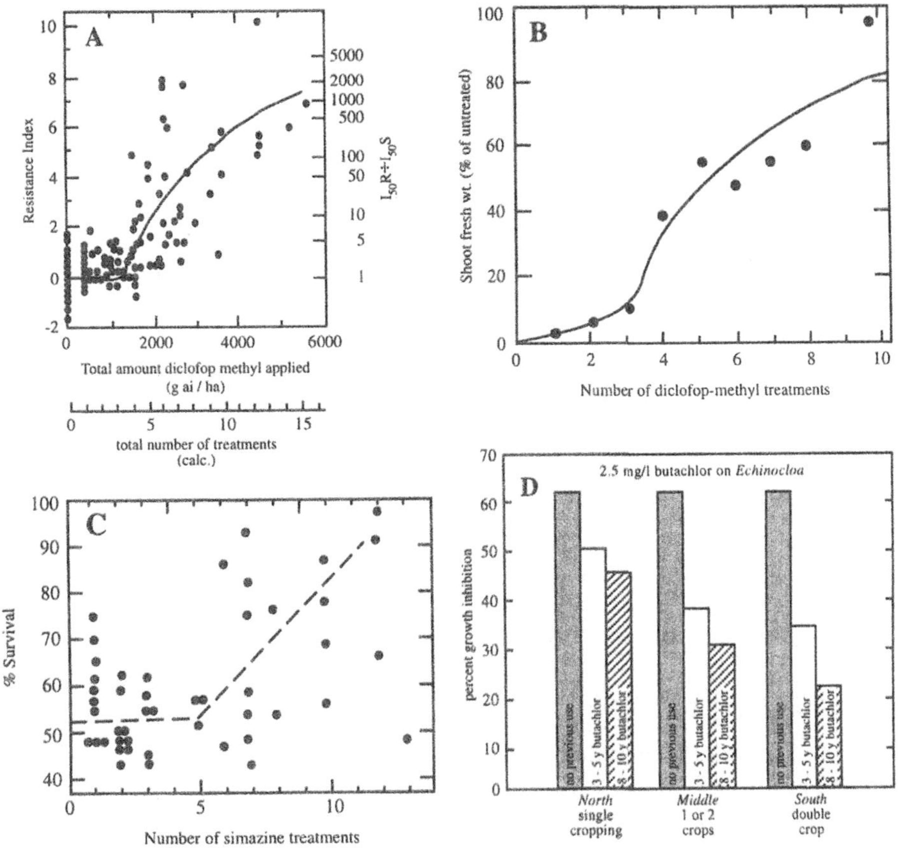

Figure 1. Incremental increase in the level of resistance in repeatedly treated populations. The lines were drawn by this author for demonstration. A. *Lolium rigidum* treated annually with 375 g/ha of diclofop-methyl in South Australia. The resistance index is the difference in log between the I_{50} of resistant and susceptible populations. Modified and redrawn from Heap (1988). B. *Lolium rigidum* treated with ACCase inhibitors in western Australia with a fixed rate of herbicide. Redrawn from Gill and Diggle (1993). C. Increased resistance of *Senecio vulgaris* to simazine as a result of repeated treatments. The last treatment of each population was with 0.7 kg/ha simazine. Drawn from data in Figures 3 and 4 in Holliday and Putwain (1980). D. *Echinochloa crus-galli* resistant to butachlor appeared as a function of treatment history. Redrawn from Huang and Lin (1993).

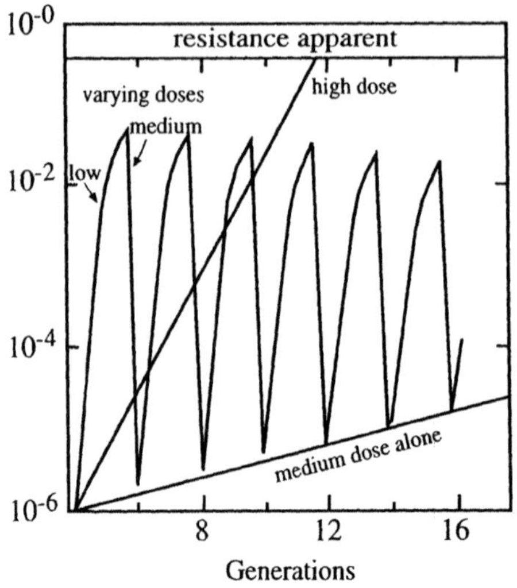

Figure 2. Modeled effects of varying low with medium doses on the enrichment of resistance. The effect of high dose, with high selection pressure and medium dose with lower selection pressure (straight lines) are plotted from the equations in Gressel and Segel (1978). The lines for 3 low doses interspersed with one medium dose were modified from those generated by the courtesy of Marc Mangel, U. California, Davis, Source: Gressel (1995), by permission.

There are times that mixtures that do not wholly meet the above criteria seem to have worked effectively. A sulfonylurea herbicide used alone in California as the only herbicide towards some weeds in rice was abandoned after four years of use, as resistance became too widespread (cf. Gressel and Baltazar, 1995). Conversely, for many years c.a. 1.6 Mha of rice in Japan received a "one-shot" application of a sulfonylurea herbicide in mixture with other herbicides, i.e. over 80% of the area in continuous paddy. Resistance has yet to be reported, despite the number of years of use. This could be for a few reasons: (a) all the sulfonylureas are in a mixture with another herbicide or herbicides. This should not be sufficient as clearly not all the mixing partners met all the above criteria for all weeds, but they may "weaken" ALS resistant weeds allowing the rice to compete with them into submission; (b) The use of the sulfonylureas in Japan replaced other effective herbicides while the levels of the most problematic weeds were still low. In California, the sulfonylurea was introduced to control large infestations of weeds that no herbicide in use at the time was controlling. Despite the "one shot" pre or early post emergence treatment, ca. 270,000 ha in Japan receive a treatment of paraquat, diquat, glyphosate, or glufosinate after rice harvest because the paddies are too weedy. Presumably all three factors act together, but complacency about long term future of ALS inhibitors in Japan would be ill advised. The criteria and models for mixtures do not yet adequately consider the situation where the mixture does not control a weed species but weakens it to a point where it no longer competes with the crop. Some such mixtures are described in Holmberg (1995).

Resistance due to enhanced metabolism of the pesticide can be overcome by synergists that block catabolism of the pesticide. Such synergists are gaining in use to prevent insecticide resistance, but not yet to suppress herbicide resistance, or to reduce the dose levels needed to control the hardest to kill weeds, despite the suggestion that this might be useful (Gressel, 1990). Resistance due to enhanced levels of a cytochrome P450 monooxygenase should be an ideal target. These specific enzymes often vary considerably among plant species vis a vis the ability to be inhibited by the same compounds. Thus, crop weed selectivity could be achieved. Additionally, these enzymes are often inhibited by compounds similar to the azole fungicides, and there is a good possibility that the compounds needed exist from fungicide screening.

Are There Resistance Prone Weed Species?

Just as there is a high mutation frequency to alleles conferring resistance to some herbicides, one can ask if there are some weed species that inherently evolve resistance more quickly than others. Epidemiology and cataloging list *Lolium* spp., *Amaranthus* spp., *Chenopodium* spp., and *Kochia scoparia* as weeds that have evolved resistance to many herbicide in many places.

This could mean that they inherently have high mutation rates, but that has yet to be demonstrated and should be checked. They could also be species with more metabolic systems to degrade xenobiotics, but that would not explain target site resistances that have evolved.

The most probable cause of their propensity towards quickly-evolving resistant populations probably comes from being copious seed producers. If one species has a million weed seeds per meter and another a million per hectare, there will be more resistant mutants around of the former, if both have the same mutation frequencies. Sex among plants is a must for polygenic resistance alleles to become enriched in populations; if a partner is rooted too far away, then pollen will be less likely to find a live pistil, thus the ability for at least some outcrossing, and copious pollen production, also contribute to the propensity to evolve resistance.

The message from this is clear; if one is about to introduce a resistance prone herbicide, it should be done where weed seed populations have been first lowered by other means.

Regulation and Resistance

Chemical producers correctly view the farmers as important stakeholders in the long term success of their products. Unfortunately for the stakeholders, shorter term considerations of their stockholders too often take precedence. Resistance takes years to evolve, even in the most rapid of cases, and stockholders demands are for quarterly gratification. Thus, resistance management takes a back seat.

There comes a time when someone must decide to regulate in favor of a stakeholder based on long term considerations. Indeed, there can be more than one stakeholder, as seen with DDT. Agricultural overuse brought DDT resistance to both agricultural pests and to mosquitoes. Mosquito control needed only a small fraction of the amount and was insufficient to select for resistance, where the agricultural use selected for resistant mosquitoes. Only the Sri Lanka government decided to intervene in favor of its whole populace of stakeholders over the agricultural stakeholders and stopped agricultural DDT use, saving many thousands of lives.

Its a wild world with herbicides. For example, despite special industry resistance action committee meetings over a long period, no recommendations (not even nonbinding ones) were ever formulated for triazine use. No action was ever taken to stop road-side and railroad use, which led to resistant weeds rolling into fields, and herbicide running into water. No binding recommendations have ever been made with ALS or ACCase inhibitors. Surely there are areas where these excellent but resistance prone herbicides should be restricted to certain crops where no alternatives are available.

The national and local extension, as well as registration services have clearly not met their duties to step in and recommend, or where necessary regulate the use of herbicides in manners that will delay resistance. Indeed, for example, the U.S. government for many years promoted monoculture through a system that provided supports only when a farmer grew the same crop in the same field; management that promotes resistance.

Some countries have come up with advisory lists of herbicide groupings to be avoided to prevent mainly target site resistances, and have the herbicides labeled with these groupings. These are heavily flawed and some modifications, additions, and risk categories are suggested in Table 1. The categories of risk within a group varies based on the rate at which a compound is used (the selection pressure), its persistence, and whether it is being used in a resistance delaying mixture. For example, the PSII inhibiting cyanazine should be a lower category than its much more persistent relative, atrazine. The risk category of atrazine should go down when mixed with alachlor. Until such information becomes rationalized internationally, and until both herbicide common name and this information appear in all advertising material, as well as labels, farmers will not be able to intelligently plan which rotations to use until it is too late. Not only should such information be presented, there should be clear explanations about how to lower the risks associated with each compound in local growing systems. The view has been "keep them stupid" instead of "keep them sophisticated". Interestingly, some of the clearest, most accurate, and erudite information is getting to the farmers from the popular agricultural press (e.g. Holmberg, 1995) and not from extension, academia or industry sources.

To my mind academia, extension services, regulators and industry have not been doing enough for the stakeholders and the long term stockholders. It is important to educate drivers not to speed, but human nature requires licensing drivers, enacting and

Table 1. Proposed rational resistance groupings and risk classings of herbicides.

Proposed code	Common cross-resistance[a]	Risk class[f]	prefix or suffix	Group(s)	
				Canadian	Australian
AC	acetyl-CoA carboxylase	M	-fop, -dim	1	A
AS	acetolactate synthase	VH-M	-sulfuron, -iman	2	B
TB	tubulin	M	-alin	3	D
AX	auxin types	L	(many)	4	I
P-II	photosystem II	L-M	-zine,-uron,-acil,-nil,-pham and others	5,6,7,14,etc	C
CL	chloroacetamides[b]	L	-aclor and others		
TH	thiocarbamates[b,c]	L	-ollate and others		
EP	EPSP synthase	L	-sate, (fosametine)	9	E/H?
P-I	photosystem I[d]	L	-quat, but not difenzoquat	10	M
PX	protoporphyrinogen oxidase	L	-fen and others	11	L
GN	glutamine synthase	L	(glufosinate, bialaphos)	12	G
CB	carbamates	VL	(various)	13	N
CLi	cellulose biosynthesis[b]	L	(isoxaben)		
CLd	cellulose biosynthesis[b]	L	(dichlobenil/chlorthiamid)		
PD	phytoene desaturase	L	-on, but not pyrazon		
WG	wheat graminicides	M	(all)		
DS	dihydropteroate synthase	VL	(asulam)		
OR	other	VL	(amitrole, dalapon and others)		

[a]Can be at target, via common structure or other commonality; [b] no cases of common structure based resistance known yet; [c]cross resistance between triallate and difenzoquat; [d] known cross resistances not at target site; [e] cross resistances are known between all graminicides selective in wheat; [f] risk classes (based on past history) VH, very high (2-4 years), M, moderate (5-10 years), L, low (>10 years), where the time refers to first case in fields; VL, very low (>10 years and unknown in the field or laboratory with eukaryotic plants or algae).

enforcing speed limits, and it is important to prevent making and selling cars that can grossly exceed the speed limits. Similarly, it is necessary to educate the farmer about resistance management, as well as to regulate how, where and when products are used. There have been potentially disastrous calls to leave only single products for each crop/ weed situation. Here too understanding is necessary.

Bean Counting Is not Regulation

Too many people have become enamored with counting instead of thinking. People want farmers to use less herbicide. The use of 30 g/ha of an ALS inhibitor that will still be biologically active next year is considered by the bean counters to be far less than 1000 g/ha of a contact herbicide that is biologically and chemically not here in a few days. Such absurdities will bring more resistance. The middle ground of rotating the two, or mixing them at lower rates, may keep the ALS inhibitor around longer, and less used over a longer period.

The Major Need - More Novel Systems of Integrated Management

Integration is sophistication - and good weed management. We need more ways to integrate. Industry has stood alone for "stand alone" products, i.e. products that do not need mixing partners, that do not need protectants (=safeners = antidotes), and that do not need synergists. The best stand alone products were thought not to need a rotation to different herbicides the following year. Hopefully, that concept is changing. Time and again it has been shown that crop rotations, especially among crops with different growth morphologies, phenologies etc. reduce weed load and increase weed intercompetition and diversity (such that no one weed is likely to get out of hand). It was discussed above how synergies and mixtures can add to resistance management. It was also described how varying rates of the same chemical might delay resistance. Other mixed strategies should be considered. Preemergence use of highly persistent chemicals has been a prescription for evolution of resistance. Are such herbicides needed on the whole field? Many weeds do disproportionate damage when present during crop germination; this is widely documented with grasses in wheat but refers only to the weeds near the crop. Consider band application at sowing, and later cultivation or herbicide application if needed. This will "lower" herbicide use (and selection pressure for resistance) and also the amount of herbicide that can make its way to water supplies is decreased. With biotechnology derived, target-site resistant crops, having a very high level of resistance one can go one step further; treat only the crop seed, obviating need for spraying. Some herbicides should spread sufficiently to provide a weed free zone around the germinating crop. This has been shown to keep *Striga* from attaching to ALS-resistant maize (Ransom *et al.*, 1995).

Resistance management must force us to think beyond herbicides; Europeans are designing huge, light-weight tine-harrows that can be pulled rapidly over fields. They could be competitive with herbicides in cost effectiveness. Surely if only this is used, the ecosystem will fill with harrow-resistant species.

Innovation is therefore needed in chemistry, mechanics, agronomy, regulation as well as ways of integrative thinking. The innovations must be integrative as well as proven to be cost-effective (if possible), or without a viable alternative. One need look to see what innovations farmers have incorporated in the last 50 years to realize how sophisticated they are and how they are given too little credit for their ever- increasing sophistication in management. Sophisticated people must make new tools available to the farmers so that the excellent chemicals we now have, and those that will be discovered can be used judiciously for the long-term profits to all. Occam has been oversimplified; what he really said was: "in solving a problem, terms, concepts, assumptions, etc., must not be multiplied beyond necessity". Herbicide resistance necessitates KISS: Keep It Sophisticated, Smarty.

Acknowledgements

The author's research activities, which are hoped will lead to long-term, judicious methods of weed control are supported by grants from BARD, GIARA and AID. The author has the Gilbert de Botton Chair of Plant Sciences.

References

Gill, G. and Diggle, A.J. (1993) Herbicide resistance in annual ryegrass in the cropping belt of western Australia -a case-study for integrated weed management. *In: Integrated Weed Management for Sustainable Agriculture* (R.K. Malik, ed.), 203-208. Indian Society Weed Science, Hisar, India.

Gressel, J. (1990) Synergizing herbicides. *Review Weed Science* **5**: 49-82.

Gressel, J. (1995) Catch 22 - Mutually exclusive strategies for delaying/preventing polygenically vs. monogenically inherited resistances. *In: Options 2000* (N.N. Ragsdale, P.C. Kearney and J.R. Plimmer, eds.), 330-349. American Chemical Society, Washington D.C.

Gressel, J. and Baltazar, A. (1995) Herbicide resistance in rice: status, causes, and prevention. *In: Weed Management in Rice* (K.-U. Kim, ed.). FAO, Rome. In press.

Gressel, J. and Segel, L.A. (1978) The paucity of genetic adaptive resistance of plants to herbicides: possible biological reasons and implications. *Journal Theoretical Biology* **75**: 349-371.

Heap, I. (1988) Resistance to herbicides in annual ryegrass (*Lolium rigidum*). *Ph.D. Thesis*. Waite Agricultural Institute, University of Adelaide.

Herms, D.A. and Mattson, W.J. (1992) The dilemma of plants, to grow or to defend. *Quarterly Review Biology* **67**: 283-335.

Holliday, R.J. and Putwain, P.D. (1980) Evolution of herbicide resistance in *Senecio vulgaris*; variation in susceptibility to simazine between and within populations. *Journal Applied Ecology* **17**: 799-808.

Holmberg, M. (1995) Multiple modes. *Successful Farming*, January, pp.48-49.

Huang, B.Q. and Lin, S.X. (1993) Study on the resistance of barnyardgrass to butachlor in paddy fields in China (in Chinese, English abstract). *Journal South China Agricultural University* **14**: 103-108.

LeBaron, H. M. and Gressel, J. (eds.) (1982) *Herbicide Resistance in Plants*. Wiley, New York.

Ransom, J.K., Odhiambo, G.D., Kisumu, K. and Gressel, J. (1995) Seed dressing maize with imidazolinone herbicides to control *Striga hermonthica* (Del.) Benth. *Weed Science Society American Abstracts* **35**: 15.

Powles S.B. and Holtum, J.A.M. (eds.) (1994) *Herbicide Resistance in Plants: Biology and Biochemistry.* CRC Press Inc., Boca Raton.

Wrubel, R.P. and Gressel, J. (1994) Are herbicide mixtures useful for delaying the rapid evolution of resistance? A case study. *Weed Technology* **8**: 635-648.

PART I

HERBICIDE RESISTANT WEEDS

CHAPTER 2
HERBICIDE-RESISTANT WEEDS IN EUROPE: AGRICULTURAL, PHYSIOLOGICAL AND BIOCHEMICAL ASPECTS

R. DE PRADO, N. LOPEZ-MARTINEZ and R. GIMENEZ-ESPINOSA
Departamento Química Agrícola y Edafología, E. T. S. I. A.M., Universidad de Córdoba, Apdo. 3048, 14080 Córdoba, Spain

Abstract

In Europe, cultivated land is mainly given over to corn (11.4 Mhas), wheat and barley (40.75 Mhas) and orchards (40.5 Mhas). For a long time, the continuous use of conventional herbicides such as triazines, phenylureas, aryloxyphenoxypropanoates, cyclohexanodiones, sulfonylureas and imidazolinones has been made but without any rotation of crops and herbicides. Now, 35 broadleaf weeds and 17 grass weeds have been identified as being resistant to various herbicides. The first herbicide-resistant biotype was a triazine-resistant *Amaranthus retroflexus* found in Austria in 1973. The proposition of the mutation of the target site as a resistance mechanism was first made in France in 1978, using chlorophyll fluorescence and Hill reaction measurements in a triazine-resistant *Chenopodium album* biotype. In 1987, the same researchers showed that the resistant *C. album* biotype had Ser264 to Gly mutation in the *psb*A gene. Triazine-resistant broadleaf weeds such as *Amaranthus bouchoni, A. hybridus, A. retroflexus, C. album* and *Erigeron canadensis* showed cross-resistance to some phenylureas, carbamates and uracils in Hungary. The use of bipyridyl herbicides in orchards has selected resistant biotypes of *Epilobium ciliatum, E. canadensis, Erigeron bonariensis, Poa annua* and *Lolium perenne*. However, various resistance mechanisms to paraquat have been proposed, according to the weed species. ALS inhibitors have been widely used in Europe, and their repeated use on crops leads to the rapid appearance of resistance. However, only three sulfonylureas-resistant biotypes of the *Stellaria media, Alisma plantago-aquatica,* and *Papaver rhoeas* species have been detected in Denmark, Portugal and Spain, respectively. In the case of grass weed, eight triazine-resistant biotypes have been detected: five of these showed resistance due to the target site mutation and the other three showed natural glutathione-atrazine formation such as a non-toxic compound. Resistance to chlorotoluron and/or diclofop-methyl due to enhanced detoxification has been confirmed in *Alopecurus myosuroides, Apera spica-venti, Bromus tectorum* and *Lolium rigidum*. Some of these biotypes have presented cross- and multiple-resistance to other herbicides due to a system such as Cytochrome P450. Recently, two cases of resistance to acetyl-CoA carboxylase (ACCase) herbicides have been detected: a

Lolium rigidum in France and an *Avena fatua* in the U.K. The tolerance of *L. multiflorum* to aryloxyphenoxypropanoates (APP) is due to the presence of a tolerant form of ACCase. With respect to other herbicides, different *Echinochloa crus-galli* populations resistant to propanil and quinclorac have appeared in rice fields in Greece and Spain, respectively. The resistance mechanism to propanil is due to a detoxification via hydrolysis catalyzed by aryl acylamidase. The tolerance to quinclorac is due to the non-stimulation of the ethylene biosynthesis in resistant *Echinochloa* spp. Combatting broadleaf resistance is not difficult in crops like corn, rice and orchards in general, where crop rotation and/or a pre- and post-emergence mixture of herbicides can be used. Cross- and multiple-resistant grass weeds are more difficult to combat and require an urgent solution because more resistant new grass weeds are appearing every year.

Introduction

Europe has a total surface area of about ten million square kilometers, of which about twenty five per cent corresponds to arable land. Cultivated land in Europe is mainly devoted to a monoculture of cereals (corn, wheat and barley) with fifty million hectares, and orchards, with forty million hectares. From the Second World War to date, an extensive number of herbicides belonging to different chemical families (aryloxyphenoxypropanoates, cyclohexanediones, dinitroanilines, imidazolinones, phenylureas, sulfonylureas and triazines, principally) have been designed and used for weed control. Both extensive monoculture and herbicide treatment, among other practices, have notably increased agricultural productivity from 1950 to 1980. However, the continuous use of the same herbicide for many years, together with other practices and factors, have originated the appearance of the so-called herbicide-resistant, cross-resistant and multiple-resistant phenomena and, hence, the selection of weed biotypes within a particular species that survived herbicide doses which control their former relatives (Gressel, 1991; Maxwell and Mortimer, 1994). The first case of a herbicide-resistant weed biotype described in Europe dates from 1973 and corresponds to an *Amaranthus retroflexus* resistant to triazines collected in corn fields in Austria (LeBaron, 1991).

Herbicide-Resistant Weeds. Distribution

To date, 35 broadleaf and 17 grass weeds resistant to different families of herbicides have been reported in the European Continent. According to the cases described, France (24 species), Spain (21 species), Switzerland (13 species) and Germany (12 species) are the countries in which this phenomenon is the most extensive. Greece and Portugal, on the other hand, have only one broadleaf weed resistant to propanil and bensulfuron, respectively (Figure 1).

The 52 herbicide-resistant weeds found belong to 12 different botanical families, the *Poaceae* (17 biotypes), the *Amaranthaceae* and *Compositae* (8 biotypes each), the *Polygonaceae* (5 biotypes), the *Caryophyllaceae* and *Chenopodiaceae* (4 biotypes each), *Onagraceae* (2 biotypes) and *Alismataceae, Ambrosiaceae, Papaveraceae, Solanaceae* and *Umbelliferae* (1 biotype each) are the most widely represented.

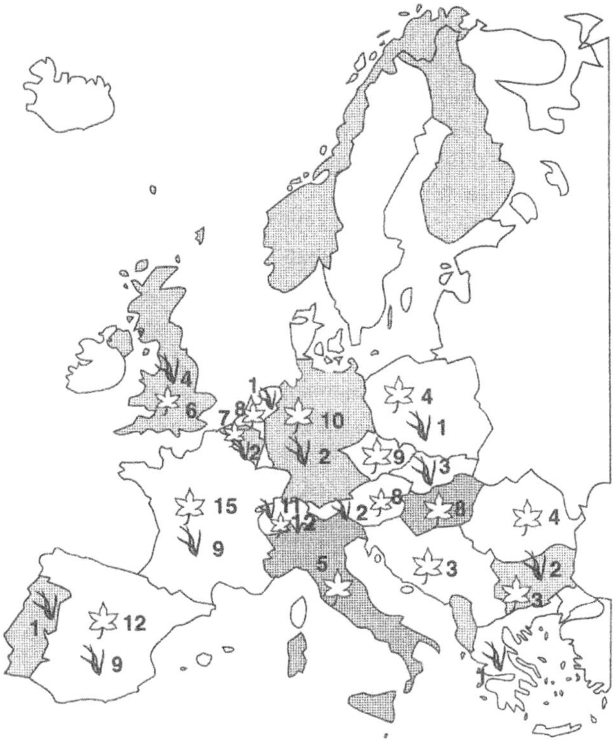

Figure 1. Distribution of resistant grass (🌾) and broadleaf (✿) weeds in Europe

In grass weeds, resistance to triazines is the most common example of herbicide resistance, with 12 cases described (Figure 2A). Resistance to paraquat has been identified in two biotypes. To date, resistance to phenylureas has been reported in four European weed biotypes. Three of them, *Apera spica-venti, Alopecurus myosuroides* and *Bromus tectorum* have only been found in Europe (Mayor and Maillard, 1996; Moss and Cussans, 1991; De Prado *et al.*, 1991, 1996), while others such as *Lolium rigidum*, previously detected in Australia, were also found in Spain (Heap, 1991; Medina, 1992) . Diclofop-methyl resistant biotypes are some of the most well known cases of herbicide resistance. This resistance has been mainly observed in four European biotypes (*A. myosuroides, Avena fatua, L. multiflorum* and *L. rigidum)* coming from winter wheat fields treated with graminicide herbicides (Cornes *et al.*, 1996; Moss, 1995; Gasquez *et al.*, 1996; Medina, 1992). *Echinochloa crus-galli* biotypes resistant to propanil and quinclorac, two herbicides commonly used in rice fields, have also been found in Europe. In 1986, the first European propanil-resistant biotype appeared in Greece and, recently, a quinclorac-resistant *Echinochloa crus-galli* biotype has been detected in Spain (Giannopolitis and Vassilou, 1989; López-Martínez *et al.*, 1995).

Regarding broadleaf weeds, triazine resistance is also the most prominent case in Europe, where 35 broadleaf triazine-resistant biotypes have been selected in different fields treated with these herbicides (Figure 2B). Some of them have also shown cross- or multiple-resistance to herbicides such as diuron, linuron, isoproturon, phenmediphan, bromacil and bromoxynil. Resistance to paraquat has been reported in a couple of biotypes coming from fruit tree orchards and vineyards which have been continuously treated with this herbicide. These biotypes, *Conyza canadensis* and *Epilobium ciliatum*, are also cross-resistant to atrazine (Pölos *et al.*, 1987; Clay, 1989). There are very few cases of resistance to auxin analog herbicides, and only resistant biotypes of *Stellaria media* and *Cirsium arvensis* detected in Hungary and England (Lutman and Lovegrove, 1985; Solymosi *et al.*, 1987). However, recently, a *Papaver rhoeas* biotype resistant to 2,4-D has been found in Northern Spain (Taberner *et al.*, 1996). Right now, resistance to sulfonylureas has been reported in three biotypes of *Stellaria media, Alisma plantago-aquatica* and *Papaver rhoeas* in Denmark, Portugal and Spain (Andreasen *et al.*, 1996; Calha *et al.*, 1996; Taberner and De Prado, unpublished data).

Mechanisms of Herbicide Resistance

Photosystem II Inhibitor Herbicides

As it has been mentioned, triazine resistance is very well represented in an ample number of species in Europe, including *Chenopodium album, Solanum nigrum, Senecio vulgaris, Poa annua, Amaranthus retroflexus, Conyza canadensis* and *Echinochloa crus-galli.* Other species, such as *Amaranthus albus, Bidens tripartita* and *Polygonum hydropiper* are only found in specific areas.

The number of broadleaf biotypes resistant to atrazine is three-fold greater than grass weed biotypes. This fact can be explained by a larger number of dicot weeds present in corn fields, where triazine herbicides are primarily used (Gasquez and Darmency, 1989). Some grass weeds are naturally resistant, as are some species of the *Panicoideae* subfamily (*Setaria faberi* and *Setaria viridis*; Jensen *et al.*, 1977). They detoxify triazines via conjugation (Edwards, this volume; De Prado *et al.*, 1995). Natural resistance via herbicide detoxification may delay the appearance of herbicide resistance when resistance is attributed to an altered site of action (Giménez-Espinosa and De Prado, unpublished data).

Resistance mechanism by a herbicide-insensitive target site (the D1 protein) was first proposed in 1978 by Ducruet and Gasquez on *C. album*, using chlorophyll fluorescence and Hill reaction analysis. This hypothesis was later confirmed by the same researchers after sequencing the *psb*A gene in the wild and resistant biotypes. The mutation, serine for glycine, at position 264 of the *psb*A gene product was found in the triazine-resistant biotype (Bettini *et al.*, 1987).

With respect to phenylurea resistance, a classic example is the selection of resistant biotypes upon continuous treatment of chlorotoluron. The first resistant grass

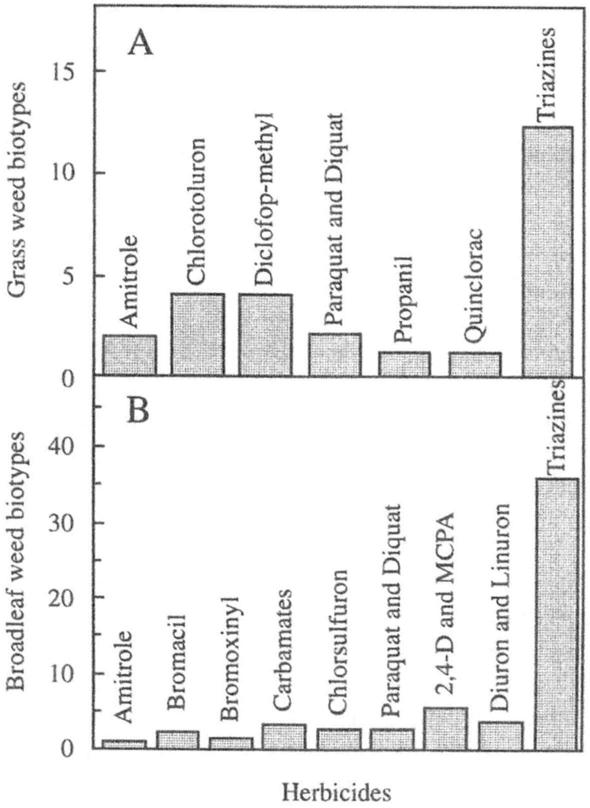

Figure 2. Number of biotypes of grass weeds (A) and broadleaf weeds (B) resistant to herbicides appearing in Europe.

biotype reported was an *Alopecurus myosuroides* biotype found in Germany (Niemann and Pestemer, 1984). Later on, similar biotypes appeared in England, where they are widespread, and, finally, in Northern Spain and France. The last chlorotoluron resistant biotypes observed are those of *Apera spica-venti*, found in Switzerland and *Lolium rigidum* and *Bromus tectorum*, found in Spain. Although the wild *Bromus tectorum* biotype is naturally tolerant to some phenylureas, the resistant biotype came from fields which were treated with chlorotoluron and isoproturon. Resistance was 3-fold greater than the non-resistant biotype (De Prado *et al.*, 1996).

Another typical case of chlorotoluron resistance is the case of *Lolium rigidum*. This is probably the weed in which herbicide resistance has been most widely studied, mainly in Australia (Hall *et al.*, 1994). In 1992, different *Lolium rigidum* populations resistant to chlorotoluron and diclofop-methyl were identified in Northern Spain. These biotypes showed levels of chlorotoluron resistance six-fold greater than those of the susceptible biotypes. Although susceptible and resistant biotypes metabolized chlorotoluron, higher levels of non-toxic chlorotoluron metabolites were earlier observed in resistant individuals. Experiments using ABT, a well known P450 monooxygenase inhibitor, indicated the important role of these enzymes in chlorotoluron metabolism. The two chlorotoluron resistant biotypes found in the same area in Northern Spain, *Lolium rigidum* and *Alopecurus myosuroides,* presented a similar behaviour with respect to chlorotoluron metabolism (Menéndez *et al.*, 1996, and this volume).

Different *Echinochloa crus-galli* populations resistant to propanil have appeared in rice fields in Greece. Increased resistance is associated with the repeated use of propanil. The mechanism of resistance to propanil is due to detoxification via hydrolysis, catalysed by an aryl-acylamidase enzyme. This enzyme is highly susceptible to carbaryl and many phosphorate insecticides, and the tolerance to propanil can be completely eliminated (Yih *et al.*, 1968; Giannopolitis and Vassilou, 1989).

Photosystem I Disrupting Herbicides

Paraquat resistance is one of the most interesting cases of herbicide resistance. Paraquat treatment, mainly in orchards, has caused the appearance of a few resistant biotypes: broadleaf species such as *Conyza canadensis* and *Epilobium ciliatum,* and grass weed species like *Lolium perenne* and *Poa annua*. The factor of resistance is highly variable, depending on the species, age of the plant and population, and it is greatly influenced by the methodology used for its determination (Preston, 1994). Mechanisms of resistance to paraquat in these biotypes have been studied using the radiolabeled herbicide. Results obtained with resistant *Conyza spp.* suggest that reduced translocation and/or sequestration of the herbicide could explain, at least in part, the resistant character of these biotypes (Tanaka *et al.*, 1986; Vaughn and Fuerst, 1985). As a consequence of paraquat action, a high concentration of very toxic oxygen species is produced. The induction of enzymes implicated in the detoxification of such species has also been suggested to explain resistance to paraquat (Harper and Harvey, 1978; Shaaltiel and Gressel, 1986).

Acetyl-CoA Carboxylase Inhibitor Herbicides

It has been firmly established that resistance to aryloxyphenoxypropanoates (APP) can be developed in grass weeds in a short period of time. In Europe four different diclofop-resistant biotypes have been detected, all of them came from winter wheat fields where graminicides have been continuously used. These resistant biotypes belong to the following species: *Lolium rigidum* (found in Spain), *Alopecurus myosuroides* and *Lolium multiflorum* (found in England, Spain and France, respectively) and *Avena fatua* (found in England). Enzyme inhibition studies have revealed that the ACCase activity from resistant *Lolium multiflorum* was 39- fold more resistant than the susceptible biotype. In this case, the resistance is due to a reduced ACCase sensitivity to the herbicide (Gasquez *et al.*, 1996).

Two resistant *L. rigidum* biotypes have been detected: one of them harvested from winter wheat fields continuously treated with diclofop-methyl plus chlorotoluron (DCT); and the second one harvested from different fields treated only with graminicide herbicides (DT). Metabolism studies showed that all the biotypes found are able to produce the conjugate. However, the amount of conjugate formed in the DCT biotype was 4-fold greater than the susceptible and the DT biotypes. Three peaks were found when plants were treated with radiolabeled herbicide on the leaves. These peaks corresponded to diclofop-methyl, diclofop acid and diclofop-conjugate. The diclofop-methyl was rapidly metabolized to the acid form of the herbicide via esterase activity followed by a diclofop-conjugation. On the other hand, ABT inhibited the metabolism of diclofop-methyl in all biotypes of *Lolium rigidum*. The resistance mechanism in *Alopecurus myosuroides* and *Lolium rigidum* is due to metabolism via arylhydroxylation, which could be catalyzed by Cytochrome P45O monooxygenases (De Prado and Menéndez, 1997; Menéndez *et al.*, 1996b). However, other mechanisms such as recovery plasma membrane depolarization in the resistant biotypes may explain this resistance (our unpublished data).

Acetolactate Synthase Inhibiting Herbicides

Acetolactate synthase (ALS) inhibitors have been widely used in Europe. The repeated use of these herbicides on crops has led to the rapid appearance of resistance in America. Only three broadleaf resistant biotypes such as *Stellaria media, Alisma plantago-aquatica* and *Papaver rhoeas* have been detected in Denmark, Portugal and Spain, respectively. The mechanism of resistance to the sulfonylurea in these weeds is due to the presence of a mutate form of ALS as has been proposed for other broadleaf biotypes (Saari *et al.*, 1994). Regarding grass weeds, the repeated use of sulfonylureas and imidazolinones has not yet selected resistant biotypes. However, some populations of *A. myosuroides* and *L. rigidum* have shown cross- and multiple-resistance to these herbicides (Moss and Cussans, 1991; De Prado and Menéndez, 1996). Cotterman and Saari (1992) reported that the resistance mechanism in *L. rigidum* to chlorsulfuron was due to the phenyl-O-glucoside formation as a major metabolite, which is inactive against ALS inhibitors.

Auxin Analog Herbicides

These chemicals (2,4-D and MCPA) were the first selective herbicides used to control dicot in cereal crops. Right now, hundreds of these herbicides have been synthesized and many active molecules are being used to control broadleaf and grass weeds. To date, three dicot *(Stellaria media, Cirsium arvensis* and *Papaver rhoeas)* and one grass weed *(Echinochloa crus-galli)* have been detected as temp tolerant to 2,4-D and quinclorac in Europe, respectively (Coupland, 1994; De Prado, 1995; Lopez-Martinez *et al.*, 1995). Quinclorac causes the inhibition of shoot growth accompanied by chlorosis and necrosis in *E. crus-galli*. (Grossmann and Kwiatkowski, 1995). Studies carried out in *E. colonum, E. oryzoides* and *E. crus-galli* have concluded that the former species are more susceptible than the latter species. The higher susceptibility in *E. colonum* and *E. oryzoides* was due to an accumulation of endogenous HCN formed as a co-product during quinclorac-stimulated ethylene biosynthesis, while quinclorac did not increase the ethylene production in *E. crus-galli* (López-Martínez and De Prado, 1996).

Conclusions

Although numerous biotypes of resistant weeds have appeared, many of them, can be controlled by different herbicides, crop rotation or other classic techniques. However, for a few biotypes (*Alopecurus myosuroides* and *Lolium rigidum*, for example), which exhibit cross- and multiple-resistance, a more complex and integrated weed management program must be adopted. In the future more exhaustive studies are needed for a better understanding of the mechanisms of resistance.

Acknowledgements

The authors thank the CICYT for financial support (Project No. AGF93-0332).

References

Andreasen, C., Jensen, J.E., Madsen, K.H., Poulsen, G.S. and Streigbig, J.C. (1996) Herbicide resistance in weeds and crops in Denmark. *In: Proceedings International Symposium on Weed and Crop Resistance to Herbicides* (R. De Prado, J. Jorrín, L. García-Torres and G. Marshall, eds.), 175-177. University of Córdoba, Córdoba, Spain.

Bettini, P., MacNally, S., Sevignac, M., Darmency, H., Gasquez, J. and Dron, M. (1987) Atrazine resistance in *Chenopodium album*: low and high levels of resistance to the herbicide and related to the same chloroplast *psb* A gene mutation. *Plant Physiology* **84**: 1442-1446.

Calha, I.M., Machado, C. and Rocha, F. (1996). A research study on herbicide resistant weeds in portuguese fields. *In: Proceedings International Symposium on Weed and Crop Resistance to Herbicides* (R. De Prado, J. Jorrín, L. García-Torres and G. Marshall, eds.), 223-226. University of Córdoba, Córdoba, Spain.

Clay, D.V. (1989) New developments in triazine and paraquat resistance and co-resistance in weed species in England. *Proceedings British Crop Protection Conference-Weeds 1989*, 317-324.

Cornes, D.W., Mills, C.E. and Ryan, P.J. (1996) Development and management of resistance to aryloxyphenoxypropanoate herbicides in *Alopecurus myosuroides* in the UK. *In: Proceedings International Symposium on Weed and Crop Resistance to Herbicides* (R. De Prado, J. Jorrín, L. García-Torres and G. Marshall, eds.), 229-230. University of Córdoba, Córdoba, Spain.

Cotterman, J.C. and Saari, L.L. (1992) Rapid metabolic inactivation is the basis for cross-resistance to chlorsulfuron in diclofop-methyl-resistant rigid ryegras (*Lolium rigidum*) biotype SR4/84. *Pesticide Biochemistry Physiology* **43**:182-192.

Coupland, D. (1994) Resistance to the auxin analog herbicides. *In: Herbicide Resistance in Plants: Biology and Biochemistry* (S.B. Powles and J.A.M. Holtum, eds.), 171-214. Lewis Publishers, Boca Raton, FL.

De Prado, R. and Menendez, J. (1996) Management of herbicide-resistant grass weeds in Europe. *In: Proceedings Second International Weed Control Congress* (H. Brown, G.W. Cussans, M.D. Devine, S.O. Duke, C. Fernández-Quintanilla, A. Helweg, R.E. Labrada, M. Landes, P. Kudsk and J.C. Streibig, eds.) **II**, 399-405. EWRS, IWSS, Denmark.

De Prado, R. and Menendez, J. (1997) Cross-resistance and herbicide metabolism in *Alopecurus myosuroides*

Huds. *In: Regulation of Enzymatic Systems Detoxifying Xenobiotics in Plants* (K.K. Hatzios, ed.). Kluwer Academic Publishers, Doordrecht. In press.

De Prado, R. Menendez, J., Tena, M., Caseley, J. and Taberner, A. (1991) Response to substituted ureas, triazines and chloroacetanilides in a biotype of *Alopecurus myosuroides* resistant to chlorotoluron. *Proceedings British Crop Protection Conference-Weeds 1991,* 1065-1070.

De Prado, R., Romera, E. and Menendez, J. (1995) Atrazine detoxification in *Panicum dichotomiflorum* and target site *Polygonum lapathifolium. Pesticide Biochemistry Physiology* **52**: 1-11.

De Prado, R., Romera, E. and Menendez, J. (1996) Chlorotoluron resistance in a *Bromus tectorum* L. biotype is due to enhanced detoxification processes. *In: Proceedings International Symposium on Weed and Crop Resistance to Herbicides* (R. De Prado, J. Jorrín, L. García-Torres and G. Marshall, eds.), 62-64. University of Córdoba, Córdoba, Spain.

Ducruet, J.M. and Gasquez, J. (1978). Observation sur la fluorescence sur feuille entière et mise en évidence de la rèsistance chloropastique à l'atrazine chez *Chenopodium album* L. et *Poa annua* L. *Chemosphere* **7**: 691-696.

Gasquez, J. and Darmency, H. (1989) Herbicide resistance in France. *In: Importance and Perspectives on Herbicide Resistant Weeds* (R. Cavalloro and G. Noyé, eds.), 81-94. Office for Official Publications of the European Communities, Brussels, Luxembourg.

Gasquez, J., Gimenez-Espinosa, R., Gronwald, J., De Prado, R. (1996). Resistance to aryloxyphenoxypropanoate herbicides in a *Lolium multiflorum* biotype in France. *In: Proceedings International Symposium on Weed and Crop Resistance to Herbicides* (R. De Prado, J. Jorrín, L. García-Torres and G. Marshall, eds.), 78-79. University of Córdoba, Córdoba, Spain.

Giannopolitis, C.N. and Vassiliou, G. (1989) The *Echinochloa crus-galli* complex in rice-morphological variants and tolerance to propanil in Greece. *In: Importance and Perspectives on Herbicide-Resistant Weeds* (R. Cavalloro and G. Noyé, eds.), 23-28. Office for Official Publications of the European Communities, Brussels, Luxembourg.

Gressel, J. (1991) Why we get resistance? It can be prevented or delayed. *In: Herbicide Resistance in Weeds and Crops* (J.C. Caseley, G.W. Cussans and R. K. Atkin, eds.) 1-26. Butterworth-Heinemann, Oxford.

Grossmann, K. and Kwiatkowski, J. (1995) Evidence for a causative role of cyanide, derived from ethylene biosynthesis, in the herbicidal mode of action of quinclorac in barnyardgrass. *Pesticide Biochemistry Physiology* **51**: 150-160.

Hall, L.M., Holtum, J.A.M. and Powles, S.B. (1994) Mechanism responsible for cross resistance and multiple resistance. *In: Herbicide Resistance in Plants: Biology and Biochemistry* (S.B. Powles and J.A.M. Holtum, eds.), 243-262. Lewis Publishers, Boca Raton, FL.

Harper, D.B. and Harvey, B.M.R. (1978) Mechanism of paraquat tolerance in perennial ryegrass. II. Role of superoxide dismutase, catalase and peroxidase. *Plant Cell Environment* **1**: 221-215.

Heap, I.M. (1991). Resistance to herbicides in annual ryegrass (*Lolium rigidum*) in Australia. *In: Herbicide Resistance in Weeds and Crops* (J.C. Caseley, G.W. Cussans and R. K. Atkin, eds.) 57-66. Butterworth Heinemann, Oxford.

Jensen, K.I.N., Stephenson, G.R. and Hunt, L.A. (1977) Detoxification of atrazine in three gramineae subfamilies. *Weed Science* **25**: 212-220.

LeBaron, H.M. (1991) Distribution and seriousness of herbicide-resistant weed infestation worldwide. *In: Herbicide Resistance in Weeds and Crops* (J.C. Caseley, G.W. Cussans and R. K. Atkin, eds.), 27-44. Butterworth-Heinemann, Oxford.

López-Martínez, N., De Prado, R., Finch, R.P. and Marshall, G. (1995) A molecular assessment of genetic diversity in *Echinochloa* spp. *In: Proceedings British Crop Protection Conference-Weeds 1995,* 445-450.

López-Martínez, N. and De Prado, R. (1996) Fate of quinclorac in resistant *Echinochloa cruss-galli. In: Proceedings Second International Weed Control Congress* (H. Brown, G.W. Cussans, M.D.

Devine, S.O. Duke, C. Fernández-Quintanilla, A. Helweg, R.E. Labrada, M. Landes, P. Kudsk and J.C. Streibig, eds.) **II**, 535-540. EWRS, IWSS, Denmark.

Lutman, P.J.V. and Lovegrove, A.W. (1985) Variation in tolerance of *Galium aparine* (cleavers) and *Stellaria media* (chickweed) to mecoprop. *Proceedings British Crop Protection Conference-Weeds 1985*, 411-418.

Mayor, J.P., Maillard, A. (1996) A new phenylurea resistant grass weed in Switzerland: *Apera spica-venti. In: Proceedings International Symposium on Weed and Crop Resistance to Herbicides* (R. De Prado, J. Jorrín, L. García-Torres and G. Marshall, eds.), 26-29. University of Córdoba, Córdoba, Spain.

Menéndez, J., Taberner, A., De Prado, J.L. and De Prado, R. (1996a) Resistance mechanisms in chlorotoluron-resistant biotypes of *Lolium rigidum* Gaudin. *In: Proceedings International Symposium on Weed and Crop Resistance to Herbicides* (R. De Prado, J. Jorrín, L. García-Torres and G. Marshall, eds.), 99-100. University of Córdoba, Córdoba, Spain.

Menéndez, J., De Prado, J.L. and De Prado, R. (1996b). Diclofop-methyl metabolism in resistant biotypes of *Lolium rigidum. In: Proceedings International Symposium on Weed and Crop Resistance to Herbicides* (R. De Prado, J. Jorrín, L. García-Torres and G. Marshall, eds.), 97-98. University of Córdoba, Córdoba, Spain.

Maxwell B. D. and Mortimer A. M. (1994) Selection for herbicide resistance. *In: Herbicide Resistance in Plants: Biology and Biochemistry* (S.B. Powles and J.A.M. Holtum eds.), 1-25. Lewis Publishers, Boca Raton, FL.

Medina, M.A., De Prado, J.L., Taberner, A., Romera, E., Menendez, J. and De Prado, R. (1992) Tolerance of *Lolium rigidum* to diclofop-methyl and isoproturon. *Proceedings 9th International Conference on Weed Biology*, **1**: 741-746. Dijon 1992.

Moss, S.R. and Cussans, G.W. (1991) The development of herbicide resistant populations of *Alopecurus myosuroides* (lackgrass) in England. *In: Herbicide Resistance in Weeds and Crops* (J.C. Caseley, G.W. Cussans and R. K. Atkin, eds.), 445-456. Butterworth-Heinemann, Oxford.

Moss, S.R. (1995) Techniques for determining herbicide resistance. *Proceedings British Crop Protection Conference-Weeds 1995*, 547-556.

Niemans, P. and Pestemer, W. (1984) Resistenz verschiedener Herkunfte von Acker-Fuchsschwanz (*Alopecurus myosuroides*) gegenuber Herbizidbehandlungen. *Nachrichtenblatt des Deutschen Pflanzenschutzdienstes* **36**: 113-118.

Pölos, E., Mikulas, J., Szigeti, Z., Laskay G. and Lehoczki E. (1987) Cross-resistance to paraquat and atrazine in *Conyza canadensis*. *Proceeding British Crop Protection Conference-Weeds* 1987, 909-916.

Preston, C. (1994) Resistance to photosystem I disrupting herbicides. *In: Herbicide Resistance in Plants: Biology and Biochemistry* (S.B. Powles and J.A.M. Holtum eds.), 61-82. Lewis Publishers, Boca Raton, FL.

Saari, L.L., Cotterman, J.C. and Thill, D.C. (1994) Resistance to acetolactate synthase inhibiting herbicides. *In: Herbicide Resistance in Plants: Biology and Biochemistry* (S.B. Powles and J.A.M. Holtum eds.), 83-139. Lewis Publishers, Boca Raton, FL.

Shaaltiel, Y. and Gressel, J. (1986) Multienzyme oxygen radical detoxifying system correlated with paraquat resistance in *Conyza bonariensis*. *Pesticide Biochemistry Physiology* **26**: 22-28.

Solymosi, P., Kostyal, Z. and Gimesi, A. (1987) Resistance of *Cirsium arvense* (L.) to phenoxyacetic acid derivatives. *Növénnyvedelem*, **23**: 301-305.

Taberner, A., Menendez, J., De Prado, J.L. and De Prado, R. (1996) Weed resistance in Catalonia. *In: Proceedings International Symposium on Weed and Crop Resistance to Herbicides* (R. De Prado, J. Jorrín, L. García-Torres and G. Marshall, eds.), 34-36. University of Córdoba, Córdoba, Spain.

Tanaka, Y., Chisaka, H. and Saka, M. (1986) Movement of paraquat in resistant and susceptible biotypes of *E. philadelphicus* and *E. canadensis*. *Physiologia Plantarum* **66**: 605-608.

Vaughn, K.C. and Fuerst, E.P. (1985) Structural and physiological studies of paraquat-resistant *Conyza*. *Pesticide Biochemistry Physiology* **24**: 86-94.

Yih, R.Y., McRae, D.H. and Wilson, H.F. (1968) Mechanism of selective action of 3',4'-dichloropropioacetanilide. *Plant Physiology* **43**: 1291-1296.

CHAPTER 3
HERBICIDE RESISTANCE IN NORTH AMERICA: HISTORY, CIRCUMSTANCES OF DEVELOPMENT AND CURRENT SITUATION

D.L. SHANER
American Cyanamid Co., P. O. Box 400, Princeton, NJ, USA 08543-0400

Abstract

The history of the selection of herbicide resistant populations in North America dates back to 1962 with the report of a *Daucus carota* population that was resistant to 2,4-D along a Canadian roadside. However, this resistant population did not spread nor has resistance to auxin type herbicide become a major problem. The discovery of triazine resistant *Senecio vulgaris* in Washington State in 1968 was the beginning of a more widespread problem. Triazine resistance has spread throughout the maize producing area of North America primarily where there is little crop rotation and/or herbicide combinations. In 1973 dinitroaniline resistant *Eleusine indica* populations were reported in cotton producing areas of the southeastern U. S. As in the case of triazine resistance, dinitroaniline resistance arose where the same herbicides were used with little or no use of other modes of action. In the 1980's resistance to ACCase inhibitors and the ALS inhibitors in Canada and the U. S. was uncovered. In the case of the ALS inhibitors, this resistance was discovered within 5 years after introduction of the sulfonylureas. In most, if not all, instances of resistance, the farmers used the same mode of action herbicides on the same piece of land over an extended period of time with no other methods or herbicides used for weed control. In the 1990's resistance to some of the older herbicides, such as triallate and propanil, as well as the newer, more potent herbicides, began to appear. Today resistant populations to many different classes of herbicides can be found throughout North America. Several different academic and industrial groups have been established and are working to educate growers and others on weed management techniques which will minimize the selection of herbicide resistant populations as well as manage resistant populations after they are selected.

Introduction

The use of selective herbicides in crop production revolutionized weed control and played a major role in increasing agricultural productivity and efficiency. This new age in weed management began in the mid-1940's with the discovery and development of 2,4-D and its analogs. This also presaged the intense effort by numerous companies to discover and develop other classes of herbicides. Weed management in North America depends on the availability and continued efficacy of these herbicides.

However, as with all methods, chemical weed control as a single means of weed control has its weaknesses. One of these weaknesses is the selection of resistant populations, the subject of this symposium. Herbicides exert a selective pressure on the weed population which can give rise to new biotypes or species that are no longer controlled. Once this occurs, that particular biotype or species will have to be managed by some other means, whether it's a different herbicide or some other weed control practices. These changes can occur very rapidly. In Craft and Harvey's review of weed control in 1949, they state that farmers in California were beginning to complain about the increase in wild oat (*Avena fatua*) infestations in cereals that were treated with 2,4-D. These complaints began within 4 years after the introduction of this new, revolutionary form of weed control.

In 1982 Bandeen *et al.* (1982) published a review of herbicide resistance development in North America. The purpose of this paper is to briefly review the development of resistance to herbicides in North America, (primarily Canada and the United States) concentrating on resistance that has developed since 1982, to look at the circumstances under which resistance has occurred and to examine the current situation in herbicide resistance in North America and what is being done to manage it. Although herbicide resistance is viewed as a potentially serious problem, it can be managed in a way that will allow the continued usefulness of herbicides as effective weed management tools.

Definitions

In order to discuss herbicide resistance it is necessary to define the terms being used. There is some controversy in the literature over the correct definitions for resistance, cross resistance and multiple resistance. For this review I will define "resistance" as "The inherited ability of a weed population to withstand a herbicide at its use rate brought about by a genetic change within the population due to selection by the herbicide." "Cross resistance" will be used to describe cases in which a weed population is resistant to two or more herbicides due to a single mechanism of resistance (Moss and Rubin, 1993). "Multiple resistance" describes those cases where the resistant plant has two or more distinct resistance mechanisms (Moss and Rubin, 1993).

Development of Herbicide Resistance in North America

One of the earliest recorded cases of resistance was in 1957 when Whitehead and Switzer (1963) discovered isolated populations of *Daucus carota* along roadsides in Ontario which were no longer sensitive to 2,4-D. Comparative studies showed that these populations were also cross resistant to 2,4-DB, and MCPB, but not to 2,4,5-T (Whitehead and Switzer, 1963). However, this resistant biotype did not spread nor was there development of widespread resistance to the auxin-type herbicides in North America. In 1990 Heap and

Morrison (1992) found populations of *Sinapsis arvense* in western Canada which were resistant to 2,4-D and/or MCPA. These populations were in fields which had been treated with auxin-like herbicides for over 20 years.

The first major case of resistance that subsequently was detected over a large part of North America was to triazine herbicides. In the mid-1960's a population of *Senecio vulgaris* infesting an ornamental nursery in Olympia, Washington was no longer controlled by high rates of simazine after 10 years of continuous use (Ryan, 1970). Greenhouse studies by Ryan (1970) and by Radosevich and Appleby (1973) confirmed the resistance. Throughout the 1970s and 1980s triazine resistance was documented in many other species (LeBaron, 1991). LeBaron (1991) reported that triazine resistance occurred in 33 states in the U. S. and 4 provinces of Canada. Resistance was particularly widespread in Wisconsin, Nebraska, Virginia, Maryland, Delaware and New York in the U. S. and in Ontario in Canada (Ritter, 1989).

In the early 1970s dinitroaniline resistant goosegrass (*Eleusine indica*) populations were discovered in the northeastern coastal plain region of South Carolina in cotton fields where trifluralin had been used on an annual basis for 10 years (Mudge *et al.*, 1984). This biotype was resistant to all dinitroanilines and subsequent research by Vaughn and others (Smeda and Vaughn, 1994) showed that the resistance was due to an alteration of the tubulin protein, the target site for the dinitroanilines. In 1988 Morrison *et al* (1989) reported the confirmation of dinitroaniline resistant *Setaria viridis* populations in Manitoba. These resistant biotypes occurred in areas where trifluralin had been used over a 20 year period in both cereals and oilseed crops. A dinitroaniline resistant biotype of *Amaranthus palmerii* was confirmed in 1991 from cotton fields that had received 24 consecutive years of trifluralin (Gossett *et al,* 1992). Both the *S. viridis* and *A. palmerii* were resistant to other dinitroanilines (Beckie and Morrison, 1993; Gossett *et al.*, 1992).

In 1988 organic arsenical resistant biotypes of *Xanthium strumarium* were discovered in three adjoining counties in South Carolina's ten-county cotton growing region (Haigler *et al.*, 1988). The farmers in this area used DSMA and MSMA on approximately 70% of the cotton with an average of 1.5 applications annually. In this particular area the cotton had received one of these two herbicides annually for about 15 years (Haigler *et al.*, 1988). Since then MSMA resistant biotypes of *X. strumarium* have been documented in other cotton growing areas in the southeastern U. S (Wills *et al.*, 1992). The mechanism of resistance of these biotypes has not yet been determined.

In 1987 paraquat resistant biotypes of *Solanum americanum* were reported in tomato producing areas of Florida (Stall *et al.*, 1987). These biotypes were selected after multiple annual applications of post-directed paraquat between the rows of tomatoes.

Aryloxyphenoxypropanoate (APP) and cyclohexanedione (CD) herbicides came on the market in North America in the late 1970s and early 1980s. Herbicides in both of these classes kill grasses by inhibiting acetyl coenzyme A carboxylase (ACCase) the first committed step in fatty acid biosynthesis (Devine and Shimabukuro, 1994). In 1988 Stanger and Appleby (1989) confirmed resistance to diclofop in two populations of *Lolium multiflorum* from wheat producing areas of Oregon. These fields had received annual applications of diclofop for at least 7 years. APP-resistant populations of *A. fatua* were discovered in Manitoba and Saskatchewan in 1990 (Heap *et al.*, 1993). These populations came from fields which were treated with an ACCase inhibiting herbicide for 9 years and they were resistant to herbicides from both the APP and CD families. In 1991 a biotype of *S. viridis* was identified in Manitoba which was highly resistant to APP and CD herbicides (Heap and Morrison, 1992). Fluazifop and quizalofop resistant populations of *Sorghum halepense* in Mississippi were confirmed in 1992 (Barrentine *et al.*, 1992). The mechanism of resistance appears to be due to an alteration in the ACCase enzyme (Marles *et al.*, 1993).

Sulfonylureas, imidazolinones, triazolopyrimidine sulfonamides and pyrimidyl oxybenzoates are classes of herbicides that inhibit acetolactate synthase (ALS), the first enzyme in the branched chain amino acid pathway (Saari *et al.*, 1994). These herbicides are some of the most effective compounds in terms of spectrum and percentage weed control currently on the market. Resistant biotypes of several weed species evolved within 6 years after introduction of this herbicide class. The first recorded case was in Idaho. In 1987 Thill *et al.* (1991) discovered a population of chlorsulfuron-resistant *Lactuca serriola* in a wheat field which had received applications of the herbicide at a 7 to 14 month interval from 1982. In 1989 sulfonylurea resistant populations of *Stellaria media, Kochia scoparia,* and *Salsola iberica* were found in Alberta and Saskatchewan, Canada and in Colorado, Kansas, Montana, North Dakota, South Dakota, Texas, Washington, as well as Idaho (Saari *et al.*, 1994). In 1991 a population of *X. strumarium* in Mississippi was confirmed to be resistant to imazaquin in a soybean field that had received multiple applications of the herbicide over a 4 year period (Schmitzer *et al.*, 1993). In 1992 bensulfuron resistant populations of *Sagittaria montevidensis* and *Cyperus difformis* were found in rice growing areas of California (Saari *et al.*, 1994). The mechanism of resistance was found to be due to an alteration in ALS, the target site for these herbicides in all the cases where the mechanism was determined (Saari *et al.*, 1994). ALS inhibitor resistant biotypes have also been found in *Amaranthus* spp., and *Lolium perenne* in the United States. Cross resistance among the different classes of ALS inhibitors has not been consistently found (Saari *et al.*, 1994).

In 1990 propanil resistant biotypes of *Echinochloa crus-galli* were discovered in rice producing areas of Arkansas in fields that had received this herbicide for over 30 years (Baltazar and Smith, 1994). The mechanism of resistance of some of these biotypes appears to be due to an increase in the ability of the resistant biotypes to detoxify the herbicide (Hoagland, personal communication).

Triallate resistant biotypes of *A. fatua* were reported by O'Donovan *et al.* (1994) in 1992 in wheat fields in Saskatchewan that had received annual applications for 20 years. Interestingly, these populations were also resistant to difenzoquat (O'Donovan *et al.*,

1994). Triallate resistant *A. fatua* populations have also been found in Montana (O'Donovan *et al.*, 1994).

Factors Leading to Resistance

Table 1 lists cases where resistant biotypes have been selected by 7 different herbicides. Close examination of these cases reveal certain common factors. One of the common factors has been the continuous use of a herbicide or herbicides sharing the same mode of action as the primary method of weed control for an extended period of time (Table 1), although the time required before resistance evolved depended on the herbicide. Lack of crop rotation was not a factor if the same mode of action herbicide was used in each crop, which was the case for the selection of resistance to ACCase inhibitors in Canada (Devine and Shimabukuro, 1994).

Another common factor where herbicide resistant biotypes have evolved is the role of the seed bank in the evolution of resistance. Resistance developed in species which have a relatively rapid turnover of their soil seed bank. This turnover was either a characteristic of species, such as kochia, or was aided by the cultural practices. One of these practices is the absence of cultivation to control weeds, particularly moldboard plowing. The lack of tillage may result in a more rapid turnover of the seed bank population. Work by Ball (1992) showed that the number of seeds in the upper 2.5 cm of soil increases after chisel plowing compared to moldboard plowing. The lack of moldboard plowing has been suggested as one of the reasons black grass (*Alopecurus myosuroides*) has become such a serious weed in England (Moss, 1980). Another practice that seems to be related to the development of triazine resistance is the application of manure to fields which will add seed back to the field and may partially mitigate the effect of crop rotation (Ritter, 1989).

A third common factor is that the spread of resistant populations appears to be due to independent selection events and not from spread from a common source. (Guttieri *et al.*, 1992) determined the change in the ALS gene in 8 different populations of sulfonylurea-resistant kochia and found 6 different mutations at the same site of the ALS gene, indicating that each population was from a separate selection event.

Management of Resistance in North America

The selection of herbicide resistant weed biotypes is a potentially serious, but manageable problem. Although there are many different geographic locations containing herbicide resistant weeds, the total percentage of land infested with resistance is still minor. Triazine resistance is the most common and widespread of all herbicide resistant biotypes and yet out of the approximately 30 million hectares of maize grown each year in the U. S, only 1 million hectares, or less than 4% of the total maize growing area, contain triazine resistant weeds (LeBaron, 1991). In addition, triazines continue to be widely used, even in areas with high instances of resistance. Wisconsin has over 400,000 hectares which are

Table 1. Development of resistance to various herbicides.

Herbicide	Year of Introduction	Year of First Case of Resistance	Years of Herbicide Use	Species	Agricultural System	Reference
2,4-D	1945	1957		*Daucus carota*	Roadside	Whitehead *et al., 1962*
Atrazine	1958	1968	10	*Senecio vulgaris*	Conifer Plantation	Ryan, 1970
Trifluralin	1963	1973	10	*Eleusine indica*	Continuous Cotton	Mudge *et al.*, 1984
Diclofop	1977	1984	7	*Avena fatua*	Wheat-Rapeseed Rotation	Devine *et al.*, 1994
Chlorsulfuron	1982	1987	6	*Lactuca serriola*	Continuous Wheat	Thill *et al.*, 1990
Triallate	1962	1987	26	*Avena fatua*	Continuous Barley	O'Donovan *et al.*, 1994
Imazaquin	1986	1992	6	*Xhantium strumarium*	Continuous Soybean	Schmitzer *et al.*, 1993

infested with triazine resistant populations (LeBaron, 1991), but 80% of this area still receives an annual application of a triazine herbicide because it is still effective on other weed species in the area (Anonymous, 1990). The resistant biotypes are treated with other herbicides.

In fact, using combinations of herbicides with different modes of action combined with tillage, crop rotation and other management practices is one of the best ways to prevent resistance from developing as well as managing resistance after it occurs. (Stephenson *et al.*, 1990) contrasted the weed control practices of two areas in Ontario, one of which had high instances of triazine resistance (TR) and the other with little triazine resistance. He found that in the area with high levels of infestation of triazine resistant biotypes over 90% of the corn is grown for silage; 40% of the corn is grown on a continuous basis; the highest rate of atrazine was used (4 kg/ha) to control perennial grasses; there was little use of post-emergence treatments; manure is put back on more than 95% of the land and more than 50% of the corn was custom harvested. By contrast, in the area with low infestations of TR biotypes, the farmers grew most of the corn for grain. Atrazine was used on less of the acreage and over 90% of the area received a non-triazine herbicide. Late applications of these herbicides with drop nozzles was common. In addition there was less continuous corn; manure was applied to less than one percent of the land and less than 20% of the corn was custom harvested. In other words, the area which used mixtures of herbicides, practiced crop rotation and minimized the spread of resistant seed through manure or custom combining had the fewest triazine resistant biotypes. This is also the reason why triazine resistance is such a problem in Wisconsin, but not in other parts of the maize producing area of the Midwest (Knake, 1992). Thus, we have the means to manage weeds that will not lead to widespread resistance.

The primary need in implementing these means is educating the farmer in effective weed management practices. Companies which manufacture and sell herbicides in North America have taken an active role in this educational process. Two working groups, the ALS/AHAS Inhibitor Resistance Working Group (AIRWG) and the ACCase Inhibitor Resistance Working Group (ACCRWG) were established in the late 1980s-early 1990s. These groups recommend technical strategies to prevent, delay, or manage herbicide resistance and monitor the subsequent implementation of these strategies. In addition, the groups establish a direct collaboration on monitoring programs and guide the activities of various national sub-groups. These groups, in cooperation with the Herbicide Resistance Action Committee, have funded research on monitoring the level of resistance to ALS inhibitors and ACCase inhibitors in northwestern United States and in western Canada. They also interact with herbicide resistance committees of various academic organizations in the United States and Canada. All of these organizations are working to educate growers on integrated weed management systems which will prevent herbicide resistance from becoming a serious problem.

Most weed control publications in the U. S. and Canada include sections on herbicide resistant weed development and management, and there are active programs to increase farmers' awareness of the issues. In addition, there have been numerous symposia on herbicide resistance at regional and nationwide weed science meetings.

For most farmers, however, herbicide resistance is not an issue that has high priority. A 1989 survey conducted by the Wisconsin Agricultural Statistics Service (Anonymous, 1990) found that 87% of the farmers surveyed used atrazine and 69% of them had used it for 10 years or more. Although 44% of the farmers had decreased the total acres treated with atrazine from 1984 to 1989, the primary reason was a concern over carryover problems rather than resistance. Interestingly, the primary reason for increasing atrazine active ingredient per acre was due to weed resistance. In another survey conducted by (Stoller *et al.*, 1993) soybean farmers ranked development of strategies for delaying and/or preventing the selection of resistant weeds 7th in 14 categories; the development of more economical weed control ranked first. Even when resistance has been well documented in an area, farmers are reluctant to change their weed control programs. In Manitoba, there has been a major effort by weed scientists in the university, extension and government to educate farmers about the need to integrate different weed control practices to prevent the spread of resistance to ACCase inhibitors, ALS inhibitors, dinitroanilines, triallate and difenzoquat. This effort has successfully raised the awareness of many farmers on the potential problem of herbicide resistance, however only 40% of the farmers who are aware of the problem have changed their weed control practices. These results indicate that farmers are more concerned in using the most effective economical weed control program than with implementing a resistance management program in the short term.

Many of the newer herbicides currently being developed or introduced share the same mode of action. These new herbicides should be part of an integrated weed management program that will maximize their utility while minimizing their impact on resistance. The challenge facing the weed management industry in North America is developing programs which are easily adapted, economical and efficacious and that will also manage resistance. For any program to be successful, however, it will have to be accepted by the farmers.

The problem of herbicide resistance in North America is still relatively small and easily manageable. With proper education and innovative programs it can remain this way.

References

Anonymous (1990) 1989 Pesticide Use Survey: Lower Wisconsin River Valley. *Wisconsin Agricultural Statistic Service*, April 27, 1990.

Anonymous (1993) Statistical Abstract of the United States. *The National Data Book, U. S. Department of Commerce.*

Ball, D.A. (1992) Weed seedbank response to tillage, herbicides and crop rotation sequences. *Weed Science* **40**: 654-659.

Bandeen, J.D., Stephenson, G.R. and Cowett. E.R. (1982) Discovery and distribution of herbicide resistant weeds in North America. *In: Herbicide Resistance in Plants* (H. M. LeBaron and J. Gressel, eds.), 9-19. John Wiley and Sons, New York.

Baltazar, A.M. and Smith Jr, R.J. (1994) Propanil-resistant barnyardgrass (*Echinochloa crus-galli)* control in rice (*Oryza sativa*). *Weed Technology* **8**: 576-581.

Barrentine, W.L., Snipes, C.E. and Smeda, R.J. (1992) Herbicide resistance confirmed in johnsongrass biotypes. *MAFES Research Report 17* No. 5.

Beckie, H.J. and Morrison, I.N. (1993) Effect of ethafluralin and other herbicides on trifluralin-resistant green foxtail (*Setaria viridis*). *Weed Technology* **7**: 6-14.

Crafts, A.S. and Harvey, W.A. (1949) Weed Control. *Advances Agronomy* **1**: 289-320.

Devine, M.D. and Shimabukuro, R.H. (1994) Resistance to acetyl coenzyme A carboxylase inhibiting herbicides. *In:* Herbicide *Resistance in Plants: Biology and Biochemistry* (S.B. Powles and J.A.M. Holtum, eds.), 141-170. CRC Press, Inc. Boca Raton, Florida.

Gossett, B.J., Murdock, E.C. and Toler, J.E. (1992) Resistance of Palmer amaranth (*Amarnathus palmeri*) to the dinitroaniline herbicides. *Weed Technology* **6**: 587-591.

Guttieri, M.J., Eberlein, C.V., Mallory-Smith, C.A., Thill, D.C. and Hoffman, D.L. (1992) DNA sequence variation in domain A of the acetolactate synthase genes of herbicide-resistant and -susceptible weed biotypes. *Weed Science* **40**: 670-676.

Haigler, W.E., Gossett, B.J., Harris, J.R. and Toler, J.E. (1988) Resistance of common cocklebur (*Xanthium strumarium*) to the organic arsenical herbicides. *Weed Science* **36:** 24-27.

Heap, I.M. and Morrison, I.N. (1992) Resistance to auxin-type herbicides in wild mustard (*Sinapsis arvense* L.) populations in western Canada. *Abstracts Weed Science Society of America,* **32**: 55.

Heap, I.M., Murray, B.G., Loepky, H.A. and Morrison, I.N. (1993) Resistance to aryloxyprhenoxypropionate and cyclohexanedione herbicides in wild oat (*Avena fatua*). *Weed Science* **41**: 232-238.

Knake, E.L. (1992) Technology transfer for herbicide-tolerant weeds and herbicide-tolerant crops. *Weed Technology* **6**: 662-664.

LeBaron, H.M. (1991) Distribution and seriousness of herbicide-resistant weed infestations worldwide. *In: Herbicide Resistance in Weeds and Crops* (J.C. Caseley, G.W. Cussans and R.K. Atkin, eds.), 27-55. Butterworth Heinemann, Oxford.

Marles, M.A.S., Devine, M.D. and Hall, J.C. (1993) Herbicide resistance in *Seteria viridis* conferred by a less sensitive form of acetyl coenzyme A carboxylase. *Pesticide Biochemistry Physiology* **46**:7-14.

Morrison, I.N., Todd, B.G. and Nawolsky, K.M. (1989) Confirmation of trifluralin-resistant green foxtail (*Setaria viridis*) in Manitoba. *Weed Technology* **3**: 544-551.

Moss, S.R. (1980) The agro-ecology and control of black-grass, *Alopecurus myosuroides* Huds., in modern ceral growing systems. *ADAS Quarterly Review* **38**: 170-191.

Moss, S.R. and Rubin, B. (1993) Herbicide-resistant weeds: a worldwide perspective. *Journal Agricultural Science* **120**: 141-148.

Mudge, L.C., Gossett, B.J. and Murphy, T.R. (1984) Resistance of goosegrass (*Eleusine indica*) to dinitroaniline herbicides. *Weed Science* **32**: 591-594.

O'Donovan, J.T., Sharma, M.P., Harker, K.N., Maurice, D., Baig, M.N. and Blackshaw, R.E. (1994) Wild oat (*Avena fatua*) populations resistant to triallate are also resistant to difenzoquat. *Weed Science* **42**: 195-199.

Radosevich, S.R. and Appleby, A.P. (1973) Relative susceptibility of two common groundsel (*Senecio vulgaris* L.) biotypes to six 2-triazines. *Agronomy Journal* **65**: 553-555.

Ritter, R.L. (1989) Understanding herbicide resistance in weeds. Sandoz Crop Protection Corp., Des Plaines, IL 60018.

Ryan, G.F. (1970) Resistance of common groundsel to simazine and atrazine. *Weed Science* **18**: 614-616.

Saari, L.L., Cotterman J.C. and Thill, D.C. (1994) Resistance to acetolactate synthase inhibiting herbicides.

In: Herbicide *Resistance in Plants: Biology and Biochemistry* (S.B. Powles and J. A.M. Holtum, eds.), 83-140. CRC Press, Inc. Boca Raton, Florida.

Schmitzer, P.R., Eilers, R.J. and Cseke, C. (1993) Lack of cross resistance of imazaquin-resistant *Xanthium strumarium* acetolactate synthase to flumetsulam and chlorimuron. *Plant Physiology* **103**: 281-283.

Smeda, R.J. and Vaughn, K.C. (1994) Resistance to dinitroaniline herbicides. *In:* Herbicide *Resistance in Plants: Biology and Biochemistry* (S. B. Powles and J. A. M. Holtum, eds.), 215-228. CRC Press, Inc. Boca Raton, Florida.

Stall, W.M., Kostewicz, S.R. and Brown, R.L. (1987) Reduction in control of common nightshade (*Solanum americanum*, Mill.) by paraquat due to copper fungicides. *Proceedings Florida State Horticultural Society* **100**: 222-224.

Stanger, C.E. and Appleby, A.P. (1989) Italian ryegrass (*Lolium multiflorum*) accessions tolerant to diclofop. *Weed Science* **37**: 350-352.

Stephenson, G.R., Dykstra, M.D., McLaren, R.D. and Hamil, A.S. (1990) Agronomic practices influencing triazine-resistant weed distribution in Ontario. *Weed Technology* **4**: 199-207.

Stoller, E.W., Wax, L.M. and Alm, D.M. (1993) Survey results on environmental issues and weed science research priorities within the corn belt. *Weed Technology* **7**: 763-770.

Thill, D.C., Mallory-Smith, C.A., Saari, L.L., Cotterman, J.C. and Primiani, M.M. (1991) Sulfonylurea herbicide-resistant weeds: discovery, distribution, biology, mechanism and management. *In: Herbicide Resistance in Weeds and Crops* (J.C. Caseley, G.W. Cussans and R.K. Atkin, eds.), 115-129. Butterworth Heinemann, Oxford.

Whitehead, C.W. and Switzer, C.M. (1963) The differential response of strains of wild carrot to 2,4-D and related herbicides. *Canadian Journal Plant Sciences* **43**: 255-262.

Wills, G.D., Byrd, J.D. and Hurst, H.R. (1992) Herbicide resistant and tolerant weeds. *Proceedings South Weed Science Society* **45:** 43.

CHAPTER 4
HERBICIDE RESISTANCE OUTSIDE NORTH AMERICA AND EUROPE: CAUSES AND SIGNIFICANCE

B. RUBIN
Department of Field Crops, Vegetables and Genetics, Faculty of Agriculture, the Hebrew University of Jerusalem, Rehovot 76100, Israel

Abstract

The increased awareness of scientists and farmers have led to the discovery of a large number of weeds conferring herbicide resistance traits throughout the world. Since 1975, when the first triazine-resistant grass weed *Brachypodium distachyon* was observed in Israel, more than 10 grasses and broad-leaves weeds were identified. Triazine resistance is closely associated with minimum tillage, mono-herbicide and high rates of application (i.e., in roadsides and orchards). In Australia, New Zealand and Japan however, the number of triazine resistant weeds is very much smaller compared to that found in Europe and North America, probably due to a limited use of atrazine. Resistance to ALS inhibiting herbicides has evolved very rapidly following repeated use of sulfonylureas and imizadolines in crops and roadsides, indicating high frequency of the resistance trait in the weed populations. Resistance due to either an altered ALS or enhanced detoxification of the herbicide or both, was confirmed in Australia, Israel, and Costa Rica. In spite of the apparent lack of selection pressure, paraquat (bipyridyls) resistance has evolved in Egypt, Japan, Australia, Kenya, Malaysia and the Caribbean Islands. Paraquat resistance is associated with orchards and plantations where repeated applications impose strong selection pressure on the weed population. Resistance to the graminicides which inhibit acetyl coenzyme A carboxilase (ACCase), evolved in Australia , Malaysia, Costa Rica, and Israel. In certain cases, insensitive ACCase was identified. In rice production in Costa Rica, Colombia, China and Egypt, continuous application of either propanil, butachlor, or benthiocarb resulted in the evolution of resistant biotypes of *Echinochloa* spp. Most populations exhibit enhanced capacity to detoxify the herbicide. Multiple resistance was detected in certain biotypes of *Lolium rigidum* in Australia which confers resistance not only to the challenging herbicide but virtually to most herbicide groups, making control extremely difficult. Multiple resistance is attributed to a common detoxification system such as cytochrome P450. The large number of weed populations conferring resistance is closely related to the extent of chemical weed control and lack of herbicide and crop rotations. The need for feasible management practices to combat this detrimental phenomenon is urgent.

Introduction

Since the identification of the first triazine-resistant weed in the US two decades ago, a dramatic increase in the number of herbicide-resistant weeds was observed. These changes in the flora along with the extended awareness of weed scientists and farmers, further increased the number of new weeds that confer resistance to herbicides throughout the world. Although the majority of cases were discovered in the US (Shaner, this volume) and Europe (De Prado *et al.*, this volume), the number of reports coming from other regions of the world is constantly increasing. For instance, the number of herbicide-resistant weeds reported during the last 6 years in Australia has increased more than three fold (Powles and Howat, 1990; Preston, 1995, personal communication). A similar trend has been observed in Israel (Rubin, 1991). Although no reports on herbicide resistance are available from a good part of the world (e. g. the majority of Asia, Africa and South American countries), more than 50 weed populations, grasses and broad-leaves, annuals and perennials are already known to be herbicide-resistant outside Europe and the US.

A clear linkage could be noticed between the level of herbicide use and the number of reported cases of resistance. Herbicide resistance is closely associated with major crops (e.g. wheat, and corn), monoculture (including perennial crops and plantations) or/and repeated use of the same herbicide or herbicides acting on the same target site (Rubin, 1991). Both in small farms and minor crops, where hand weeding and cultivation are the dominant methods for weed control, there are small chances for resistance to evolve.

The terms used in this article are based on those suggested by Rubin (1991) and Moss and Rubin (1993) . "Resistance" will be used where population has acquired an inheritable capacity to withstand herbicide(s) applied at or above field rate. "Cross-resistance" describes plants which confer resistance to several herbicides acting at the same site or detoxified by the same enzyme system. "Multiple resistance" will be used for weed populations that are resistant to several chemicals acting at different sites or detoxified by a different enzyme system.

The aim of this paper is to review the present situation regarding the evolution of herbicide-resistant weeds and their spread outside Europe and North America. I will also try out the processes involved in their evolution as well as the significance of these weeds in agriculture.

Mechanisms of Herbicide Resistance

Resistance due to an altered target site in the regions was reported mainly for triazines, ALS inhibitors and to a lesser extent- ACCase inhibitors. The inherited modification of the herbicide binding site generally results in lack of weed control even at high doses of the challenging herbicide as well as various degrees of cross resistance to other herbicides acting at the same site. However, there are certain resistance cases such as that found in

New Zealand for phenoxy herbicides (Popay *et al.*, 1991; Bourdot and Harvey, 1992) where the mechanism has not yet been elucidated.

Triazines

Cohen and Nir observed in 1975 that high rates of triazines herbicides failed to control several grass weeds such as *Brachypodium distachyon* in Israeli roadsides (Cohen and Nir, 1979). At present, at least 11 different grasses and broad leaf weeds were discovered in Israel and characterized as resistant due to an altered binding site at the chloroplast (Table 1).

Table 1: Occurrence and distribution of triazine-resistant weed populations, grasses and broadleaves, outside North America and Europe.

Species	Location	Reference
Alopecurus myosuroides	Israel	Rubin *et al.*, 1985
Brachypodium distachyon	Israel	Cohen & Nir, 1979
Crypsis schoenoides	Israel	Rubin *et al.*, 1995
Lolium rigidum	Australia	Powles & Howat, 1990
Lolium rigidum	Israel	Rubin *et al.*, 1985
Lopochloa smyrnacea	Israel	Yaacoby, 1983
Phalaris paradoxa	Israel	Rubin *et al.*, 1985
Poa annua	Japan	Le Baron, 1991
Polygonon monspeliensis	Israel	Yaacoby, 1983
Amaranthus blitoides	Israel	Schonfeld *et al.*, 1988
Amaranthus hybridus	Israel	Schonfeld *et al.*, 1988
Chenopodium album	New Zealand	Rahman *et al.*, 1983
Conyza canadensis	Israel	Lior *et al.*, 1994
Plantago lagopus	Israel	Lior *et al.*, 1994
Polygonum persicaria	New Zealand	Rahman & Patterson, 1987

In most cases, triazine-resistant weed populations evolved in locations where mono-herbicide and high rates regularly applied (e.g. roadsides and orchards). In Australia (Powles and Howat, 1990), New Zealand (Rahman *et al.*, 1983; Rahman and Patterson, 1987) and Japan (Le Baron, 1991) however, the number of triazine-resistant weeds is much smaller compared to that found in Europe and North America, probably due to the limited use of triazine. In spite of the wide use of triazines in Central and South American countries such as Mexico and Brazil, the author is not aware of any relevant publication reporting triazine resistance evolution in these parts of the world.

ALS Inhibiting Herbicides

The worldwide use of ALS inhibiting herbicides has increased dramatically since their first introduction into the market in early 1980. At present, herbicides of four major groups (sulfonylureas, imizadoline, triazolopyrimidins, and pyrimidinylthiobenzoates) that inhibit the enzyme ALS are available on the market (Saari *et al.*, 1994). Low application rate, wide spectrum weed control, good crop selectivity, and low mammalian toxicity have made the ALS inhibitors an important component of the weed management in many annual and perennial crops, as well as in forestry and non-cultivated land. However, resistance to ALS inhibitors has evolved very rapidly following repeated use of sulfonylureas and imizadolines in crops and roadsides, indicating high frequency of the resistance trait in the weed populations. Resistance was confirmed in six different weed species in Australia, at least five weeds in Israel, and two weeds in Costa Rica (Table 2).

The resistance mechanism has been elucidated only in part of them as yet. Several weeds confer resistance due to enhanced degradation of the herbicide which is probably mediated by a cytochrome P450 monooxygenase catalysis (Hall *et al.*, 1994). A mutated target site, plants having a mutated enzyme, may exhibit different levels and patterns of cross resistance, depending on the type and location of the mutation (Saari, this volume; Saari *et al.*, 1994; Shaner, 1991). For example, on the whole plants basis, the resistant (R) biotype of *A. retroflexus* is highly resistant to sulfonylureas (X30-40), moderately resistant to flumetsulam (X20), and only about ten fold more resistant than the sensitive (S) biotype to imidazolinines (Sibony and Rubin, 1996).

Recently, R. populations of the parasitic weed *Cuscuta campestris* were detected while growing on R plants of *A. blitoides* (Sibony and Rubin, 1996). Our preliminary results suggest that the observed resistance is due to insensitive ALS. Additional weed species suspected as resistant to ALS inhibitors in Israel are now under investigation (Rubin and Sibony, unpublished data).

Bipyridyls

The bipyridyls, paraquat and diquat are non-selective, pre-emergent herbicides, which act rapidly and control a wide range of weeds, but lack any significant soil activity. Although widely used all over the world for almost two decades, the reported number of resistance to these herbicides is limited. In spite of the apparent absence of selection pressure, paraquat resistance was detected in several countries, mainly in perennial crops, orchards and plantations where repeated application of herbicides is practiced. In Australia, paraquat resistance was confirmed in *Hordeum glaucum* (Powles, 1986) and *H. leporinum* (Tucker and Powles, 1991), *Arototheca calendula* (Powles *et al.*,1989), and *Vulpia bromoides* (Purba *et al.*, 1993). In banana plantations in the Caribbean Islands, *Commelina elegans* was no longer controlled by the recommended rate of paraquat forcing the farmers to use alternatives suchs as asulam (Parham, 1991, personal communication). In Egypt (LeBaron, 1991), Japan (Matsunaka and Itoh, 1991), and Malaysia (Itoh *et al.*, 1992), several populations of *Erygeron* spp. (=*Conyza* spp) conferring various degrees of resistance to paraquat have evolved. In addition, in Malaysia paraquat-resistant biotypes of *Solanum nigrum, Crassocephalum crepoidiodes, Amaranthus lividus* and *Conyza sumatrensis*, were discovered in potato and sweet pear fields thar were treated twice annually with paraquat (Itoh *et al.*,

Table 2. Occurrence and distribution of weed populations resistant to ALS inhibitors, grasses and broad leaves, outside North America and Europe.

Weed species	Selecting herbicide	Location	Resistance Mechanism	Reference
*Amaranthus blitoides***	Sulfumeturon	Israel	Resistant ALS	Rubin *et al.*, 1992
Amaranthus retroflexus	Sulfumeturon	Israel	Resistant ALS	Rubin *et al.*, 1992
Brasatoa tournefortii	Chlorsulfuron	Australia	Resistant ALS	Preston, 1995*
*Conyza bonaerensis***	Sulfumetorum	Israel	Unknown	Lior et al., 1994
Conyza canadensis	Sulfumeturon	Israel	Unknown	Lior *et al.*, 1994
Cuscuta campestris	Sulfumeturon	Israel	Resistant ALS	Sibony *et al.*, 1995
Cyperus difformis	Bensulfuron	Australia	Unknown	Preston, 1995*
Eleusine indica	Imazapyr	Costa Rica	Unknown	Valverde *et al.*, 1993
Ixophorus unisetum	Imazapyr	Costa Rica	Unknown	Valverde *et al.*, 1993
*Lolium rigidum***	Chlorsulfuron	Australia	Resistant ALS	Christopher *et al.*,1992
*Lolium rigidum***	Diclofop	Australia	Enhanced Metabolism	Burnet *el al.*, 1994
Polygonum convolvulus	Chlorsulfuron	Australia	Unknown	Adkins *et al.*, 1996
Sisymbrim orientale	??	Australia	Unknown	Preston, 1995*
Sonchus oleraceus	Chlorsulfuron	Australia	Resistant ALS	Preston, 1995*

* Personal communication; ** Multiple resistant populations.

1992). Paraquat resistant *Youngia japonica* biotype was also found in mulberry fields in Japan (Matsunaka and Itoh, 1991), and *Parthenium hysterophorus* in Kenya (LeBaron, 1991). It should be noted that in most cases, paraquat-resistant biotypes were much less resistant to diquat (Dodge, 1991). Preston (1994), recently reviewed in detail the mechanism proposed for paraquat (and diquat) resistance. Limited paraquat translocation in resistant biotypes as compared to the wild type, as well as its sequestration away from its target site, leading to reduced uptake into leaf cells were considered as a possible mechanism of resistance. In addition, an increased activity of enzymes involved in detoxification of the active oxygen species produced by the herbicide, was found in some paraquat-resistant plants. The possibility that more than one resistance mechanism acts simultaneously at the same plant, is not ruled out (Preston, 1994).

ACCase Inhibitors

The worldwide situation in resistance to acetyl coenzyme A carboxylase (ACCase) inhibiting herbicides was recently reviewed (Devine and Shimabukuro, 1994). Resistance was first found in Australia in 1982, in *Lolium rigidum* after four years of diclofop-methyl use in wheat (Heap and Knight, 1982). Since then, large numbers of farms and enormous acreage were reported to be infested with biotypes of *L. rigidum* in Australia. Many R populations exhibit various degrees and patterns of cross- and multiple-resistance to various herbicides, hence causing serious practical difficulties in weed management. Several mechanisms proposed as responsible for the resistance were reviewed by Hall and her co-workers (1994). Altered target site (insensitive ACCase), which bestow different patterns of cross resistance to many aryloxyphenoxipropanoate (APP) and CHD (cyclohexadione) herbicides was found in *L. rigidum* populations that were exposed for several years to annual applications of either diclofop-methyl or sethoxydim. Enhanced metabolism of ACCse inhibiting herbicides is considered to be of minor importance, unlike that of the substituted ureas and ALS inhibitors. Diclofop as well as other APPs (and to lower extent CHDs), causes depolarization of the plasma membrane potential in both resistant and susceptible *L. rigidum*, in unknown mechanism. However, some R biotypes are able to restore the transmembrane potentials following removal of the APP or CHD herbicide. The link of this phenomenon the so-called membrane recovery response, to plant resistance, is not completely clear, and is associated mainly with plants which do not confer resistance due to insensitive ACCase. (Devine and Shimabukuro, 1994; Hall *et al.*, 1994).

Diclofop-methyl resistance was also detected in *Avena fatua* in an Australian wheat-lupine field that was treated annually six times with the herbicide (Mansooji *et al.*, 1992). A biotype of *A. sterilis* that exhibited an outstanding resistance to a wide range of APP and CHD herbicides, was found in a clover field in Australiain 1989. This biotype was not controlled by two consecutive haloxyfop treatments (Mansooji *et al.*, 1992). For eight years (1981- 1988), this field was under crop rotation of wheat, legume and canola as well as herbicide rotation of which one was triallate and one glyphosate. Although both *Avena* biotypes were exposed to a similar number of previous treatments with APPs, only *A.*

sterilis confers insensitive ACCase (Maneechote *et al.*, 1994), indicating that different mechanisms might be responsible for the resistance in each *Avena* population.

Recently (Zarka and Rubin, 1996), a resistant population of *Phalaris minor* was discovered, in a wheat monoculture (more than 20 years) field in Israel. The field was treated once a year with diclofop (1976-1988), fenoxaprop+safener (1989-1991) and tralkoxydim (1992). In spite of the fact that diclofop was repeatedly applied for 12 years, the eruption of resistance occurred only after fenoxaprop was used. Furthermore, as compared to the susceptible population, the R biotype was 30 times more resistant to fenoxaprop and only 3 time more resistant to diclofop. No cross resistance was observed with APPs such as haloxyfop, fluazifop and quizalofop, whereas some resistance was detected to sethoxydim but not to tralkoxydim. The infestation of *P. minor* has increased with time in the original location, but no further spread was so far reported. The mechanism for this resistance has not yet been elucidated.

Several populations of *E. colona* were detected in irrigated rice fields from the Central Pacific region of Costa Rica that have shown resistance to fenoxaprop following six years of use (Riches *et al.*, 1996; Valverde *et al.*, 1996). Although the resistance mechanism is still under study, the selection pressure imposed by the herbicide, and the pattern of cross resistance to other APPs and CHDs suggest that the R biotype confers resistant ACCase.

Resistance to ACCase inhibiting herbicides was also detected in *Eleusine indica* in Malaysia in 1989, following repeated application of fluazifop, which confers resistance mainly to APPs and much lower to CHDs (Leach *et al.*, 1993; Marshall *et al.*, 1994). ACCase isolated from the R biotypes exhibited high resistance to fluazifop, fenoxaprop and sethoxydim with I_{50} of >500, 25 and 77 mM, respectively, but not to clethodim (Leach *et al.*, 1995). Based on observation made on the growth and survival of F1 populations, Marshall and his co-workers (1994) concluded that the trait for graminicide resistance is not strongly heritable in *Eleusine indica* and its spread is unlikely to develop.

Enhanced Degradadation

Wheat is grown in North-West India (Haryana, Pinjab and Westerns Uttar Pradesh), mainly as monoculture or in wheat-rice rotation. *Phalaris minor* is the major annual grass weed that infests wheat in this area, where phenylurea herbicides such as isoproturon, methabenzothiazuron and metoxuron were used (Malik and Singh, 1993). Repeated use of isoproturon for 15 years resulted in the evolution of several *P. minor* populations which were resistant to the recommended rate of herbicide (Singh and Malik, 1993). Some of the populations exhibited cross-resistance to diclofop-methyl (Malik and Singh, 1995). The nature of photosynthesis inhibitions and their differential recovery among biotypes indicates that resistance may be due to enhanced metabolism of isoproturon (Singh *et al.*, 1995, 1996). Further studies are needed, but it seems likely that, unless special measures are employed, we will face a new case of multiple resistance similar to that developed in *Alopecurus myosoroides* (Moss, this volume) and *Lolium rigidum* (Tardil *et al.*, this volume).

Multiple resistance is evident when a population confers resistance to several chemicals acting at different sites or detoxified at different enzymes (Rubin, 1991; Moss and Rubin, 1993). Intensive studies have been carried out by Powles and his research group in Australia on the nature and significance of multiple resistance in *L. rigidum* (Powles, *et al.*, this volume) to various combinations of APPs, CHDs, ALS inhibitors, dinitoanilines, substituted ureas and carbamates (Hall *et al.*, 1994).

Propanil and thiobencarb (=benthiocarb) are frequently used for the control of *Echinochloa colona* and *E. crus-galli*, the major weeds in rice production worldwide. In many rice fields in Colombia (Fischer *et al.*, 1993), and Costa Rica (Valverde *et al.*, 1996), where propanil has been heavily used for decades, *E. Colona* was inconsistently controlled by the recommended rates of propanil. Propanil-resistance in *E. colona* was attributed mainly to rapid detoxification of the herbicide via hydrolysis to the 2,4-dichloroaniline product, catalyzed by aryl acylamidase. During the early growth of the weed (2 to 6 leaves), the enzyme activity extracted from the R biotypes as well as the herbicide metabolism, are much higher than that found in the S biotypes (Leah *et al., 1996).* As mentioned above, the same group has identified other population of *E. colona* which may confer ACCase-resistance as well (Riches *et al.*, 1996).

Resistance to butachlor and thiobencarb was detected in *E. crus-galli* in China (Huang *et al.*, 1994). It was suggested that resistance to butachlor and thiobencarb is due to an increased activity of amid hydrolase and MFO (mixed function oxidases), respectively. When appropriate inhibitors were used in mixture with the herbicide, the resistance was abolished. Thiobencarb resistance in *E. crus-galli* was recently reported in rice production in Egypt.

Concluding Remarks

The observed increase in the number of weed populations conferring resistance to herbicides is closely related to the extent of chemical weed control and lack of tillage, mono-herbicide and monoculture. In order to prevent further spread of this detrimental phenomenon in the region, there is a need for management practices attainable by small farmers. In spite of the increase in awareness to the problem throughout the world, strategy planning and action groups to tackle the problem were established only in the developed part of the world. In the region covered in this article, (apart from Australia), there are no known reports of official programs that attempt to combat resistance. In most countries, we have no estimate of the economic damage caused by herbicide resistance and its magnitude. The role that governments, the scientific community and the industry can and should play in educating the end users, has not yet been accomplished. The apparent trend to introduce herbicide-resistant crops to combat troublesome weeds (e.g, parasitic weeds), will inevitably lead to the evolution of more herbicide-resistant species, particularly in the developing part of the world. To better combat herbicide-resistant weeds, new and old ideas regarding breeding, tillage, herbicide mixtures and rotations of crops and herbicides should be re-examined and fitted to practical and operative weed management practices.

Acknowledgements

I would like to thank Drs. C. Preston, S. Singh and T. Yaacoby for providing me with unpublished data.

References

Adkins, S.W., Boersma, M., Walker, S.R. and Robinson, G. (1996) Herbicide-resistant weeds from the North-East Belt of Australia. *In: Proceedings International Symposium on Weed and Crop Resistance to Herbicides* (R. De Prado, J. Jorrín, L. Garcia-Torres and G. Marshall, eds.), 11-13. University of Córdoba, Córdoba, Spain.

Bourdot, G. and Harvey, I. (1992) A potential mycoherbicide for the control of phenoxy- herbicide resistant giant buttercup (*Ranunclus acris*) in dairy pastures. *In: Proceedings 1st. International Weed Control Congress* (R. G. Richardson, ed.), 94. Weed Science Society, Victoria, Melbourne.

Burnet, M.W., Hart, Q., Holtum, J.A.M. and Powles, S.B. (1994) Resistance to nine herbicide classes in a population of rigid ryegrass (*Lolium rigidum*). *Weed Science* **42**: 369-377.

Christopher, J.T., Powles, S.B. and Holtum, J.A.M. (1992) Resistance to acetolactate synthase inhibiting herbicides in annual ryegrass (*Lolium rigidum*) involves at least two mechanism. *Plant Physiology* **100**: 1909-1913.

Cohen, M. and Nir, A. (1979) Escape of grass weeds following triazine herbicides treatment. *Proceedings 7th Conference Weed Science Society*, 14. Israel, Rehovot (in Hebrew).

Devine, M. D. and Shimabukuro, R.H. (1994) Resistance to acetyl coenzyme A carboxylase inhibiting hebicide. In: *Herbicide Resistance in Plants: Biology and Biochemistry* (S.B. Powles and J.A.M. Holtum, eds.), 141-169. Lewis Published, CRC Press Inc. Boca Raton.

Dodge, A. D. (1991) Mechanism of paraquat tolerance. In: *Herbicide Resistance in Weeds and Crops* (J.C. Caseley, G. Cussans and R. Atkin, eds.), 165-175. Butterworths-Heinemann, UK.

Fischer, A.J., Granados, E. and Trujillo, D. (1993) Propanil resistance in populations of junglerice (*Echinochloa colona*) in Colombian rice fields. *Weed Science* **41**: 201-206.

Hall, L.M., Holtum, A.M. and Powles, S.B. (1994) Mechanism responsible for cross resistance and multiple resistance. In: *Herbicide Resistance in Plants: Biology and Biochemistry* (S.B. Powles and J.A.M. Holtum, eds.), 243-261. Lewis Publishers, CRC Press. Inc. Boca Raton.

Heap, J. and Knight, R. (1982) A population of ryegrass tolerant to the herbicide diclofop- methyl. *Journal Australian Institute Agricultural Sciences* **48**: 156-157.

Huang, B.Q., Lin, S.X., Xizo, Z.Y. and Luo, Y.P. (1994) Study on mechanism and countermeasure of the resistance barnyardgrass. *In: Proceedings 5th. Weed Science Conference,* 85-88. China, Kunming.

Itoh, K, Azmi, M. and Ahmad, A. (1992) Paraquat resistance in *Solanun nigrum, Crassocephalum crepotdiodes, Amaranthus lividus* and *Conyza sumatrensis* in Malaysia. *In: Proceedings 1st. International Weed Control Congress.* (R.G Richardson, ed.), 224-228. Weed Science Society, Victoria, Melbourne.

Leach, G.E., Devine, M.D., Kirkwood, R.C. and Marshall, G. (1995) Target enzyme-based resistance to acetyl-coenzyme A carboxylase inhibitors in *Eleusine indica. Pesticide Biochemistry Physiology* **51**:129-136.

Leach, G.E., Kirkwood, R.C. and Marshall, G. (1993) The basis of resistance displayed to fluazifop-butyl by biotypes of *Elusine indica. In: Brighton Crop Protection Conference-Weeds* **1**: 201- 206.

Leah, J.M., Caseley, J.C., Riches, C.R. and Vaverde, B. (1996) Mechanisms of propanil resistance in *Echinochloa colona. In: Proceedings International Symposium on Weed and Crop Resistance to Herbicides* (R. De Prado, J. Jorrín, L. Garcia-Torres and G. Marshall, eds.), 95-96. University of Córdoba, Córdoba, Spain.

LeBaron, H.M. (1991) Distribution and seriuosness of herbicide-resistant weeds infestations worlwide. In: *Herbicide Resistance in Weeds and Crops* (J.C. Caseley, G. Cussans and R. Atkin, eds.), 27-43. Butterworths-Heinemann, UK.

Lior, E., Kigel, H. and Rubin, B. (1994) Variation in response of *Conyza* species to atrazine, diuron and sulfometuron. *Phytoparasitica* **22**: 134.

Malik, R.K. and Singh, S. (1993) Evolving strategies for herbicide use in wheat: Resistance and integrated weeds management. *Proceedings International Symposium Indian Society Weed Science* **1**: 225-238. Hisar.

Malik, E. and Singh, S. (1995) Littleseed canarygrass (*Phalaris minor*) resistance to isoproturon in India. *Weed Technology* **9**:413-425.

Manneechote, C.J., Holtum, A.M. and Powles, S.B. (1994) Resistant acetyl-coA carboxylase is the mechanism of herbicide resistance in a biotype of *Avena steriis* ssp. *Ludiviciana. Plant Cell Physiology* **35**: 627-635.

Mansooji, A.M., Holtum, J.A., Boutssalis, P., Matthews, M. and Powles, S.B. (1992) Resistance to aryloxyphenoxypropanoate herbicides in two wild oat species (*Avena fatua and Avena sterlis* ssp. *Ludiviciana.). Weed Science* **40**: 599-605.

Marshall, G., Kirkwood, R.C. and Leach, G.E. (1994) Comparative studies on graminicide-resistant and susceptible biotypes of *Eleusine indica. Weed Research* **34**: 177-185.

Matsunaka, S. and Itoh, K. (1991) Paraquat resistance in Japan. In: *Herbicide Resistance in Weeds and Crops* (J.C. Caseley, G. Cussans and R. Atkin, eds.), 77-86. Butterwoths- Heinemann, UK.

Moss, S.R and Rubin, B. (1993) Herbicide resistant weeds: A worldwide perspective. *Journal Agricultural Sciences* **120**: 141-148.

Parham, M. (1991) ICI agrochemicals, Jealott´s Hill Research Atation, Bracknell, Berkshire, UK.

Popay, A.I., Bourdot, G.W., Harrington, K.C. and Rahman, E. (1991) Herbicide resistance in weeds in New Zealand.*In*: *Herbicide Resistance in Weeds and Crops* (J.C. Caseley, G. Cussans and R. Atkin eds.), 77-86. Butterwoths-Heinemann, UK.

Powles, S.B. (1986) Appearance of a biotype of the weed, *Hordeum glaucum* Steud., resistant to the herbicide paraquat. *Weed Research* **26**: 167-172.

Powles, S.B. and Howat, D. (1990) Herbicide resistant-weed in Australia. *Weed Technology* **4**: 178-185.

Powles, S.B., Tucker, E.S. and Morgan, T.R. (1989) A capeweed (*Arototheca calendula*) biotype in Australia resistant to bipyridyl herbicides. *Weed Science* **37**: 60-62.

Preston, C. (1994) Resistance to phosynthesis I disrupting herbicides. In: *Herbicide Resistance in Plants: Biology and Biochemistry* (S.B. Powles and J.A.M. Holtum, eds.), 61-109. Lewis Publishers, CRC Press Inc. Boca Ratón.

Preston, C. (1995) Waite Agricultural Research Institute, University od Adelaide, Glen Osmond, South Australia. Personal Communication.

Purba, E., Preston, C. and Powles, S.B. (1993) Paraquat resistance in a biotype of *Vulpia bromoides* (L.) S.F. Gray. *Weed Research* **33**: 409-413.

Rahman, A., James, T.K. and Mortimer, J. (1983) Control of atrazine-resistant fathem in maize. *In: Proceedings 36th New Zealand Weed and Pest Control Conference*, 229-232.

Rahman, A. and Patterson, T.M. (1987) *Polygonum persicaria* triazine-resistant biotype. *In: Proceedings 40th New Zealand Weed and Pest Control Conference,* 186-188.

Riches, C.R., Caseley, J.C., Valverde, B.E. and Down, V.M. (1996) Resistance of *Echinochloa colona* (L.) link to ACCase inhibiting herbicides. *In: Proceedings International Symposium on Weed and*

Crop Resistance to Herbicides (R. De Prado, J. Jorrín, L. Garcia-Torres and G. Marshall, eds.), 14-16. University of Córdoba, Córdoba, Spain.

Rubin, B. (1991) Herbicide resistance in weeds and crops, progress and prospects. *In: Herbicide Resistance in Weeds and Crops* (J. C. Caseley, G. Cussans and R. Akin, eds.), 387-414. Butterwoth-Heinemann, UK.

Rubin, B. and Sibony, M. (1995) The Hebrew University of Jerusalem. Rehovot, Israel. Unpublished data.

Rubin, B., Sibony, M. and Levy, O. (1995) The Hebrew University of Jerusalem, Rehovot, Israel. Unpublished data.

Rubin, B., Sibony, M., Benyamini, Y. and Danino, Y. (1992) Resistance to sulfonylurea herbicides in redroot pigweed (*Amaranthus retroflexus L.*). *WSSA Abstracts* **32**: 66.

Rubin, B., Yaacobi, T. and Schonfeld, M. (1985) Triazine resistant grass weeds: Cross resistance with wheat herbicide, a possible threat to cereal crops. *British Crop Protection Conference-Weeds* **3**: 1171-1178.

Saari, L.L., Cotterman, J.C. and Thill, D.C. (1994) Resistance to acetolactate synthase inhibiting herbicides. In: *Herbicide Resistance in Plants: Biology and Biochemistry* (S. B. Powles and J.A.M. Holtum, eds.), 83-139. Lewis Published, CRC Press Inc. Boca Raton.

Schonfeld, M., Yaacoby, T., Benyamini, Y., Michael, O. and Rubin, B. (1988) Phosynthetic capacity of triazine resistant weeds. *EWRS Symposium Factors Affecting Herbicidal Activity and Selectivity*, 361-366. Wageningen.

Shaner, D.L. (1991) Mechanisms of resistance to acetolactate synthase/ acetohydroxyacid synthase inhibitors. *In: Herbicide Resistance in Weeds and Crops* (J.C. Caseley, G. Cussans, and R. Atkin, eds.), 387-414. Butterworths-Heinemann, UK.

Sibony, M., Lior, E., Fisher, A. and Rubin, B. (1995) Resistance of *Amaranthus* spp., horseweed (*Conyza canadensis*) and field dodder (*Cuscuta campestris*) to acetolactate synthase (ALS) inhibitors. *WSSA Abstracts* **35**: 87.

Sibony, M. and Rubin, B. (1996) Multiple resistance to ALS inhibitors and triazine herbicides in *Amaranthus* species. *In: Proceedings International Symposium on Weed and Crop Resistance to Herbicides* (R. De Prado, J. Jorrín, L. Garcia-Torres and G. Marshall, eds.), 112-114. University of Córdoba, Córdoba, Spain.

Singh, S. and Malik, R.K. (1993) Resistance of some *Phalaris minor* Retz. biotypes to isoproturon but not to pemdimethalin. *Proceedings Weed Science Indian Society* **2**: 125- 130. Hisar.

Singh, S., Kirkwood, R.C. and Marshall, G. (1995) Evaluation of isoproturon-resistant *Phalaris minor* against diclofop-methyl and fenoxaprop-P-ethyl. Test of Agrochemicals and Cultivars. *Annals Applied Biology* **125** (Suppl.).

Singh, S., Kirkwood, R.C. and Marshall, G. (1996) The effect of isoproturon on phtosynthesis of resistant and susceptible biotypes of *Phalaris minor* Retz. *In: Proceedings International Symposium on Weed and Crop Resistance to Herbicides* (R. De Prado, J. Jorrín, L. Garcia-Torres and G. Marshall, eds.), 115-117. University of Córdoba, Córdoba, Spain.

Tucker, E.S. and Powles, S.B. (1991) A biotype of hare barley (*Hordeum leporinum*) resistant to parquat and diquat. *Weed Science* **39**: 159-162.

Valverde, B.E., Chaves, L., González, J. and Garita, I. (1993) Field-evolved imazapyr resistance in*Ixophorus unisetum* and *Eleusine indica* in Costa Rica. *Brighton Crop Protection Conference Weeds* **3**: 1189-1194.

Valverde, B.E., Garita, I., Riches, C.R., Caseley, J.C. and Chacón, L.A. (1996) Integrated management of propanil-resistant *Echinochloa colona* in dryland rice in Costa Rica. *In: Proceedings International Symposium on Weed and Crop Resistance to Herbicides* (R. De Prado, J. Jorrín, L. Garcia-Torres and G. Marshall, eds.), 264-265. University of Córdoba, Córdoba, Spain.

Yaacoby, T. (1983) Ministry of Agriculture, Plant Protection and Inspection Services, Bet- Dagan, Israel. Unpublished data.

Yaacoby, T., Schonfeld, M. and Rubin, B. (1986) Characteristics of atrazine resistant biotypes of three grass weeds. *Weed Science* **34**: 181-185.

Zarka, S. and Rubin, B. (1996) Evolution of fenoxaprop-resistant *Phalaris minor* population in wheat monoculture. *In: Proceedings International Symposium on Weed and Crop Resistance to Herbicides* (R. De Prado, J. Jorrín, L. Garcia-Torres and G. Marshall, eds.), 39-41. University of Córdoba, Córdoba, Spain.

PART II

HERBICIDE RESISTANCE MECHANISMS

CHAPTER 5
RESISTANCE TO PS II INHIBITOR HERBICIDES

J. W. GRONWALD
Plant Science Research Unit, USDA-ARS and Department of Agronomy and Plant Genetics, University of Minnesota, St. Paul, Minnesota 55108 USA

Introduction

In 1970, Ryan (Ryan, 1970) reported that a population of *Senecio vulgaris* (L.) was not controlled by recommended rates of simazine in a conifer nursery in Washington state (USA). Since this first report, resistance to photosystem II (PS II) inhibitor herbicides has become widespread. In terms of numbers of species, triazine resistance is the most prevalent type of herbicide resistance found in weeds (Holt and LeBaron, 1990; LeBaron, 1991; LeBaron and MacFarland, 1990). The most recent worldwide survey of herbicide resistance, which was conducted in 1989, indicated that there were 57 weed species with biotypes resistant to triazine herbicides (LeBaron, 1991). In the majority of cases, resistance to PS II-inhibiting herbicides is due to a modification at the target site, the D1 protein of the PS II reaction center (Trebst, 1991). Since the mid-1980's, there have been increasing reports of resistance to PS II inhibitors due to enhanced herbicide detoxification (Burnet *et al.*, 1993a,b; Gronwald *et al.*, 1989; Hoagland *et al.*, this volume; Kemp *et al.*, 1990; Leah *et al.*, 1994).

Resistance Due to Altered Target Site

It has been known for over 30 years that the mechanism of action of PS II inhibitor herbicides involves inhibition of electron transport on the reducing side of PS II (Gronwald, 1994). Our current understanding of the PS II reaction center in higher plants is based largely on knowledge gained from the elucidation of the 3-D structure of the purple photosynthetic bacterium (*Rhodopseudomonas viridis*) reaction center by X-ray crystallography (Deisenhofer and Michel, 1989). A simplified schematic of the PS II reaction center complex of higher plants is illustrated in Figure 1. Two integral membrane proteins, Dl and D2, form the reaction center. Associated with the D1/D2 heterodimer are bound redox components that are involved in electron transfer. The reaction center catalyzes the light-dependent oxidation of water and the transfer of electrons to Q_B, an exchangeable plastoquinone that binds in a niche on the Dl protein. PS II inhibitor herbicides, such as atrazine, are herbicidal because they have structural determinants that allow them to bind in the Q_B-binding niche, displacing plastoquinone, and thereby blocking electron transfer from Q_A to Q_B (Gronwald, 1994; Trebst, 1991).

Research conducted with triazine-resistant weeds during the 1980's clearly established that target site-based resistance is due to a mutation of the chloroplast *psbA* gene that encodes the Dl protein (Trebst, 1991). In all cases of target site resistance in weeds where the *psbA* gene has been sequenced, resistance is due to a point mutation resulting in the substitution of Gly for Ser at residue 264 (Gronwald, 1994; Trebst, 1991). This modification greatly reduces the affinity of atrazine at the Q_B-binding site since binding affinity is strongly dependent on hydrogen bonding with the hydroxyl side chain of Ser 264 (Trebst, 1991). The Ser 264 to Gly mutation does not confer cross-resistance to diuron, a urea herbicide, but leads to negative cross-resistance (i.e. increased sensitivity) to dinoseb, a phenol herbicide (Trebst, 1991). Other mutations of the *psbA* gene conferring resistance to PS II herbicides have been selected in algae, *cyanobacterium*, and cell cultures of higher plants. Eight different amino acid substitutions that confer resistance have been identified in or near the Q_B-binding niche (Gronwald 1994; Trebst 1991). Depending on the residue altered, different levels and patterns of cross resistance to PS II inhibitor herbicides are observed.

There is considerable evidence that the Ser 264 to Gly mutation of the *psbA* gene reduces plant fitness. Triazine-resistant biotypes exhibit reduced carbon dioxide fixation, quantum yield, and seed and biomass production (Holt and Thill, 1994).

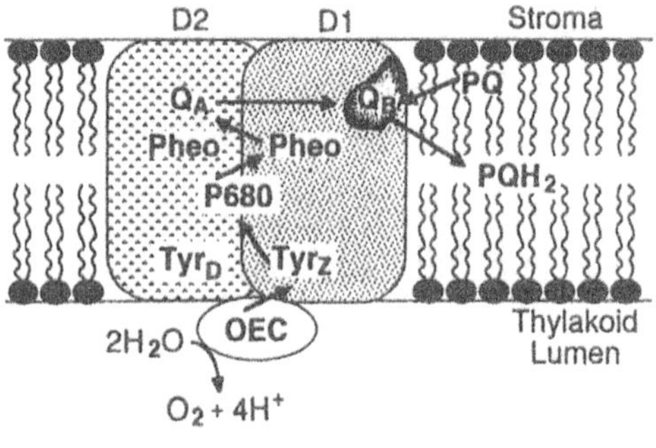

Figure 1: Simplified schematic of the photosystem II reaction center complex (Gronwald, 1994). Arrows indicate the direction of electron flow: Abbreviations: OEC, oxygen-evolving complex; Tyr_Z, tyrosine residue 161 that acts as electron donor to P680; P680, reaction center chlorophyll a dimer; Pheo, pheophytin; Q_A. plastoquinone tightly bound to the D2 protein; Q_B, exchangeable plastoquine bound to the D1 protein; PQ, plastoquinone; PQH_2, plastohydroquinone.

However, there is a lack of agreement concerning the specific mechanism. It is well established that the Ser 264 to Gly mutation reduces the rate of electron transfer between Q_A and Q_B (Bowes et al., 1980; Gronwald, 1994; Jansen and Pfister, 1990; Jursinic and Pearcy, 1988), but it is not clear how this relates to the reduction in photosynthetic capacity and relative fitness. There are conflicting reports as to whether the Ser 264 to Gly mutation reduces whole chain electron transport rate in isolated chloroplasts (Gronwald, 1994; Jansen and Pfister, 1990; Jursinic and Pearcy, 1988).

It has also been proposed that the altered kinetics of Q_A to Q_B electron transfer associated with the Ser 264 to Gly mutation increases the susceptibility of PS II to photoinhibition (i.e. photodamage) (Hart and Stemler, 1990). The Dl protein normally undergoes a continuous cycle of degradation and resynthesis in the light because of photodamage that occurs during normal operation of the PS II reaction center (Barber and Andersson, 1992; Kyle, 1987). It is postulated that the slower rate of electron transfer from Q_A to Q_B. due to the Ser 264 to Gly mutation, increases radical formation in the Q_B-binding niche which in turn increases the turnover of the D1 protein (Hart and Stemler, 1990). Increased susceptibility to photoinhibition would reduce fitness, particularly under high light environments, because of the high energy cost associated with turnover of the D1 protein (Barber and Andersson, 1992).

Resistance Due to Enhanced Detoxification

Since the mid-1980's, there have been increasing reports of resistance to PS II inhibitor herbicides in weeds due to enhanced herbicide detoxification (Burnet *et al.*, 1993a,b; Gronwald 1994; Kemp *et al.*, 1990; Leah *et al.*, 1994; Hoagland *et al.*, this volume). In most cases, it appears that the resistant weeds are mimicking the mechanism that confers herbicide resistance to the crop. In cotton and tobacco, resistance to the PS II inhibitor bromoxynil was engineered through transformation with a bacterial enzyme (nitralase) that detoxifies the herbicide (Gronwald, 1994; Stalker *et al.*, 1988).

Glutathione S-transferase

In 1984, an *Abutilon theophrasti* Medic. biotype resistant to simazine and atrazine was found in a maize field in Maryland (USA) where these herbicides had been applied for several years (Andersen and Gronwald, 1987). Compared to an atrazine-susceptible *Abutilon* biotype from Minnesota, the Maryland biotype is tenfold more resistant to atrazine (Gronwald *et al.*, 1989). Atrazine-resistant biotypes of *Abutilon theophrasti* have also been found in Wisconsin (USA) (Gray *et al.*, 1995). For both the Maryland and Wisconsin biotypes, resistance is due to an enhanced capacity to detoxify atrazine via glutathione conjugation in leaves and stems (Gray *et al.*, 1995; Gronwald *et al.*, 1989). In the case of the atrazine-resistant Maryland biotype, the enhanced rate of conjugation is due to an increase in activity of two glutathione S-transferase isozymes that exhibit activity with atrazine (Anderson and Gronwald, 1991). This trait is controlled by a single nuclear gene exhibiting partial dominance (Anderson and Gronwald, 1987).

Cytochrome P450

Substituted ureas have been used for over twenty years for selective weed control in the wheat-growing regions of Europe (Moss and Cussans, 1991). During the past decade, there have been increasing reports of chlortoluron resistance in biotypes of *Alopecurus myosuroides* Huds. This type of resistance has been found in wheat-growing regions of the U.K. (Moss and Cussans, 1991), Germany (Niemann and Pestemer 1984), The Netherlands (Oorschot and Leeuwen, 1992), and Spain (Menendez *et al.*, 1994). The mechanism of resistance has been partially characterized in the biotype "Peldon" found in England. Compared to susceptible *Alopecurus myosuroides* biotypes, "Peldon" exhibits an enhanced capacity to detoxify chlortoluron via N-dealkylation and ring-methyl hydroxylation (Kemp and Caseley, 1991; Kemp *et al.*, 1990). This is the same mechanism that confers chlortoluron resistance in wheat. Experiments conducted with the cytochrome P450 inhibitor, 1-aminobenzotriazole (ABT), have provided indirect evidence that the enhanced capacity for chlortoluron metabolism is due to cytochrome P450 (Kemp *et al.*, 1990). ABT blocks the enhanced rate of chlortoluron metabolism in the resistant biotype and causes it to become susceptible. Preliminary studies suggest that chlortoluron resistance is a nuclear, polygenic trait that results in no obvious decrease in fitness (Moss and Cussans, 1991).

In the wheat-growing regions of Australia, resistance to s-triazines (simazine and atrazine) and the substituted phenylurea chlortoluron has appeared in biotypes of *Lolium rigidum* Gaud. (Powles *et al.*, 1990). The mechanism of resistance has been investigated in two biotypes, WLR 2 and VLR 69. For both biotypes, resistance to simazine and chlortoluron is due to an enhanced rate of herbicide metabolism via *N*-dealkylation (Burnet, 1992; Burnet *et al.*, 1993a,b). Indirect evidence, based largely on the synergistic effect of cytochrome P450 inhibitors on chlortoluron efficacy in the resistant biotypes, suggests that the enhanced oxidative capacity is mediated by cytochrome P450 (Burnet 1992; Burnet *et al.*, 1993a,b). However, the involvement of cytochrome P450 has yet to be demonstrated in vitro. Preliminary studies suggest that resistance to s-triazines and substituted phenylureas in WLR 2 and VLR 69 is a polygenic, nuclear trait that results in little or no reduction in fitness (Burnet, 1992).

Aryl Acylamidase

During the past 30 years, propanil has been widely used to control *Echinochloa crus-galli* in rice-growing regions of Arkansas (USA) (Baltazar and Smith, 1994; Carey, *et al.,* 1995). In 1989, poor control of *Echinochloa crus-galli* by recommended field rates of propanil (3.4-5.6 kg/ha) was reported in one field in Arkansas (Carey *et al.,* 1995). A survey conducted in 1991-1992 indicated that propanil-resistant *Echinochloa crus-galli* biotypes were found in 16 of 38 rice-producing counties in Arkansas (Carey *et al.,* 1995). The level of resistance varied among biotypes but was less than 20-fold (Carey *et al.,* 1995). Propanil resistance in *Echinochloa crus-galli* has also been reported in rice-growing regions of Mississippi (Carey *et al.,* 1995), Louisiana (Carey *et al.,* 1995), Texas (Carey *et al.,* 1995), and Greece (Giannopolitis and Vassiliou, 1989). The mechanism of propanil resistance has been characterized in one biotype of *Echinochloa crus-galli* found in Arkansas. In this biotype, resistance is due to enhanced activity of aryl acylamidase, an enzyme that catalyzes the metabolism of propanil to 3,4-dichloroaniline, and the same enzyme that confers propanil resistance in rice (Hoagland *et al.*, this volume). Additional evidence that resistance is due to enhanced aryl acylamidase activity is the synergism observed when the insecticide carbaryl (l-naphthyl-N-methylcarbamate), an inhibitor of aryl acylamidase, is applied to the resistant biotype in combination with propanil (Hoagland *et al.*, this volume) .

Propanil resistance has also appeared in biotypes of *Echinochloa colona* (L.) Link found in Texas, USA (Carey *et al.*, 1995), Costa Rica (Garro *et al.*, 1991), and Columbia, South America (Fischer *et al.*, 1993). As expected, resistance was found in rice fields with a long history of propanil use. The level of resistance, which varies among biotypes, is in the range of two to eight-fold. Propanil resistance for one *Echinochloa colona* biotype found in Columbia, South America, is due to an enhanced capacity for detoxification via aryl acylamidase (Leah *et al.*, 1994). Preliminary studies indicated a trend toward lower reproductive fitness in propanil-resistant *Echinochloa colona* biotypes (Fischer *et al.*, 1993).

Nitralase

The tools of biotechnology have been used to confer resistance to bromoxynil in tobacco and cotton (Stalker *et al.*, 1988). A gene encoding nitralase, an enzyme that detoxifies bromoxyml, was isolated and cloned from a soil-borne microorganism (*Klebsiella ozaenae*). The cloned gene was put under the control of a plant promoter and used to transform tobacco and cotton (Gronwald, 1994; Stalker *et al.*, 1988). Bromoxynil-resistant cotton, developed by Calgene, Inc., Davis, CA (USA), was commercially available in 1995.

References

Andersen, R.N. and Gronwald, J.W. (1987) Noncytoplasmic inheritance of atrazine tolerance in velvetleaf (*Abutilon theophrasti*). *Weed Science* **35**: 496-498.

Anderson, M.P. and Gronwald, J.W. (1991) Atrazine resistance in a velvetleaf (*Abutilon theophrast*) biotype due to enhanced glutathione S-transferase activity. *Plant Physio*logy **96**: 104- 109.

Baltazar, A.M. and Smith, R.J. Jr. (1994) Propanil-resistant barnyardgrass (*Echinochloa crus-galli*) control in rice (*Oryza sativa*). *Weed Technology* **8**: 576-581.

Barber, J. and Andersson, B. (1992) Too much of a good thing: Light can be bad for photosynthesis. *TIBS* **17**: 61-66.

Bowes, J., Crofts, A.R. and Arntzen, C.J. (1980) Redox reactions on the reducing side of photosystem II in chloroplasts with altered herbicide binding properties. *Archives Biochemistry Biophysic* **200**: 303-308.

Burnet, M.W.M. (1992) Mechanisms of herbicide resistance in *Lolium rigidum. Ph.D. Thesis*, University of Adelaide, Adelaide, Australia.

Burnet, M.W.M., Loveys, B.R., Holtum, J.A.M. and Powles, S.B. (1993a) Increased detoxification is a mechanism of simazine resistance in *Lolium rigidum. Pesticide Biochemistry Physiology* **46**: 207-218.

Burnet, M.W.M., Loveys, B.R., Holtum, J.A.M. and Powles, S.B. (1993b) A mechanism of chlorotoluron resistance in *Lolium rigidum. Planta* **190**: 182-189.

Carey III, V.F, Hoagland, R.E. and Talbert, R.E. (1995) Verification and distribution of propanil-resistant barnyardgrass [*Echinochloa crus-galli* (L.) Beauv.] in Arkansas. *Weed Technology* **9**: 366-372.

Deisenhofer, J. and Michel, H. (1989) The photosynthetic reaction center from the purple bacterium *Rhodopseudomonas viridis*. *Science* **245**: 1463-1473.

Fischer, A.J., Granados, E. and Trujillo, D. (1993) Propanil tolerance in populations of junglerice (*Echinochloa colona*) in Columbian rice fields. *Weed Science* **41**: 201-206.

Garro, J.E., de la Cruz, R. and Shannon, P.J. (1991) Propanil resistance in *Echinochloa colona* populations with different herbicide use histories. In: *Brighton Crop Protechon Conference -Weeds* **8A**-11:1079-1083.

Giannopolitis, C.N. and Vassiliou G. (1989) Propanil tolerance in *Echinochloa crus-galli* (L.) Beauv. *Tropical Pest Management* **35**: 6-7.

Gray, J.A., Balke, N.E. and Stoltenberg, D.E. (1995) Enhanced detoxification is the mechanism of atrazine resistance in a Wisconsin velvetleaf *(Abutilon theophrasti* Medic.) biotype. *WSSA Abstracts* **35**: 66.

Gronwald, J.W. (1994) Resistance to photosystem II inhibiting herbicides. *In: Herbicide Resistance in Plants: Biology and Biochemistry* (S.B. Powles and J.A.M. Holtum, eds.), 27-60. Lewis Publishers, Boca Raton, FL.

Gronwald, J.W., Andersen, R.N. and Yee, C. (1989) Atrazine resistance in velvetleaf (*Abutilon theophrasti*) due to enhanced atrazine detoxification. *Pesticide Biochemistry Physiology* **34**: 149-163.

Hart, J.J. and Stemler, A. (1990) High light-induced reduction and low light-enhanced recovery of photon yield in triazine-resistant *Brassica napus* L. *Plant Physiology* **94**: 1301-1307.

Holt, J.S. and LeBaron, H.M. (1990) Significance and distribution of herbicide resistance. *Weed Technology* **4**: 141-149.

Holt, J.S. and Thill, D.C. (1994) Growht and productivity of resistant plants. *In: Herbicide Resistance in Plants: Biology and Biochemistry* (S.B. Powles and J.A.M. Holtum, eds.), 299-316. Lewis Publishers, Boca Raton, FL.

Jansen, M.A.K. and Pfister, K. (1990) Conserved kinetics at the reducing side of reaction-center II in photosynthetic organisms: Changed kinetics in triazine-resistant weeds. *Zeischrift Naturforschung* **45c**: 441-445.

Jursinic, P.A. and Pearcy, R.W. (1988) Determination of the rate limiting step for photosynthesis in a nearly isonuclear rapeseed (*Brassica napus* L.) biotype resistant to atrazine. *Plant Physiology* **88**: 1195-1200.

Kemp, M.S. and Caseley, J.C. (1991) Synergists to combat herbicide resistance. *In: Herbicide Resistance in Weeds and Crops* (J.C. Casely, G.W. Cussans and R.K. Atkin eds.), 273-292. Butterworth-Heinemann, Oxford.

Kemp, M.S., Moss, S.R. and Thomas, T.H. (1990) Herbicide resistance in *Alopecurus mysoruoides. In: Managing Resistance to Agrochemicals: From Fundamental Research to Practical Strategies* (M.B. Green, H.M. LeBaron and W.K. Moberg, eds.), 376-393. American Chemical Society Symposium Series 421, Washington, D.C.

Kyle. D.J. (1987) The biochemical basis for photoinhibition of photosystem II. *In: Topics in Photosynthesis-Vol. 9, Photoinhibition-* (D.J. Kyle, C.B. Osmond and C.J. Artnzen, eds.), 197-226. Elsevier, Amsterdam.

Leah, J.M., Caseley, J.C., Riches, C.R. and Valverde, B. (1994) Association between elevated activity of aryl acylamidase and propanil resistance in jungle-rice, *Echinochloa colona. Pesticide Science* **42**: 271-289.

LeBaron, H.M. (1991) Distribution and seriousness of herbicide-resistant weed infestations worldwide. *In: Herbicide Resistance in Weeds and Crops* (J.C. Casely, G.W. Cussans and R.K. Atkin, eds.), 27-43. Butterworth-Heinemann, Oxford.

LeBaron, H.M. and McFarland, J. (1990) Herbicide resistance in weeds and crops: An overview and prognosis. *In: Managing Resistance to Agrochemicals: From Fundamental Research to Practical Strategies* (M.B. Green, H.M. LeBaron and W.K. Moberg, eds.), 336-352. American Chemical Society Symposium Series 421, Washington, D.C.

Menendez, J., Jorrin, J., Romera, E. and De Prado, R. (1994) Resistance to chlorotoluron of a slender foxtail (*Alopecurus myosuroides*) biotype. *Weed Science* **42**: 340-344.

Moss, S.R. and Cussans, G.W. (1991) The development of herbicide-resistant populations of *Alopecurus myosuroides* (black-grass) in England. *In: Herbicide Resistance in Weeds and Crops* (J.C. Casely, G.W. Cussans and R.K. Atkin, eds.), 45-55. Butterworth-Heinemann, Oxford.

Niemann, P. and Pestemer, W. (1984) Resistenz verschiedener herkunfte von ackerfuchsschwanz (*Alopecurus myosuroides*) gegenuber herbizidbehandlungen. *Nachrichtenblatt Deutschen Pflanzenschutzdienstes* **36**: 113-118.

Oorschot, J.L.P. and van Leeuwen, P.H. (1992). Use of fluorescence induction to diagnose resistance of *Alopecurus myosuroides* Huds. (black-grass) to chlortotoluron. *Weed Research* **32**: 473-482.

Powles, S.B., Holtum, J.A.M., Matthews, J.M. and Liljegren. D.R. (1990) Herbicide cross-resistance in annual ryegrass (*Lolium rigidum* Gaud): The search for a mechanism. *In: Managing Resistance to Agrochemicals: From Fundamental Research to Practical Strategies* (M.B. Green, H.M. LeBaron and W.K. Moberg, eds.), 394-406. American Chemical Society Symposium Series 421, Washington, D.C.

Ryan, G.F. (1970) Resistance of common groundsel to simazine and atrazine. *Weed Science* **18**: 614-616.

Stalker, D.M., McBride, K.E. and Malyj, L.D. (1988) Herbicide resistance in transgenic plants expressing a bacterial detoxif1cation gene. *Science* **242**: 419-423.

Trebst, A. (1991) The molecular basis of resistance of photosystem II herbicides. *In: Herbicide Resistance in Weeds and Crops* (J.C. Casely, G.W. Cussans and R.K. Atkin, eds.), 145-164. Butterworth-Heinemann, Oxford.

Product names are necessary to report factually on available data; however, the USDA neither guarantees nor warrants the standard of the product to the exclusion of others that may also be suitable.

CHAPTER 6
TARGET-SITE BASED RESISTANCE TO ACCASE INHIBITORS

M.D. DEVINE
Department of Crop Science and Plant Ecology, University of Saskatchewan, Saskatoon, Saskatchewan, Canada S7N OWO

Introduction

The enzyme acetyl-coenzyme A carboxylase (ACCase; E.C. 6.4.1.2) is the target site of two major groups of synthetic herbicides, the aryloxyphenoxypropanoates (AOPP) and cyclohexanodiones (CHD). ACCase is a key enzyme in the biosynthesis of acyl lipids, catalyzing the addition of HCO_3^- to acetate to form malonate, a precursor of fatty acid biosynthesis. This is a two-step reaction involving the carboxylation of a biotin group on the enzyme, followed by a carboxytransfer in which the carboxy group is transferred to acetate (in the form of acetyl-CoA) to form malonyl-CoA (Scheme 1).

$$\text{E-biotin} + HCO_3^- + \text{ATP} \rightarrow \text{E-biotin-}CO_2 + \text{ADP}$$
$$\text{E-biotin-}CO_2 + \text{acetyl-CoA} \rightarrow \text{E-biotin} + \text{malonyl-CoA}$$

Scheme 1. Diagramatic representation of the reaction catalyzed by ACCase. E-biotin = enzyme linked to the biotin prosthetic group. The second step (carboxytransferase) is inhibited by AOPP and CHD herbicides.

A wide variety of AOPP and CHD herbicides inhibit fatty acid biosynthesis by inhibiting ACCase (Burton *et al.*, 1987; Focke and Lichtenthaler, 1987; Rendina and Felts, 1988; Rendina *et al.*, 1988; Secor and Cseke, 1988). Kinetic analysis suggests that the herbicides interfere with the carboxytransferase reaction; inhibition with herbicides is non-competitive with respect to the three substrates MgATP, HCO_3^- and acetyl-CoA, although "nearly competitive" vs. acetyl-CoA (Burton *et al.*, 1991; Rendina *et al.*, 1990). It has been speculated that the herbicides interfere with a hydrophobic region of the protein, perhaps corresponding to the acetyl-CoA binding domain. In general, ACCase from grasses is sensitive to AOPP and CHD herbicides, with I_{50} values in the range of 0.1-5 µM. In contrast, broadleaf ACCase is resistant to these herbicides, and concentrations >50 µM are usually required for 50% inhibition in *in vitro* studies. Recent evidence suggests that dicots contain two forms of the enzyme: the prokaryotic form (herbicide-insensitive) is located in the plastids, the site of fatty acid biosynthesis, whereas the eukaryotic form (herbicide-sensitive) is extra-plastidic (Konishi and Sasaki, 1994). The function of the extra-plastidic form is unknown. Grasses contain only the eukaryotic (herbicide-sensitive) form, but it is targeted to the plastids. Multiple forms of ACCase have been identified in maize, and these apparently differ in their herbicide sensitivity (Egli *et al.*, 1993). However, there is no evidence that differential sensitivities of ACCase isoforms within an individual plant play a role in herbicide resistance in weeds.

Although most grass species contain a herbicide-sensitive form of ACCase, there are two broad exceptions. Several *Festuca* species carry an insensitive form of ACCase, making them resistant to AOPP and CHD herbicides (Catanzaro *et al.*, 1993; Stoltenberg *et al.*, 1989). The second exception is in several species of herbicide resistant grass weeds (see section ACCase-Based Resistance Mechanisms).

Resistance to ACCase Inhibitors

Resistance to ACCase inhibitors was first reported soon after these herbicides were introduced (Heap and Knight, 1982; Thai *et al.*, 1985), although resistance did not appear to be a serious concern at that time. Reports became more frequent, and by the late 1980s it was evident that resistance had developed in several major weed species in many parts of the world (Devine and Shimabukuro, 1994). Although the current status is difficult to assess, resistance to ACCase inhibitors has now been identified in at least 11 species (Table 1), and is suspected in several others. In almost all cases, resistance developed after repeated use of these herbicides in intensively managed agricultural systems. Given the high level of selection pressure and the high level of effectiveness of these herbicides, it is not altogether surprising that resistance has arisen in many common weed species.

Most of the AOPP- and CHD-resistant weed biotypes characterized to date are resistant to ACCase inhibitors, but not to herbicides with other modes of action. However, they are not equally resistant to all ACCase inhibitors, and show differences in the levels of resistance and patterns of cross-resistance to different AOPP- and CHD- herbicides. In addition, some AOPP- and CHD-resistant weed biotypes are cross resistant to other classes of herbicides. In these, resistance can be conferred by single or multiple mechanisms, including target-site mutations and other mechanisms (enhanced herbicide detoxification, etc.). Non-target-site based resistance mechanisms are discussed in more detail in other Chapters in this volume and elsewhere (e.g., Devine and Shimabukuro, 1994; Hall *et al.*, 1994).

Table 1. Weed species in which resistance to aryloxyphenoxypropanoate and cyclohexanedione herbicides has developed.

Species	Source
Alopecurus myosuroides	U.K., W. Europe
Avena fatua	Canada, U.S.A., Australia
Avena sterilis,	Australia,U.K.
Digitaria sanguinalis	U.S.A.
Echinochloa colona	Costa Rica
Eleusine indica	Malasya
Lolium multiflorum	U.S.A., U.K., France
Lolium rigidum	Australia, Spain
Setaria viridis	Canada
Setaria faberi	U.S.A.
Sorghum halepense	U.S.A.

ACCase-Based Resistance Mechanisms

In the majority of AOPP- and CHD-resistant weed biotypes analyzed to date, resistance is conferred by an ACCase alteration that results in reduced herbicide sensitivity. As indicated previously, ACCase from most grasses is sensitive to AOPP and CHD herbicides, with I_{50} values in the range 0.1 to 5 µM, depending on the species and herbicide. ACCase from resistant biotypes is less sensitive to herbicides. Typical ACCase inhibition data from susceptible and resistant weed biotypes are shown in Figure 1. In this case, ACCase from the resistant biotype of *E. indica* was very resistant to fluazifop (I_{50} >500 µM), but much less resistant to clethodim (I_{50} = 6.6 µM) (Leach *et al.*, 1995).

ACCase inhibition data from eight different weed species, including several different resistant biotypes within some species, are summarized in Table 2. Although it is not always valid to compare data derived from different species, using slightly different experimental protocols, it is apparent that several different forms of herbicide-resistant ACCase are represented in Table 2. For example, the ACCase from the resistant *S. viridis* biotype UM8 is highly resistant to all AOPP and CHD herbicides tested; in contrast, the ACCase from the resistant *E. indica* biotype is extremely resistant to fluazifop, but carries only a low level of resistance to clethodim. However, the *E. indica* biotype is clearly resistant to sethoxydim (R/S I_{50} = 20), whereas the ACCase from the resistant *L. multiflorum* and *A. fatua* (UM33) biotypes is almost completely susceptible to CHD herbicides.

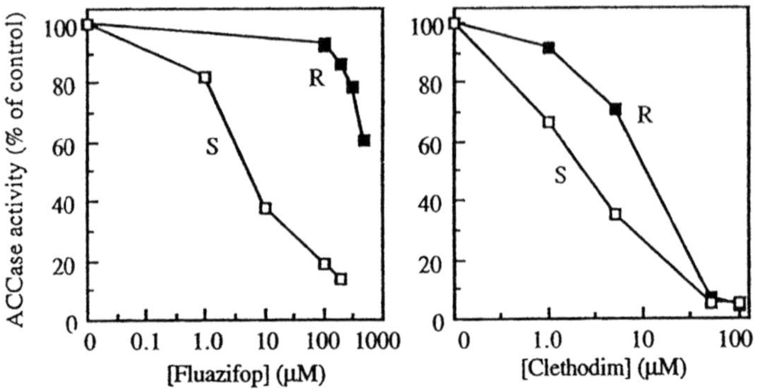

Figure 1. Inhibition of ACCase from susceptible and resistant biotypes of *E. indica* by fluazifop an clethodim. Data from Leach *et al.* (1995); used with permission.

In most instances, the level of resistance to particular inhibitors at the whole plant level correlates quite closely with resistance at the target enzyme level. For example, the resistant *E. indica* biotype which is very resistant to fluazifop, shows an intermediate level of resistance to sethoxydim and fenoxaprop, but only a very low level of resistance to clethodim (Marshall *et al.*, 1994); the enzyme inhibition results show a similar pattern (Leach *et al.* 1995; Figure 1 and Table 2).

There has been limited comparative kinetic analysis of herbicide-resistant and susceptible forms of ACCase from grasses. In *L. multiflorum*, the Km for acetyl-CoA was equal in the resistant and sensitive forms of the enzyme (Evenson *et al.*, 1994). However, in *A. fatua* the Km for acetyl-CoA was slightly lower in the sensitive than the resistant (UMl) enzyme (Km values 80 and 115 μM, respectively; Marles and Devine, unpubl. results). Although this difference is small and unlikely to affect the catalytic function of the enzyme, the altered affinity for acetyl-CoA in UM1 ACCase suggests that the resistance mutation may affect acetyl-CoA binding in addition to herbicide binding. This is not unexpected, since the results of competitive inhibition experiments suggest that the herbicides interact at or near the acetyl-CoA binding site on the enzyme (Burton *et al.*, 1991; Rendina *et al.*, 1990). Not surprisingly, herbicide binding was also different in the resistant and sensitive forms of ACCase in the above studies. In *L. multiflorum*, inhibition by diclofop was non-competitive in both cases, but changed from linear in the sensitive enzyme to S-hyperbolic-I-hyperbolic in the resistant enzyme (Evenson *et al.*, 1994). Diclofop inhibition was mixed, non-competitive in both the resistant and sensitive ACCase forms from *A. fatua*, as was sethoxydim inhibition in the sensitive form (Marles and Devine, unpublished results).

However, sethoxydim inhibition of the resistant ACCase from *A. fatua* was uncompetitive. The change in the inhibition characteristics may explain why this resistant biotype (UM1) can survive treatment with exceptionally high doses of sethoxydim (>150-fold those required to kill the sensitive biotype; Heap *et al.*, 1993), although UM1 ACCase shows only 16-fold resistance compared to the sensitive enzyme.

It appears from the above data that resistance to ACCase inhibitors can be conferred by several different mutations, each conferring a different level of resistance and pattern of cross-resistance to AOPP and CHD herbicides. This is analogous to the situation with s-triazine and sulfonylurea herbicides, in which molecular analysis has shown that different point mutations in the genes coding for the target proteins (the D- 1 protein in the photosystem II reaction centre and acetolactate synthase, respectively; Devine *et al.*, 1993) confer different resistant phenotypes.

The mutations conferring resistance to ACCase inhibitors have not yet been identified. However, recent advances in characterizing the ACCase gene (e.g., Gornicki and Haselkorn, 1993; Schulte *et al.*, 1994) will open the door to further analysis of the gene in both crop and weed species. Ultimately, this will allow identification of the mutations responsible for reduced herbicide sensitivity in weeds.

Summary and Future Prognostications

ACCase-based target site resistance has been confirmed in eight weed species to date, and it is likely to be identified in others. In most cases the ACCase mutations confer a substantial level of resistance to most AOPP and CHD herbicides; instances in which there is no cross-resistance to certain ACCase inhibitors are the exception rather than the rule. This precludes the use of any other ACCase inhibitor for the control of the resistant biotypes, necessitating the use of herbicides with other modes of action or alternative weed control methods.

The incidence of resistance to ACCase inhibitors is increasing rapidily and, in the majority of cases analyzed, resistance is conferred by target site mutations. Since ACCase inhibitor resistance based on target site alterations is controlled by single gene and inherited as a dominant or semi-dominant trait (Betts *et al.*, 1992; Morrison *et al.*, 1992; Parker *et al.*, 1990), it is likely that it will continue to increase in frequency in weed populations subjected to high selection pressure from AOPP or CHD herbicides.

Future research in this area is likely to proceed in several areas. First, additional ACCase mutants will be analyzed to obtain a more complete picture of the range and diversity of resistant ACCase phenotypes. Second, mutations conferring AOPP and CHD

Table 2. Summary of I_{50} ratios (R/S) for ACCase from herbicide-resistant (R) and -susceptible (S) biotypes of various weeds.

Specie/ Biotype	I_{50} ratios (R/S)								
	Sethox[1]	Tralk	Cleth	Fenox	Dic	Quiz	Fluaz	Halox	Ref[2]
A. fatua									
UM1	16	3.2	-	4.3	6.2	-	-	-	1
UM33	1.3	2.5	-	3.3	1.8	-	-	-	1
A. sterilis									
SAS1	8	6	-	-	52	-	-	-	2
L. rigidum									
SLR31 (R2)	26	14	-	-	6	-	55	15	3
SLR3	7.8	>9.5	-	-	>37	-	>2.9	20	4
VLR69	1.2	1.2	-	-	31	-	4.3	15	5
WLR96	>1.6	6.0	-	-	>217	-	>6.9	>56	6
L. multiflorum									
	0.9	-	1.5	-	28	10	-	9	7
S. viridis									
UM8	50	31	31	48	47	60	-	-	8
UM131	>33	-	8	-	-	-	21	-	9
E. indica									
	-	20	4	25	-	-	>90	-	10
S. halepense									
Leland	4.7	-	1.7	-	6.7	-	15	-	11
Natchez	2.1	-	1.9	-	3.9	-	30	-	11
A. myosuroides									
Lincs	3.3	>4.5	-	>17	5.7	-	>59	-	12

1. Abbreviations: Sethox, sethoxydim; Tral, tralkoxydim; Cleth, clethodim; Fenox, fenoxaprop; Dic, diclofop; Quiz, quizalofop; Fluaz, fluazifop; Halox, haloxyfop. 2. references: 1, Marles *et al.*, in prep; 2, Maneechote *et al.*, 1994; 3, Tardif and Powles, 1994; 4, Tardif *et al.*, 1993; 5, Preston *et al.*, 1996; 6, Tardif *et al.*, 1996; 7, Gronwald *et al.*, 1992; 8, Marles *et al.*, 1993; 9, Leach and Devine, unpubl.; 10, Leach *et al.*, 1995; 11, Marles and Devine, unpubl.; 12, Hall, unpubl.

herbicide resistance will be identified by molecular techniques. Third, based on the above, the structure of the herbicide-binding region of the protein may be predicted, and inhibitor interactions with sensitive and resistant ACCase forms may be modelled, as has been achieved previously for triazine herbicides and the photosystem II binding site (Tietjen *et al.,* 1991; Trebst, 1987). This will provide (a) a more complete understanding of the biochemical basis of resistance and (b) possible leads for the development of inhibitors of both the sensitive and resistant forms of the enzyme.

Acknowledgements

The author acknowledges the contributions of M.A.S. Marles, G.E. Leach and A. Shukla to this research. Appreciation is also expressed to F.J. Tardif, C. Preston, S.B. Powles and L.M. Hall for sharing unpublished data on ACCase mutants from Australia and the U.K. Financial support from the Natural Sciences and Engineering Research Council of Canada and the Herbicide Resistance Action Committee is gratefully acknowledged.

References

Betts, K.J., Ehlke, N.J., Wyse, D.L., Gronwald, J.W. and Somers, D.A. (1992) Mechanism of inheritance of diclofop resistance in Italian ryegrass (*Lolium multiflorum*). *Weed Science* **40**: 184-189.

Burton, J.D., Gronwald, J.W., Keith, R.A., Somers, D.A., Gengenbach, B.G. and Wyse, D.L. (1991) Kinetics of inhibition of acetyl-coenzyme A carboxylase by sethoxydim and haloxyfop. *Pesticide Biochemistry Physiology* **39**: 100-109.

Burton, J.D., Gronwald, J.W., Somers, D.A., Connelly, J.A., Gengenbach, B.G. and Wyse, D.L. (1987) Inhibition of plant acetyl-coenzyme A carboxylase by the herbicides sethoxydim and haloxyfop. *Biochemical Biophysical Research Communications* **148**: 1039-1044.

Catanzaro, C.J., Burton, J.D. and Skroch, W.A. (1993) Graminicide resistance of acetyl-CoA carboxylase from ornamental grasses. *Pesticide Biochemistry Physiology* **45**: 147-153.

Devine, M.D., Duke, S.O. and Fedtke, C. (1993) *Physiology of Herbicide Action*. Prentice-Hall Inc, Englewood Cliffs, NJ.

Devine, M.D. and Shimabukuro, R.H. (1994) Resistance to acetyl coenzyme A carboxylase inhibiting herbicides. *In: Herbicide Resistance in Plants: Biology and Biochemistry* (S.B. Powles and J.A.M. Holtum eds.), 141-169. Lewis Publishers, Boca Raton, FL.

Egli, M.A., Gengenbach, B.G., Gronwald, J.W., Somers, D.A. and Wyse, D.L. (1993) Characterization of maize acetyl-coenzyme A carboxylase. *Plant Physiology* **101**: 499-506.

Evenson, K.J., Gronwald, J.W. and Wyse, D.L. (1994) Purification and characterization of acetyl-coenzyme A carboxylase from diclofop-resistant and -susceptible *Lolium multiflorum. Plant Physiology* **105**: 671-680.

Focke, M. and Lichtenthaler, H.K. (1987) Inhibition of the acetyl-CoA carboxylase of barley chloroplasts by cycloxydim and sethoxydim. *Zeischrift Naturforschung* **42c**: 1361-1363.

Gornicki, P. and Haselkorn, R. (1993) Wheat acetyl-CoA carboxylase. *Plant Molecular Biology* **22**: 547-552.

Gronwald, J.W., Eberlein, C.V., Betts, K.J., Baerg, R.J., Ehlke, N.J. and Wyse, D.L. (1992) Mechanism of diclofop resistance in an Italian ryegrass (*Lolium multiflorum* Lam.) biotype. *Pesticide Biochemistry Physiology* **44**: 126-139.

Hall, L.M., Holtum, J.A.M. and Powles, S.B. (1994) Mechanisms responsible for cross resistance and multiple resistance. *In: Herbicide Resistance in Plants: Biology and Biochemistry* (S.B. Powles and J.A.M. Holtum, eds.), 243-261. Lewis Publishers, Boca Raton, FL.

Heap, I.M., Murray, B.G., Loeppky, H.A., and Morrison, I.N. (1993) Resistance to aryloxyphenoxypropionate and cyclohexanedione herbicides in wild oat (*Avena fatua*). *Weed Science* **41**: 232-238.

Heap, J. and Knight, R. (1982) A population of ryegrass tolerant to the herbicide diclofop-methyl. *Journal Australian Institute Agricultural Sciences* **48**: 156-157.

Konishi, T. and Sasaki, Y. (1994) Compartmentalization of two forms of acetyl-CoA carboxylase in plants and the origin of their tolerance toward herbicides. *Proceedings National Academy Sciences USA* **91**: 3598-3601.

Leach, G.E., Devine, M.D., Kirkwood, R.C. and Marshall, G. (1995) Target enzyme-based resistance to acetyl-coenzyme A carboxylase inhibitors in *Eleusine indica. Pesticide Biochemistry Physiology* **51**: 129-136.

Maneechote, C., Holtum, J.A.M., Preston, C. and Powles, S.B. (1994) Resistant acetyl-CoA carboxylase is a mechanism of herbicide resistance in a biotype of *Avena sterilis* spp. *ludoviciana. Plant Cell Physiol*ogy **35**: 627-235.

Marles, M.A.S., Devine, M.D. and Hall, J.C. (1993) Herbicide resistance in *Setaria viridis* conferred by a less sensitive form of acetyl coenzyme A carboxylase. *Pesticide Biochemistry Physiology* **46**: 7-14.

Marles, M.A.S., Shukla, A. and Devine, M.D. Manuscript in preparation.

Marshall, G., Kirkwood, R.C. and Leach, G.E. (1994) Comparative studies on graminicide resistant and susceptible biotypes of *Eleusine indica. Weed Research* **34**: 177-185.

Morrison, I.N., Heap, I.M. and Murray, B. (1992) Herbicide resistance in wild oat - the Canadian perspective. *Proceedings Fourth International Oat Conference*, 36-40. Adelaide, Australia.

Parker, W.B., Marshall, L.C., Burton, J.D., Somers, D.A., Wyse, D.L., Gronwald, J.W. and Gengenbach, B.G. (1990) Dominant mutations causing alterations in acetyl-coenzyme A carboxylase confer tolerance to cyclohexanedione and aryloxyphenoxypropionate herbicides in maize. *Proceedings National Academy Sci*ences USA **87**: 7175-7179.

Preston, C., Tardif, F.J., Christopher, J.T. and Powles, S.B. (1996) Multiple resistance to dissimilar herbicide chemistries in a biotype of *Lolium rigidum* dute to enhanced activity of several herbicide degrading enzymes. *Pesticide Biochemistry Physiology* **54**: 123-134.

Rendina, A.R., Craig-Kennard, A.C., Beaudoin, J.D. and Breen, M.K. (1990) Inhibition of acetyl-coenzyme A carboxylase by two classes of grass-selective herbicides. *Journal Agricultural Food Chemistry* **38**: 1282-1287.

Rendina, A.R. and Felts, J.M. (1988) Cyclohexanedione herbicides are selective and potent inhibitors of acetyl-CoA carboxylase from grasses. *Plant Physiology* **86**: 983-986.

Rendina, A.R., Felts, J.M., Beaudoin, J.D., Craig-Kennard, A.C., Look, L.L., Paraskos, S.L. and Hagenah, J.A. (1988) Kinetic characterization, stereoselectivity, and species selectivity of the inhibition of plant acetyl-CoA carboxylase by the aryloxyphenoxypropionic acid grass herbicides. *Archives Biochemistry Biophysic* **265**: 219-225.

Schulte, W., Schell, J. and Topfer, R. (1994) A gene encoding acetyl-coenzyme A carboxylase from *Brassica napus. Plant Physiology* **106**: 793-794.

Secor, J. and Cseke, C. (1988) Inhibition of acetyl-CoA carboxylase activity by haloxyfop and tralkoxydim. *Plant Physiology* **86**: 10-12.

Stoltenberg, D.E., Gronwald, J.W, Wyse, D.L., Burton, J.D., Somers, D.A. and Gengenbach, B.G. (1989) Effect of sethoxydim and haloxyfop on acetyl-cenzyme A carboxylase activity in *Festuca* species. *Weed Science* **37**: 512-516.

Tardif, F.J., Holtum, J.A.M. and Powles, S.B. (1993) Occurrence of a herbicide-resistant acetyl-cenzyme A carboxylase mutant in annual ryegrass *(Lolium rigidum)* selected by sethoxydim. *Planta* **190**: 176-181.

Tardif, F.J. and Powles, S.B. (1994) Herbicide multiple-resistance in a *Lolium rigidum* biotype is endowed by multiple mechanisms: isolation of a subset with resistant acetyl-CoA carboxylase. *Physiologia Plantarum* **91**: 488-494.

Tardif, F.J., Preston, C., Holtum, J.A.M. and Powles, S.B. (1996) Resistance to acetil-Coenzyme A carboxilase-inhibiting herbicides endowed by a single major gene encoding a resistant target site in a biotype of *Lolium rigidum*. *Australian Journal Plant Physiology* **23**: 15-23.

Thai, K.M., Jana, S. and Naylor, J.M. (1985) Variability for response to herbicides in wild oat (*Avena fatua*) populations. *Weed Science* **33**: 829-835.

Tietjen, K.G., Kluth, J.F., Andree, R., Haug, M., Lindig, M., Muller, K.H., Wroblowsky, H.J. and Trebst, A. (1991) The herbicide binding niche of photosystem II - a model. *Pesticide Science* **31**: 65-72.

Trebst, A. (1987) The three-dimensional structure of the herbicide binding niche on the reaction center polypeptides of photosystem II. *Zeischrift Naturforschung* **42c**: 742-750.

CHAPTER 7
PERTURBATION OF THE TRANSMEMBRANE PROTON GRADIENT AND RESISTANCE TO AOPP HERBICIDES

R.H. SHIMABUKURO and B.L. HOFFER
Biosciences Research Laboratory, USDA, Agricultural Research Service, State University Station, Fargo, North Dakota 58105-5674, USA

Introduction

The mechanism of action of the postemergence herbicide, diclofop-methyl (DM), and other aryloxyphenoxypropionic acid esters (AOPP) has been a topic of controversy due to the absence of definitive evidence implicating a specific mechanism. Two models have been proposed for the mode of action of diclofop-methyl. Model 1 is a single site-single effect mechanism that involves the inhibition of acetyl-CoA carboxylase (ACCase), a key enzyme in fatty acid biosynthesis. It is believed that by limiting the supply of long-chain fatty acids the biosynthesis of cellular membranes will be affected to cause lethality. With few exceptions, positive correlations exist between *in vitro* sensitivity of ACCase and whole plant resistance or susceptibility (Powles and Preston, 1995 and ref. therein). However, it is not certain that limiting the *in vivo* supply of fatty acids alone will cause the rapid lethality that has been observed in susceptible plants treated with DM.

Model 2 is a single site-multiple effects mechanism involving the collapse of the transmembrane proton gradient by DM. Such an action by DM would affect at least three major cellular processes simultaneously (Figure 1). Auxin-induced growth is immediately antagonized and active membrane transport functions cease (Shimabukuro, 1990). Evidence also suggest that depolarization of E_m (membrane potential) may lead to the catabolism of the plasmalemma by lipid peroxidation and free radical oxidation (Figure 1) (Banas *et al.*, 1993; Paliyath and Droillard, 1992; Shimabukuro and Hoffer, 1995). Therefore, both models 1 and 2 seriously affect major biochemical and biophysical cellular processes that could account for the rapid lethal action of DM on whole plants.

Whole Plant and Tissue Responses to Diclofop-Methyl and AOPP Herbicides

A mechanism for DM and other analogs should explain whole plant and tissue responses observed in field or laboratory experiments. Resistance mechanisms in tolerant species may be expected to counter the action of the herbicides and prevent or reverse the observed responses.

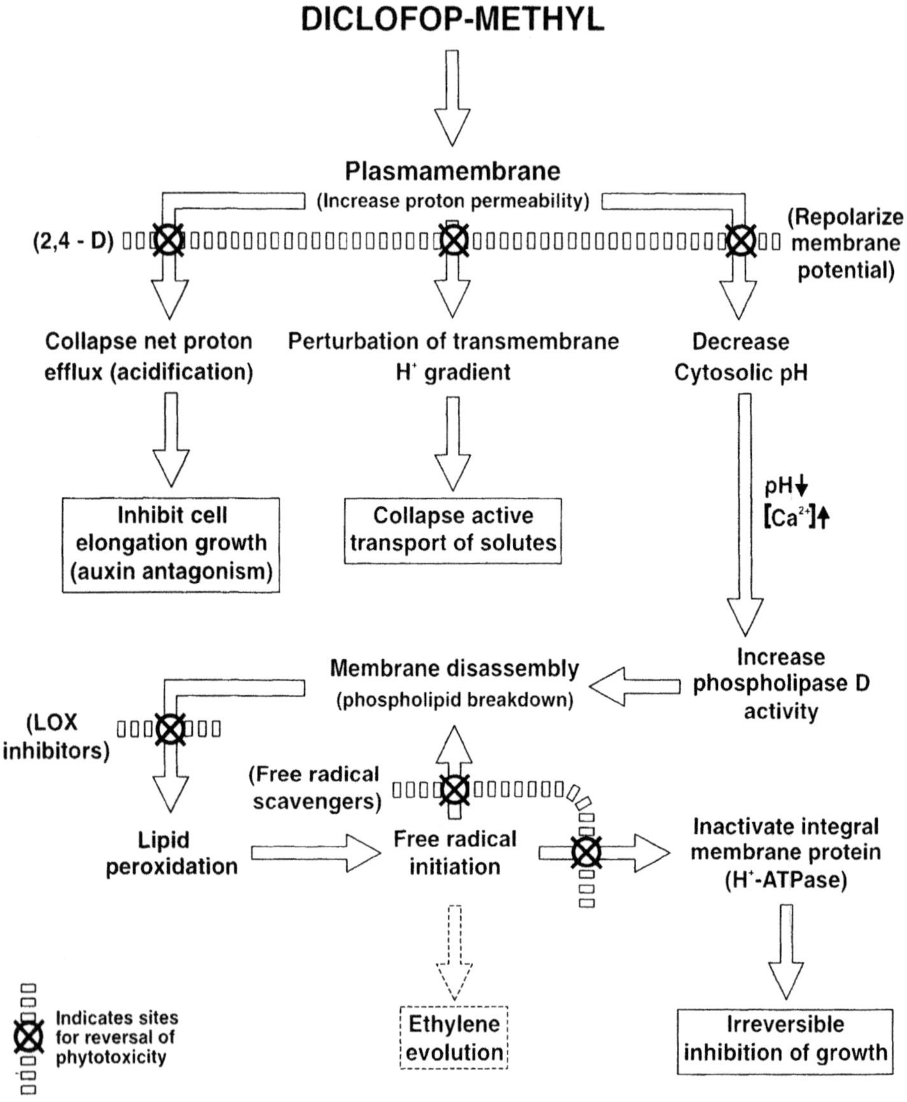

Figure 1. Multiple effects of diclofop-methyl involving lipid peroxidation and free radical oxidation.

Significant whole plant and tissue responses to DM and AOPP herbicides are: (**1**) The rate of cell growth in isolated tissues and whole plants is decreased nearly immediately upon contact and complete inhibition of growth occurs within 2-4 h following herbicide application (Shimabukuro, 1990; Donald *et al.*, 1982; Banas *et al.*, 1990). (**2**) Irreversible herbicide damage to whole plants in field applications occurs within 6-9 h following DM application (Olson and Nalewaja, 1981; Kafiz *et al.*, 1989). (**3**) Phytotoxicity of AOPP herbicides is antagonized or reversed by auxinic compounds (2,4-D, dicamba, MCPA) if applied within 9 h following herbicide application in the field (Olson and Nalewaja, 1981; Kafiz *et al* 1989; Aquero-Alvarado *et al.*, 1991). (**4**) Qualitative and quantitative changes in cellular lipid components are caused within 1-4 h with minimum inhibition of *de novo* lipid biosynthesis (Banas *et al.*, 1990, 1993; Shimabukuro and Hoffer, 1994, 1995). (**5**) Integrity of the plasmalemma as indicated by electrolyte leakage is destroyed within 12 h after DM application in susceptible wild oat (*Avena fatua*) but not in resistant wheat (*Triticum aestivum*) leaf tissues (Crowley and Prendeville, 1979). (**6**) A major integral membrane protein (H^+-ATPase) involved in auxin-induced growth, active membrane transport and modulation of cytoplasmic pH becomes nonfunctional within 3 h after DM application (Shimabukuro *et al.*, 1982, 1989). (**7**) Inhibition of growth and changes in cellular lipid profile are reversed by various lipoxygenase inhibitors and free radical scavengers (Banas *et al.*, 1993). (**8**) A significant increase in ethylene evolution is induced in susceptible monocots and a dicot within 7-12 h following herbicide application (Banas *et al.*, 1993 and unpublished).

The rapidity of whole plant and tissue responses to AOPP herbicides is probably due to a highly destructive, quick-acting mechanism that affects many seemingly unrelated cellular processes as outlined above. A mechanism that appears to account for most if not all of the observations above is the induction of senescence (Paliyath and Droillard, 1992; Leshem, 1987) resulting from model 2. This mechanism may be readily reversed by 2,4-D, lipoxygenase inhibitors and free radical scavengers (Figure 1) (Banas *et al.*, 1993; Paliyath and Droillard, 1992; Leshem, 1987; Thompson, 1988).

Senescence in Plants

Senescence is an active process triggered by external and internal signals that result in rapid destruction of cell membrane, integral membrane protein and other components through lipid peroxidation and free radical oxidation (Paliyath and Droillard, 1992; Leshem, 1987; Thompson, 1988). Mechanisms for the induction of active oxygen species may differ, but the end result is rapid cell death (Foyer *et al.*, 1994). An induction of senescence by DM, possibly through second messengers such as H^+ and Ca^{2+} (Paliyath and Droillard, 1992; Leshem, 1987; Bush, 1993), may be analogous to the localized senescence induced by elicitors of incompatible plant pathogens (Keppler and Novacky, 1987).

Senescense Sequence of Plasma Membrane

The most characteristic change in a senescing plasma membrane is a sharp decline in membrane phospholipids. The deesterification of phospholipids leads to an increase in free polyunsaturated fatty acids (PUFA) which act as primary substrates for lipoxygenase. Lipid peroxidation leads to the generation of free radicals and the whole destructive process will become self-perpetuating unless terminated by lipoxygenase inhibitors and/or enzymatic and nonenzymatic free radical scavenging systems (Paliyath and Droillard, 1992; Thompson, 1988; Foyer *et al.*, 1994; Scandalios, 1993). Auxin and 2,4-D appear to delay the onset of senescence through several possible mechanisms that may prevent the deesterification of phospholipids (Paliyath and Droillard, 1992; Leshem, 1987). A most characteristic feature of membrane senescence is the increased evolution of ethylene which is not the actual destructive or phytotoxic component in senescence (Paliyath and Droillard, 1992; Thompson, 1988; Scandalios, 1993). Rapid and extensive membrane and protein damage may be expected as a consequence of free radical oxidations (Paliyath and Droillard, 1992; Thompson, 1988).

Diclofop-Methyl and Herbicidal Effects

DM is applied postemergence and causes nearly immediate inhibition of meristematic growth in susceptible species (Shimabukuro, 1990). Most monocots are susceptible to DM and other analogs whereas all dicots are known to be resistant. In susceptible oat (*Avena sativa*) visible injury (chlorosis) appears after 48-72 h following herbicide application.

The rapid inhibition of growth and changes in cellular lipid components in wheat by haloxyfop-ethoxyethyl which acts similarly to DM were reversed by various lipoxygenase inhibitors and free radical scavengers (Banas *et al.*, 1993). This observation implicated lipid peroxidation and free radical oxidation as a major mechanism of action for DM and other analogs.

Effect on Growth and Cellular Lipid Biosynthesis by DM

The immediate decrease in the rate of IAA-induced growth in oat coleoptiles due to DM coincides with the collapse of the transmembrane proton gradient (depolarization of E_m) and the elimination of net acidification at the plasmalemma (Shimabukuro, 1990). However, complete inhibition of growth after 3 h of DM treatment appears to be due to the failure of the proton pump (H^+-ATPase) to resume proton transport activity following removal of DM and the addition of 2,4-D (Shimabukuro *et al.*, 1982, 1989). Coleoptile growth coinciding with the resumption of acidification and repolarization of E_m occurred when DM was removed and 2,4-D or IAA added prior to 3 h (Shimabukuro *et al.*, 1982, 1989; Shimabukuro and Hoffer, 1992; Holtum *et al.*, 1994). DM does not interact directly with H^+-ATPase (Shimabukuro, 1990). Therefore, the rapid inactivation of H^+-ATPase was not due to DM but probably due to oxidative damage by free radicals (Figure 1) (Paliyath and Droillard, 1992). A similar mechanism may be the cause of irreversible herbicide damage observed shortly after field applications of DM (Olson and Nalewaja, 1981; Kafiz *et al.*, 1989).

Inhibition of *de novo* lipid biosynthesis by DM and haloxyfop does not appear to be the cause of growth inhibition and changes to lipid components (Banas *et al.*, 1990,

1993; Shimabukuro and Hoffer, 1994, 1995). Growth rate of oat coleoptiles was only slightly decreased by the antibiotic, cerulenin, in contrast to the complete inhibition of growth by DM after 3 h. However, *de novo* lipid biosynthesis was inhibited 57% by cerulenin with no inhibition by DM (100 μM) (Shimabukuro and Hoffer, 1994). Growth inhibition and lipid profile changes in wheat caused by haloxyfop were reversed by various lipoxygenase inhibitors and free radical scavengers, but similar effects caused by cerulenin were not reversed with the same compounds (Banas *et al.*, 1993). Changes to cellular lipid components by DM and haloxyfop occurred with little or no inhibition of *de novo* lipid biosynthesis with [^{14}C] acetate (Banas *et al.*, 1990, 1993; Shimabukuro and Hoffer, 1994, 1995).

Cerulenin does not collapse the transmembrane proton gradient in contrast to DM and haloxyfop (Shimabukuro and Hoffer, 1992, 1994). The collapse of the proton gradient has been implicated as a possible mechanism for the initiation of senescence (Paliyath and Droillard, 1992; Leshem, 1987; Bush, 1993). Cytosolic acidification and increased Ca^{2+} concentration due to membrane depolarization by DM and haloxyfop create the conditions that activate phospholipase D or other phospholipases to initiate deesterification of membrane phospholipids to release PUFAs (Fig. 1) (Paliyath and Droillard, 1992; Leshem, 1987; Bush, 1993). The disassembly of phospholipids releases free oleic, linoleic and linolenic acids as the principal PUFAs in plant plasma membrane. The role of H^+ and Ca^{2+} as second messengers in the induction of the overall process is still uncertain (Paliyath and Droillard, 1992; Leshem, 1987; Bush, 1993). Increased levels of free PUFA with concomitant decreases in phospholipids were detected shortly after treatment of wheat with haloxyfop (Banas *et al.*, 1993). Therefore, the effects of DM and haloxyfop appear to correlate with their action on the plasmalemma to induce the senescence process and not to inhibition of fatty acid biosynthesis alone, a common mechanism shared with cerulenin.

Ethylene Evolution and Senescence

Ethylene evolution is characteristic of senescence (Paliyath and Droillard, 1992; Leshem, 1987; Thompson, 1988) and it increases in plants under oxidative stress (Scandalios, 1992). Ethylene appears to be a consequence of lipid peroxidation and free radical generation in senescence (Figure 1) (Bousquet and Thimann, 1984). Ethylene, however, is not the destructive intermediate in senescence. Evidence indicates that IAA and 2,4-D stimulate ethylene by increasing the activity of ACC synthase which catalyzes the conversion of AdoMet (SAM) to ACC (1-aminocyclopropane-1-carboxylic acid), the rate-limiting step in ethylene biosynthesis. Senescence induces the activity of ACC oxidase which catalyzes the conversion of ACC to ethylene (Zarembinski and Theologis, 1994). Therefore, it appears that 2,4-D and DM may stimulate ethylene evolution by affecting different sites in the biosynthetic pathway of ethylene. Ethylene due to IAA and 2,4-D treatments appears to participate in specific physiological functions whereas ethylene due to senescence or DM appears to be a consequence of a catabolic process (Paliyath and Droillard, 1992; Bousquet and Thimann, 1984; Zarembinski and Theologis, 1994).

Induction of Ethylene by Diclofop-methyl

DM and haloxyfop are potent inducers of ethylene production (Banas *et al.*, 1993; Shimabukuro and Hoffer, 1996). Ethylene evolution increased by 146 and 207% of control in wheat shoots after 18 and 25 h of haloxyfop treatment, respectively (Banas *et al.*, 1993).

Immature apical leaves of leafy spurge (*Euphorbia esula*), a dicotyledonous weed, were totally resistant to a foliar application of 3.0 mM DM, but the more mature leaves below the apical leaves were highly susceptible. Mature leaves were killed (severe chlorosis and desiccation) and abscission occurred within 7 days after treatment (Shimabukuro and Hoffer, 1996). Seven hours after DM application, ethylene evolution increased by 278 and 616% of untreated controls in resistant and susceptible leaves, respectively. Only 16 and 32% of absorbed [^{14}C] DM was metabolized in susceptible and resistant leaves over 24 h, respectively. The rate of cellular lipid biosynthesis *in vivo* from [^{14}C]acetate was 90% of control in both resistant and susceptible leaf discs over 4 h at 100 μM DM (Shimabukuro and Hoffer, 1996). Therefore, the response of leafy spurge leaves to DM was unrelated to metabolism and detoxification of DM or to a putative DM-insensitive ACCase. Resistance to DM by immature leafy spurge leaves may be due to an inherently greater capacity in younger tissues to counter or neutralize oxidative stress caused by DM as compared to more mature tissues (Casano *et al.*, 1994). This appears to be a common observation in senescence.

Induction of Ethylene by Diclofop-methy and 2,4-D in Susceptible Oat

Oat is susceptible to DM with complete inhibition of growth and severe chlorosis occurring within 4-5 days following application of 4.5 mM DM. Phytotoxicity of DM was completely reversed when applied in combination with 2.0, 3.0 and 4.5 mM 2,4-D. No antagonism occurred when DM was applied as a combination with 0.75 mM 2,4-D (Casano *et al.*, 1994).

Evolution of ethylene increased significantly within 7 hours following DM application at 4.5 mM to 1.5-leaf oat seedlings. DM-induced ethylene increased by 232 and 112% of untreated controls in susceptible oat and resistant wheat, respectively. The relative lack of response in wheat is due to the rapid metabolism and detoxification of DM in this species (Shimabukuro, 1990). Ethylene evolved in tissues treated with 0.75 and 2.0 mM 2,4-D was 103 and 312% of controls, respectively. Except for a distorted bending of stems, 2,4-D application was nonphytotoxic to oat. Total amount of ethylene evolved when DM (4.5 mM) was combined with 0.75 and 2.0 mM 2,4-D was 400 and 391% of controls, respectively. Therefore, total amount of ethylene evolved in DM + 2,4-D combinations was approximately the sum of ethylene induced by DM and 2,4-D (0.75 mM) alone when no antagonism occurred. Total amount of ethylene evolved when DM phytotoxicity was reversed by 2,4-D was only slightly more than the amount of ethylene induced by 2,4-D alone (312% of control at 2.0 mM 2,4-D alone vs. 391% of control in combination with 4.5 mM DM) (Shimabukuro and Hoffer, 1996). Total amount of ethylene evolved with other combinations of DM and 2,4-D (3.0 and 4.5 mM) was also only slightly more than with 2,4-D alone (Shimabukuro and Hoffer, 1996).

From these results, it was concluded that 2,4-D indirectly prevents induction of lipid peroxidation and free radical generation by DM which induces the increased production of ethylene (Figure 1) (Paliyath and Droillard, 1992; Bousquet and Thimann, 1984; Zarembinski and Theologis, 1994).

Reversal of Diclofop-Methyl Phytotoxicity by 2,4-D

IAA and 2,4-D stimulate H^+-ATPase to repolarize the E_m, reestablish net acidification at the plasmamembrane, and restore cytoplasmic pH to homeostasis (Figure 1) (Shimabukuro, 1990; Paliyath and Droillard, 1992; Shimabukuro *et al.,* 1982, 1989; Shimabukuro and Hoffer, 1992; Holtum, *et al.,* 1994). This process should interrupt the lipid peroxidation and free radical oxidation mechanism induced by DM and delay or reverse senescence (Paliyath and Droillard, 1992). However, such a reversal or antagonism is only possible if the proton pump is still functional within 3 h in tissues such as oat coleoptiles and within 6-9 h in field applications of DM (Olson and Nalewaja, 1981; Kafiz *et al.,* 1989; Shimabukuro *et al.,* 1982). The integral membrane protein, H^+-ATPase, may be vulnerable to rapid damage by free radical oxidation (Paliyath and Droillard, 1992) as indicated by its rapid inactivation in isolated tissues (Figure 1) (Shimabukuro *et al.,* 1982, 1989).

Conclusions

Two major mechanisms of action have been proposed for DM and other AOPP analogs. Model 1 should disrupt an anabolic process resulting in unknown pleiotropic effects to cause plant lethality. The resistance mechanism against model 1 may be limited to the development of DM-insensitive ACCase, the target site for the proposed mechanism. Model 2 is essentially a signal perception and transduction mechanism that induces a catabolic process resulting in severe oxidative stress. Resistance mechanisms against model 2 may include inherent differences in various enzymatic and nonenzymatic mechanisms for countering or neutralizing oxidative stress (Foyer *et al.,* 1993; Scandalios, 1993).

Several possible sites for the reversal of oxidative stress are indicated in Figure 1. Resistance to DM may occur through endogenous or exogenous (2,4-D, dicamba, MCPA) compounds that may prevent the initiation of free radicals or through inherent free radical scavenging systems (Foyer *et al.,* 1993; Scandalios, 1993) that can terminate the self-generating process. The life of a cell is likely to be a delicate balance between anabolic and catabolic processes. Both models could shift the balance in favor of catabolism to cause cell death.

Both models are not free from contradictory evidence. Resistant biotypes of rigid ryegrass (*Lolium rigidum*) and wild oat possess DM-sensitive ACCase (Powles and Preston, 1995; Holtum *et al.,* 1994). Some resistant biotypes of rigid ryegrass also show no membrane repolarization activity (Holtum *et al.,* 1994). Growth of susceptible maize roots was inhibited with little effect on E_m and the concept underlying model 2 has been questioned (DiTomaso, 1994). However, the biological significance of membrane depolarization and the interaction between DM and 2,4-D and its possible role in signal transduction (Barnwell and Cobb, 1994) support the processes discussed under model 2.

References

Aquero-Alvarado, R., Appleby, A.P. and Armstrong, D.J. (1991) Antagonism of haloxyfop activity in tall fescue (*Festuca arundinacea*) by dicamba and bentazon. *Weed Science* **39**: 1-5.

Bana,A., Johansson, I., Stenlid, G. and Stymne, S. (1990) The effect of haloxyfop-ethoxyethyl on lipid metabolism in oat and wheat shoots. *Swedish Journal Agricultual Research* **20**: 97-104.

Bana, A., Johansson, I., Stenlid, G., and Stymne, S. (1993) Free radical scavengers and inhibitors of lipoxygenases as antagonists against the herbicides haloxyfop and alloxydim. *Swedish Journal Agricultual Research* **23**: 67-75.

Barnwell, P. and Cobb, A.H. (1994) Graminicide antagonism by broadleaf weed herbicides. *Pesticide Science* **41**: 77-85.

Bousquet, J-F. and Thimann, K.V. (1984) Lipid peroxidation forms ethylene from 1-aminocyclopropane-1-carboxylic acid and may operate in leaf senescence. *Proceedings National Academy Sciences USA* **81**: 1724-1727.

Bush, D.S. (1993) Regulation of cytosolic calcium in plants. *Plant Physiology* **103**: 7-13.

Crowley, J. and Prendeville, G.N. (1979) Effect of diclofop methyl on leaf-cell membrane permeability in wild oat, barley and wheat. *Cananadian Journal Plant Sciences* **59**: 275-277.

Casano, L.M., Martin, M. and Sabater, B. (1994) Sensitivity of superoxide dismutase transcript levels and activities of oxidative stress is lower in mature-senescent than in young barley leaves. *Plant Physiology* **106**: 1033-1039.

DiTomaso, J.M. (1994) Evidence against a direct membrane effect in the mechanism of action of graminicides. *Weed Science* **42**: 302-309.

Donald, W.W., Parke, R.V. and Shimabukuro, R.H. (1982) The effects of diclofop-methyl on root growth of wild oat. *Physiologia Plantarum* **54**: 467-474.

Foyer, C.H., Lelandais, M. and Kunert, K.J. (1994) Photooxidative stress in plants. *Physiologia Plantarum* **92**: 696-717.

Holtum, J.A.M., Häusler, R.E., Devine, M.D. and Powles, S.B. (1994) Recovery of transmembrane potentials in plants resistant to aryloxyphenoxypropanoate herbicides: A phenomenon awaiting explanation. *Weed Science* **42**: 293-301.

Kafiz, B., Caussanel, J.P., Scalla, R. and Gaillardon, P. (1989) Interaction between diclofop-methyl and 2,4-D in wild oat (*Avena fatua* L.) and fate of diclofop-methyl in cultivated oat. *Weed Research* **29**: 299-305.

Keppler, L.D. and Novacky, A. (1987) The initiation of membrane lipid peroxidation during bacteria-induced hypersensitive reaction. *Physiological Molecular Plant Pathology* **30**: 233-245.

Leshem, Y.Y. (1987) Membrane phospholipid catabolism and Ca^{2+} activity in control of senescence. *Physiologia Plantarum* **69**: 551-559.

Olson, W.A. and Nalewaja, J.D. (1981) Antagonistic effects of MCPA on wild oat (*Avena fatua*) control with diclofop. *Weed Science* **29**: 566-571.

Paliyath, G. and Droillard, M.J. (1992) The mechanisms of membrane deterioration and disassembly during senescence. *Plant Physiology Biochemistry* **30**: 789-812.

Powles, S.B. and Preston C. (1995) Herbicide cross resistance and multiple resistance in plants. *The Herbicide Resistance Action Committee,* Monography No. 2.

Scandalios, J.G. (1993) Oxygen stress and superoxide dismutases. *Plant Physiology* **101**: 7-12.

Shimabukuro, R.H. (1990) Selectivity and mode of action of the postemergence herbicide diclofop-methyl. *Plant Growth Regulation Society American Quarterly* **18**: 37-54.

Shimabukuro, R.H. and Hoffer, B.L. (1992) Effect of diclofop on the membrane potentials of herbicide-resistant and -susceptible annual ryegrass root tips. *Plant Physiology* **98**: 1415-1422.

Shimabukuro, R.H. and Hoffer, B.L. (1994) Effects on transmembrane proton gradient and lipid biosynthesis in the mode of action of diclofop-methyl. *Pesticide Biochemistry Physiology* **48**: 85-97.

Shimabukuro, R.H. and Hoffer, B.L. (1995) Enantiomers of diclofop-methyl and their role in herbicide mechanism of action. *Pesticide Biochemistry Physiology* **51**: 68-82.

Shimabukuro, R.H. and Hoffer, B.L. (1996) Induction of ethylene as an indicator of senescence in the mode of action of diclofop-methyl. *Pesticide Biochemistry Physiology* **54**: 146-158.

Shimabukuro, M.A., Shimabukuro, R.H. and Walsh, W.C. (1982) The antagonism of IAA-induced hydrogen ion extrusion and coleoptile growth by diclofop-methyl. *Physiologia Plantarum* **56**: 444-452.

Shimabukuro, R.H., Walsh, W.C. and Wright, J.P. (1989): Effect of diclofop-methyl and 2,4-D on the transmembrane proton gradient: A mechanism for their antagonistic interaction. *Physiologia Plantarum* **77**: 107-114.

Thompson, J.E. (1988) The molecular basis for membrane deterioration during senescence. *In: Senescence and Aging in Plants* (L.D. Noodén and A.C. Leopold, eds.), 51-83. Academic Press.

Zarembinski, T.I. and Theologis, A. (1994) Ethylene biosynthesis and action: A case of conservation. *Plant Molecular Biology* **26**: 1579-1597.

CHAPTER 8
TARGET-SITE RESISTANCE FOR ACETOLACTATE SYNTHASE INHIBITOR HERBICIDES

L. L. SAARI and C. A. MAXWELL
DuPont Agricultural Products, Stine-Haskell Research Center, P.O. Box 30, Newark Delaware U.S.A.

Introduction

Acetolactate synthase (ALS) is the first enzyme common to the biosynthesis of the branched-chain amino acids, leucine, valine, and isoleucine. Currently, there are four commercial classes of herbicides which inhibit this enzyme including sulfonylureas, imidazolinones, triazolopyrimidines, and pyrimidinyl thiobenzoates.

Resistance to ALS-inhibitor herbicides is both problematic and useful. While resistant weeds have been selected through several years of herbicide use, resistance has also been used as a mechanism to introduce additional crop selectivity. Both consequences of resistance have been recently reviewed (Saari *et al.*, 1994; Saari and Mauvais, 1995). The predominant mechanism of resistance is associated with the modification of ALS. That is, the target site is less sensitive to inhibition by the herbicide. However, overexpression of the target site is also possible. Both target amplification and target insensitivity have been used in the development of sulfonylurea-resistant crops, but only decreased target-site sensitivity has been found in weeds that are resistant to ALS inhibitor herbicides.

In this paper, we discuss the extent of ALS-inhibitor resistance, mutations capable of conferring resistance, and the roles of decreased target-site sensitivity vs. target-site amplification.

Extent of ALS-Inhibitor Herbicide Resistance in Weeds

Resistance to ALS-inhibitor herbicides in weeds was first discovered in 1987 (Mallory-Smith *et al.*, 1990; Primiani *et al.*, 1990). Since then, over 15 weed species have become resistant with instances reported in North America, Europe, the Middle East, and Australia (Saari *et al.*, 1994; Brown and Cotterman, 1994). The majority of resistant weeds are dicotyledons, but monocotyledons and sedges are also represented.

Genes and Mutations Conferring Resistance

In laboratory studies, several mutations in the ALS gene have been identified which result in resistance. In contrast, very few ALS mutations have been identified in weeds selected in the field. This observation, in part, is due to less molecular analysis of resistant biotypes from the field. However, fewer mutations are expected from field isolates because any mutation would need to allow a level of fitness sufficient for successful competition in a plant community. Under laboratory conditions, a study of spontaneous mutations in yeast (*Saccharomyces cerevisiae*) identified 24 amino acid changes at 10 different sites which conferred resistance (Mazur and Falco, 1989; Falco *et al.*, 1989). Additional mutations in the yeast ILV2 gene were identified by site-directed mutagenesis. These experiments showed that substitution of nearly any amino acid at ala117, pro192, or trp586 resulted in a herbicide-resistant phenotype.

In tobacco (*Nicotiana tabacum*), two unlinked loci for ALS were identified (as SurA and SurB) by mutations resulting in sulfonylurea resistance (Chaleff and Bascomb, 1987; Chaleff and Ray, 1984). A pro196 to gln change occurs as a result of one mutation in the SurA gene of tobacco line C3 (Lee *et al.*, 1988). The resistant tobacco line, HRA, has two mutations in the SurB gene resulting in pro196 to ala and trp573 to leu changes (Lee *et al.*, 1988) (Table 1). Either the *SurA-C3* or the *SurB-Hra* gene, if used to transform wild-type tobacco, results in resistant plants. Tobacco expressing the *SurB-Hra* gene is resistant to both sulfonylurea and imidazolinone herbicides. In addition to tobacco, the *SurB-Hra* gene has been used to transform sugarbeet (*Beta vulgaris*) (D'Halluin *et al.*, 1992) and cotton (*Gossypium hirsutum*) (Mauvais, 1994).

A single gene, CSRl, encodes ALS in *Arabidopsis thaliana*. Several alleles have been identified as conferring resistance to ALS inhibitors (Table 1). The *csrl-l* gene differs from the wild-type gene by only one base pair resulting in a pro197 to ser change in the encoded protein (Haughn *et al.*, 1988; Mazur *et al.*, 1987). Analogous proline changes that result in resistance have been identified in ALS from yeast (Yadav *et al.*, 1986), C3 tobacco (Lee *et al.*, 1988), and canola (*Brassica napus*) (Wiersma *et al.*, 1989). Changes in pro197 (and equivalent prolines in ALS from other organisms) have often been described as resulting in phenotypes of sulfonylurea but not imidazolinone resistance. Closer scrutiny of organismic and enzymic data indicates that it is more accurate to describe the proline changes as resulting in a high degree of resistance to several sulfonylurea herbicides with a slight degree of cross resistance to some, but probably not all, imidazolinone herbicides (e.g., imazapyr) (Saari *et al.*, 1994).

The *A. thaliana csrl-l* gene has been frequently used to introduce sulfonylurea resistance into crop plants. Examples of transgenic crops include canola (Miki *et al.*, 1990), chicory (*Cichorium intybus*) (Vermeulen *et al.*, 1992), flax (*Linum usitatissimum*) (McHughen, 1989), and rice (*Oryza sativa*) (Li *et al.*, 1992). In addition, canola (Wiersma *et al.*, 1989) and maize (Zea mays) (Fromm *et al.*, 1990) genes encoding changes in analogous prolines have been used to transform canola and maize, respectively.

Table 1. Genes conferring ALS-inhibitor resistance.

Gene	Species	Amino Acid Change	Resistant to:	Reference
csrl-1	*Arabidopsis thaliana*	Pro197 to ser	Sulfonylureas, Triazolopyrimidines	1
csrl-2	*Arabidopsis thaliana*	Ser653 to asn	Imidazolinones, Pyrimidinyloxybenzoates	2
csrl-3	*Arabidopsis thaliana*	?	Triazolopyrimidines	3
csrl-4	*Arabidopsis thaliana*	Pro197 to ser and	Sulfonylureas	3
		Ser653 to asn	Imidazolinones	4
			Triazolopyrimidines	
SurA-C3	*Nicotiana tabacum*	Pro196 to gln	Sulfonylureas	5
SurB-Hra	*Nicotiana tabacum*	Pro196 to ala and Trp573 to leu	Sulfonylureas, Imidazolinones	5
ahas3r	*Brassica napus*	Trp557 to leu	Sulfonylureas, Imidazolinones, Triazolopyrimidines	6

1. Haughn *et al.*, 1988; 2. Sathasivan *et al.*, 1991; 3. Mourad *et al.*, 1993; 4. Hattori *et al.*, 1992; 5. Lee *et al.*, 1988; 6. Brandle *et al.*, 1994

Other *A. thaliana* alleles of CSR1 also confer resistance. The *A. thaliana csrl-2* gene encodes an ALS with a ser653 to asn change (Sathasivan *et al.*, 1991) resulting in much higher resistance to imidazolinone and pyrimidinyl oxybenzoate herbicides than sulfonylurea or triazolopyrimidine herbicides (Mourad and King, 1992). Another mutant allele, designated *csrl-3,* was identified after selection of M2 populations of *A. thaliana* with a triazolopyrimidine herbicide (Mourad *et al.*, 1993). The mutation(s) responsible for the resistance has not been identified but the ser653 to asn change has been ruled out. A chimeric ALS gene was constructed that contained the mutations associated with the *csrl-l* and *csrl-2* alleles (Hattori *et al.*, 1992). Tobacco plants transformed with this gene were resistant to both sulfonylurea and imidazolinone herbicides. Through intragenic recombination of *A. thaliana*, the nucleotide changes encoding both the prol97 to ser and ser 653 to asn changes were placed in one gene, *csrl-4*, which resulted in a similar phenotype (Mourad *et al.*, 1994). Tobacco was transformed with the canola ALS gene, *ahas3r*, resulting in resistance to sulfonylurea, imidazolinone, and triazolopyrimidine herbicides (Brandle *et al.*, 1994). The *ahas3r* gene encodes an ALS with a single amino acid change of trp557 to leu.

Several sulfonylurea-resistant crop lines were developed through selection of callus, microspores, or seeds with sulfonylurea herbicides (see Saari and Mauvais, 1995 for review). Mutagens such as ethyl methane sulfonate are often used to increase the frequency of mutation. Examples of resistant lines developed by this method include sugarbeet (Saunders *et al.*, 1992; Hart *et al.*, 1992, 1993), canola (Kenyon *et al.*, 1987; Swanson *et al.*, 1988; Saxena *et al.*, 1990; Tonnemaker *et al.*, 1992), soybean (*Glycine max*) (Sebastian and Chaleff, 1987; Sebastian *et al.*, 1989), lettuce (*Lactuca sativa*) (Mallory-Smith *et al.*, 1993), flax (Jordan and McHughen, 1987), birdsfoot trefoil (*Lotus corniculatus*) (Pofelis *et al.*, 1992), and tobacco (Harms *et al.*, 1992; Chaleff and Bascomb, 1987; Creason and Chaleff, 1988; Chaleff and Ray, 1984). The resistance was due to a sulfonylurea-insensitive ALS in nine of the studies, but the mechanism of resistance was not always identified. In one instance, a soybean line was selected that was sulfonylurea-resistant but retained ALS sensitivity to sulfonylurea herbicides (Sebastian and Chaleff, 1987). The resistance trait is recessively inherited in this soybean line in contrast to the dominantly or semidominantly inherited resistance observed for most of the resistant crop lines.

Mutations Identified in Resistant Weeds

In several agronomic areas, weeds that are resistant to ALS-inhibitor herbicides have appeared after three to five years of selection. The great majority of these resistant cases are due to target-site resistance rather than other mechanisms (Saari *et al.*, 1994). The rapidity of selection probably is a function of the high degree of selection pressure imposed by these herbicides rather than an unusually high frequency of resistant mutations. The frequency of finding a resistant mutant, whether selecting for resistant seeds, cells, or protoplasts, is not appreciably greater than the expected 10^{-6} frequency for mono-gene, dominant mutations (Table 2).

Table 2. Initial frequency of ALS-inhibitor resistance (Target-site).

Study	Frequency	Reference
Arabidopsis thaliana M2 seeds	10^{-9}	Haughn & Somerville,1987
Germinated *Medicago sativa* seeds	10^{-7}	Stannard, 1987
Tobacco cell culture	2.7×10^{-8}	Harms & DiMaio, 1991
Tobacco leaf protoplasts	10^{-6} to 10^{-5}	Mauvais, 1994

Analysis of mutations responsible for ALS-inhibitor resistance in weeds is limited. Guttieri *et al.* (1992) showed that single point mutations encoding proline changes (pro173) in Domain A, one of the highly-conserved regions of ALS, were associated with resistant biotypes of *Kochia scoparia, Salsola iberica,* and *Lactuca serriola* that were selected with sulfonylurea herbicides. However, one *K. scoparia* biotype in this study did not have a corresponding proline change in Domain A even though the ALS isolated from this biotype was less sensitive to ALS-inhibitors. The majority of these results are consistent with laboratory studies where *A. thaliana csrl-l* and tobacco *SurA-C3* mutants selected with sulfonylurea herbicides are resistant due to proline changes in ALS (Haughn *et al.*, 1988; Lee *et al.*, 1988).

Resistant weeds selected with imidazolinones can respond differently to ALS inhibitor herbicides than those selected with sulfonylureas. For example, a *Xanthium strumarium* biotype selected with the imidazolinone, imazaquin, is resistant to imazaquin but not to the sulfonylurea, chlorimuron ethyl, or the triazolopyrimidine, flumetsulam (Schmitzer *et al.*, 1993). While the mutation has not been identified, the phenotype is similar to *A. thaliana* harboring the *csrl-2* gene (ser653 to asn). Other *X. strumarium* biotypes selected with imidazolinones have a single point mutation in the ALS gene which encodes ALS containing a trp552 to leu change that is resistant to representatives from the sulfonylurea, imidazolinone, triazolopyrimidine, and pyrimidinyl benzoate herbicide classes (Bernasconi and Subramanian, personal communication) .

Overexpression of the Target-Site as a Result of Selection

ALS-inhibitor resistance resulting from ALS overexpression has not been identified as a mechanism of resistance in weeds. Where studied, specific activities of ALS extracted from susceptible and resistant weed biotypes are similar (Saari *et al.*, 1990, 1992; Christopher *et al.*, 1992).

Two studies have shown that resistance as a result of ALS overexpression can occur through selection. Increased ALS specific activity and increased gene expression were observed in *Daucus carota* (Caretto *et al.*, 1994) and tobacco (Harms *et al.*, 1992) selected for resistance. Both resistant lines also had a less-sensitive ALS. Step-wise selection with increasing concentrations of sulfonylurea herbicides was required to obtain lines with both target-site resistance and target amplification. In addition to selected plants, several transformed plant species overexpress ALS (e.g., Odell *et al.*, 1990).

Summary

Several mutations in the ALS gene are capable of conferring resistance to ALS-inhibitor herbicides as exemplified by studies with yeast (Mazur and Falco, 1989). In plants, changes in certain proline, serine, and tryptophan residues in ALS have been among the more frequently-studied changes. The phenotypes of plants expressing ALS with specific amino acid changes are somewhat predictable. Selection with sulfonylurea and imidazolinone herbicides may predominantly select for resistant biotypes or lines with proline and serine changes, respectively. Overexpression of the target site, ALS, has been observed in the laboratory but has not been identified as a mechanism of resistance in the field.

References

Brandle, L.E., Morrison, M.J., Hanori, J. and Miki, B.L. (1994) A comparison of two genes for sulfonylurea herbicide resistance in transgenic tobacco seedlings. *Crop Science* **34**: 226-229.

Brown, H.M. and Cotterman, J.C. (1994) Recent advances in sulfonylurea herbicides. *In: Herbicides Inhibiting Branched-Chain Amino Acid Biosynthesis* (J. Stetter, ed.). Springer-Verlag, New York.

Caretto, S., Giardina, M.C. and Nicolodi, C. (1994) Chlorsulfuron resistance in *Daucus carota* cell lines and plants: Involvement of gene amplification. *Theoretical Applied Genetics* **88**: 520-524.

Chaleff, R.S. and Bascomb, N.F. (1987) Genetic and biochemical evidence for multiple forms of acetolactate synthase in *Nicotiana tabacum. Molecular General Genetics* **210**: 33-38.

Chaleff, R.S. and Ray, T.B. (1984) Herbicide-resistant mutants from tobacco cell cultures. *Science* **223**: 1148-1151.

Christopher, J.T., Powles, S.B. and Holtum, J.A.M. (1992) Resistance to acetolactate synthase-inhibiting herbicides in annual ryegrass (*Lolium rigidum*) involves at least two mechanisms. *Plant Physiology* **100**: 1909-1913.

Creason, G.L. and Chaleff, R.S. (1988) A second mutation enhances resistance of a tobacco mutant to sulfonylurea herbicides. *Theoretical Applied Genetics* **76**: 177-182.

D'Halluin, K., Bossut, M., Bonne, E., Mazur, B., Leemans, J. and Bonerman, J. (1992) Transformation of sugarbeet (*Beta vulgaris L.*) and evaluation of herbicide resistance in transgenic plants. *Bio/ Technology* **10**: 309-314.

Falco, S.C., McDevitt, R.E., Chui, C.-F., Hartnett, M.E., Knowlton, S., Mauvais, C.J., Smith, J.K. and Mazur, B.J. (1989) Engineering herbicide-resistant acetolactate synthase. *Developmental Industrial Microbiology* **30**: 187-194.

Fromm, M.E., Morrish, F., Armstrong, C., Williams, R., Thomas, J. and Klein, T.M. (1990) Inheritance and expression of chimeric genes in the progeny of transgenic maize plants. *Bio/Technology* **8**: 833-839.

Guttieri, M.J., Eberlein, C.V., Mallory-Smith, C.A., Thill, D.C. and Hoffman, D.L. (1992) DNA sequence variation in domain A of the acetolactate synthase genes of herbicide resistant and susceptible weed biotypes. *Weed Science* **40**: 670-677.

Harms, C.T. and DiMaio, J.J. (1991) Primisulfuron herbicide-resistant tobacco cell lines. Application of fluctuation test design to *in vitro* mutant selection with plant cells. *Journal Plant Physiology* **137**: 513-519.

Harms, C.T., Armour, S.L., DiMaio, J.J., Middlesteadt, L.A., Murray, D., Negrotto, D.V., Thompson-Taylor, H., Weymann, K., Montoya, A.L., Shillito, R.D. and Jen, G.C. (1992) Herbicide resistance due to amplification of a mutant acetohydroxyacid synthase gene. *Molecular General Genetics* **233**: 427-435.

Hart, S.E., Saunders, J.W. and Penner, D. (1992) Chlorsulfuron-resistant sugarbeet: Cross-resistance and physiological basis of resistance. *Weed Science*. **40**: 378-383.

Hart, S.E., Saunders, J.W. and Penner, D. (1993) Semidominant nature of monogenic sulfonylurea herbicide resistance in sugarbeet (*Beta vulgaris*). *Weed Science* **41**: 317-324.

Hattori, J., Rutledge, R., Labbe, H., Brown, D., Sunohara, G. and Miki, B. (1992) Multiple resistance to sulfonylureas and imidazolinones conferred by an acetohydroxyacid synthase gene with separate mutations for selective resistance. *Molecular General Genetics* **232**: 167-173.

Haughn, G. and Somerville, C.R. (1987) Selection for herbicide resistance at the whole plant level. *In: Biotechnology in Agricultural Chemistry, Symposium Series* **334** (H.M. LeBaron, R.O. Mumma, R.C. Honeycutt and J.H. Duesing, eds.), 98-107. American Chemical Society, Washington, D.C.

Haughn, G.W., Smith, J., Mazur, B. and Somerville, C. (1988) Transformation with a mutant *Arabidopsis* acetolactate synthase gene renders tobacco resistant to sulfonylurea herbicides. *Molecular General Genetics*. **211**: 266-271.

Jordan, M.C. and McHughen, A. (1987) Selection for chlorsulfuron resistance in flax (*Linum usitatissimum*) cell cultures. *Journal Plant Physiology* **131**: 333-338.

Kenyon, P.D., Marshall, G. and Morrison, I.N. (1987) Selection for sulfonylurea herbicide tolerance in oilseed rape (*Brassica napus*) using microspore culture. *In: Proceedings Brighton Crop Protection Conference-Weeds*, 871-877.

Lee, K.Y., Townsend, J., Tepperman, J., Black, M., Chui, C.F., Mazur, B., Dunsmuir, P. and Bedbrook, J. (1988) The molecular basis of sulfonylurea herbicide resistance in tobacco. *EMBO Journal* **7**: 1241-1248.

Li, Z., Hayashimoto, A. and Murai, N. (1992) A sulfonylurea herbicide resistance gene from *Arabidopsis thaliana* as a new selectable marker for production of fertile transgenic rice plants. *Plant Physiology* **100**: 662-668.

Mallory-Smith, C.A., Thill, D.C. and Dial, M.J. (1990) Identification of sulfonylurea herbicide-resistant prickly lettuce (*Lactuca serriola*). *Weed Technology* **4**: 163-168.

Mallory-Smith, C., Thill, D.C. and Dial, M.J. (1993) ID-BRI: Sulfonylurea herbicide-resistant lettuce germplasm. *HortScience* **28**: 63-64.

Mauvais, C. (1994) DuPont Agricultural Products, Wilmington, DE. Personal communication.

Mazur, B.J., and Falco, S.C. (1989) The development of herbicide resistant crops, *Annual Review Plant Physiology Plant Molecular Biology* **40**: 441-470.

Mazur, B.J., Chui, C.-F. and Smith, J.K. (1987) Isolation and characterization of plant genes coding for acetolactate synthase, the target enzyme for two classes of herbicides. *Plant Physiology* **85**: 1110- 1117.

McHughen, A. (1989) *Agrobacterium* mediated transfer of chlorsulfuron resistance to commercial flax cultivars. *Plant Cell Reports* **8**: 445-449.

Miki, B.L., Labbe, H., Hattori, J. Ouellet, T. Gabbard, J., Sunohara, G., Charest, P.J. and Iyer, V.N. (1990) Transformation of *Brassica napus* canola cultivars with *Arabidopsis thaliana* acetohydroxyacid synthase genes and analysis of herbicide resistance. *Theoretical Applied Genetics* **80**: 449-458.

Mourad, G. and King, J. (1992) Effect of four classes of herbicides on growth and acetolactate-synthase activity in several variants of *Arabidopsis thaliana*. *Planta* **188**: 491-497.

Mourad, G., Pandey, B. and King, J. (1993) Isolation and genetic analysis of a triazolopyrimidine-resistant mutant of *Arabidopsis*. *Journal Heredity* **84**: 91-96.

Mourad, G., Haughn, G. and King, J. (1994) Intragenic recombination in the CSR1 locus of *Arabidopsis*. *Molecular General Genetics* **243**: 178-184.

Odell, J.T., Caimi, P.G., Yadav, N.S. and Mauvais, C.J. (1990) Comparison of increased expression of wild-type and herbicide-resistant acetolactate synthase genes in transgenic plants, and indication of posttranscriptional limitation on enzyme activity. *Plant Physiology* **94**: 1647-1654.

Pofelis, S., Le, H. and Grant, W.F. (1992) The development of sulfonylurea herbicide-resistant birdsfoot trefoil (*Lotus corniculatus*) plants from *in vitro* selection. *Theoretical Applied Genetics* **83**: 480-488.

Primiani, M.M., Conerman, J.C. and Saari, L.L. (1990) Resistance of kochia (*Kochia scoparia*) to sulfonylurea and imidazolinone herbicides. *Weed Technology* **4**: 169-172.

Saari, L.L. and Mauvais, C.J. (1995) Sulfonylurea herbicide resistant crops. *In: Herbicide Resistant Crops: Agricultural, Environmental, Economic, Regulatory, and Technical Aspects* (S. O. Duke, ed.), 127-142. Lewis Publishers, Chelsea, MI.

Saari, L.L., Cotterman, J.C. and Primiani, M.M. (1990) Mechanism of sulfonylurea herbicide resistance in the broadleaf weed, *Kochia scoparia. Plant Physiology* **93**: 55-61.

Saari, L.L., Cotterman, J.C., Smith, W.F. and Primiani, M.M. (1992) Sulfonylurea herbicide resistance in common chickweed, perennial ryegrass, and Russian thistle. *Pesticide Biochemistry Physiology* **42**: 110-118.

Saari, L.L., Cotterman, J.C. and Thill, D.C. (1994) Resistance to acetolactate synthase inhibiting herbicides. *In: Herbicide Resistance in Plants, Biology and Biochemistry* (S.B. Powles and J.A.M. Holtum, eds.), 83-139. Lewis Publishers, Boca Raton, FL.

Sathasivan, K., Haughn, G.W. and Murai, N. (1991) Molecular basis of imidazolinone herbicide resistance in *Arabidopsis thaliana* var Columbia. *Plant Physiology* **97**: 1044-1050.

Saunders, J.W., Acquaah, G., Renner, K.A. and Doley, W.P. (1992) Monogenic dominant sulfonylurea resistance in sugarbeet from somatic cell selection. *Crop Science* **32**: 1357-1360.

Saxena, P.K., Williams, D. and King, J. (1990) The selection of chlorsulfuron-resistant cell lines of independent origin from an embryogenic cell suspension culture of *Brassica napus L. Plant Science* **69**: 231-237.

Schmitzer, P.R., Eilers, R.J. and Cseke, C. (1993) Lack of cross-resistance of imazaquin-resistant *Xanthium strumarium* acetolactate synthase to flumetsulam and chlorimuron. *Plant Physiology* **103**: 281-283.

Sebastian, S.A. and Chaleff, R.S. (1987) Soybean mutants with increased tolerance for sulfonylurea herbicides. *Crop Science* **27**: 948-952.

Sebastian, S.A., Fader, G.M., Ulrich, J.F., Forney, D.R. and Chaleff, R.S. (1989) Semidominant soybean mutation for resistance to sulfonylurea herbicides. *Crop Science* **29**: 1403-1408.

Stannard, M.E. (1987) Weed control in alfalfa (*Medicago sativa L.*) grown for seed. *MS Thesis*, Montana State University, Bozeman, MT.

Swanson, E.B., Coumans, M.P., Brown, G.L., Patel, J.D. and Beversdorf, W.D. (1988) The characterization of herbicide tolerant plants in *Brassica napus.* 1. After *in vitro* selection of microspores and protoplasts. *Plant Cell Reports* **7**: 83-87.

Tonnemaker, K.A., Auld, D.L., Thill, D.C., Mallory-Smith, C.A. and Erickson, D.A. (1992) Development of sulfonylurea-resistant rapeseed using chemical mutagenesis. *Crop Science* **32**: 1387-1391.

Vermeulen, A., Vaucheret, H., Pautot, V. and Chupeau, Y. (1992) *Agrobacterium* mediated transfer of a mutant *Arabidopsis* acetolactate synthase gene confers resistance to chlorsulfuron in chicory (*Cichorium intybus L.*). *Plant Cell Reports* **11**: 243-247.

Wiersma, P.A., Schmiemann, M.G., Condie, J.A., Crosby, W.L. and Maloney, M.M. (1989) Isolation, expression and phylogenetic inheritance of an acetolactate synthase gene from *Brassica napus. Molecular General Genetics* **219**: 413-420.

Yadav, N., McDevitt, R.E., Benard, S. and Falco, S.C. (1986) Single amino acid substitutions in the enzyme acetolactate synthase confer resistance to the herbicide sulfometuron methyl. *Proceedings National Academy Sciences USA* **83**: 4418-4422.

CHAPTER 9
MECHANISM OF RESISTANCE TO DINITROALINE HERBICIDES

R.J. SMEDA and K.C.VAUGHN
USDA-ARS, Southern Weed Science Laboratory, Stoneville, MS, USA

Introduction

Dinitroaniline herbicides belong to a class of compounds whose primary mechanism of action is the depolymerization of microtubules necessary for cell division and root cell elongation, though these are conveniently known as mitotic disrupter herbicides. Many compounds in this class were discovered in the early 1960's (Probst *et al.*, 1975), and include diverse groups of compounds with a similar structure (dinitroanilines, e.g. trifluralin, pendimethalin; phosphoric amides, e.g. butamiphos, amiprophos-methyl; carbamates, e.g. propham, chlorpropham), as well as compounds with a unique structure (dithiopyr, DCPA, pronamide).

Many of the mitotic disrupters are soil applied herbicides for pre-emergence control of many weedy grass and small-seeded broadleaf species. Mitotic disrupters are used widely in row crop production (e.g. cotton, soybeans), but also are utilized for weed control in small grain, vegetable, turf, ornamental, small fruit and tree fruit production. For some of the mitotic disrupters, a high vapor pressure and susceptibility to photodecomposition reduces herbicide efficacy, unless the herbicide is incorporated in the upper soil profile (5 cm) following application. Generally, mitotic disrupters exhibit low water solubility, and thus remain in the zone where most weed seeds germinate. Herbicide is taken into plants primarily through the shoot of emerging monocots and the hypocotyl or hypocotyl hook of dicots (Parka and Soper, 1977). Mitotic disrupters are taken up by roots, but the herbicide is translocated poorly to the hypocotyl or epicotyl region (Parka and Soper, 1977). Some tolerant crop plants resist the effects of these herbicides either by sequestering the herbicide in lipid bodies and/or having enough seed reserves to grow their root through the zone of herbicide incorporation. Members of the *Umbelliferae* are especially resistant to the dinitroaniline herbicides; however these are apparently not enriched in lipids nor have especially large seeds.

Mechanism and Mode of Action

Mitotic disrupters interfere with the polymerization of free tubulin protein into microtubules. Microtubules are 25 nanometer, hollow cylinders, comprised primarily of α- and β-tubulin subunits. Because microtubules form different arrays (cortical and mitotic; pre-prophase; spindle and phragmoplast) which orchestrate various processes during the cell cycle, from defining cell shape to organizing cell plate formation, disrupting microtubules also disrupt a number of cellular processes including, and most importantly, root cell elongation. Microtubules exist in a dynamic state, with incorporation of α- and β-tubulin in the form of a heterodimer on one end ("+ end") and disassembly on the other ("- end"). In addition, the microtubules exhibit "dynamic instability", a process of spontaneous disassembly, making the half-life of a given microtubule, in general, fairly short. Gamma tubulin is present at the "end", as well as at periodic distances along the microtubule, and is believed to be necessary for microtubule nucleation. Herbicides which interfere with microtubule polymerization bind to free tubulin heterodimers, and either block the incorporation of the complex or if the complex is added to the growing "+ end", further assembly of tubulin heterodimer into microtubules is prevented (Morejohn *et al.*, 1987; Vaughn and Vaughan, 1990). The disassembly of microtubules on the "- end" continues, however, shortening and eventually leading to loss of the entire microtubule.

Post microscopic analysis of structural abnormalities induced by mitotic disrupters indicated that these compounds could be classified in their ability to disrupt different microtubule arrays and/or microtubule organization (Vaughn and Lehnen, 1991). Many of these studies were conducted on fairly tolerant crop species, so that only marginal effects of the herbicide were detected. More recent studies demonstrate that mitotic disrupters affect all microtubule arrays by inhibiting polymerization, but the effective concentration varies between compounds (Hoffman and Vaughn, 1994). Also, the concentration of herbicide necessary to disrupt formation of one array differs from that needed to disrupt another array, with spindle/phragmoplast arrays being most sensitive and cortical arrays most resistant. This might reflect in vivo differences in the stability or rate of turnover of the microtubules in these arrays. However, it is the elimination or disorganization of cortical arrays which is primarily responsible for the utility of the compounds as herbicides.

Characteristic injury of susceptible plants treated with mitotic disrupters is club shaped or swollen root tips due to cortical array disruption (Figure 1) (Hess, 1987; Vaughn, 1986). Cytologically, different effects are observed, depending on the microtubule array disrupted. When spindle or kinetochore microtubules are not produced, chromosomes in treated cells are arrested in prometaphase, resulting in cells with highly lobed tetraploid (4C) nuclei after reformation of the nuclear envelope around the mass of chromosomes (Figure 2) (Vaughn, 1986; Vaughn *et al.*, 1987). However, isodiametric expansion of root cells in the zone of elongation is the result of loss or disruption of cortical microtubules, those primarily responsible for cell shape (Vaughn and Vaughan, 1988). Swollen roots result from isodiametrically swollen cells, which would normally undergo transverse elongation (Hess, 1987). Abnormal phragmoplast arrays disrupt cell wall formation during telophase, resulting in irregularly shaped, reticulate, or incomplete cell walls (Hess, 1987; Vaughn *et al.*, 1987). Disruption of normal root development stunts susceptible plants, often resulting in plant death, perhaps due to restricted water uptake by these underdeveloped roots.

A B

Figure 1. Typical development of carrot roots following a 24 h exposure to water (A) or 10^{-5} oryzalin (B). Note swelling near the root tip.

Resistant Biotypes

Although initially developed almost three decades ago, mitotic disrupters remain widely used because they are inexpensive, effective, and numerous agronomic crops exhibit tolerance. In most crop production systems, mitotic disrupters are applied prior to crop emergence and activity persists for a period of weeks to months. Subsequent plants emerging are managed by application of herbicides with a different mechanism of action and/or cultivation. The use of multiple approaches for weed control diminishes selection of plants resistant to any one approach. In the cotton belt of the United States, the dinitroaniline herbicide trifluralin, is used as a primary herbicide for a large percentage of the acreage. The high selectivity, relatively long persistence, and repeated use have all encouraged the development of resistant biotypes.

After a decade of continuous annual application of trifluralin in cotton fields throughout northeast South Carolina, a population of goosegrass (*Eleusine indica* (L.) Gaertner) resistant to this and six additional dinitroaniline herbicides was identified (Mudge *et al.*, 1984). Subsequent laboratory studies determined the level of resistance (as measured

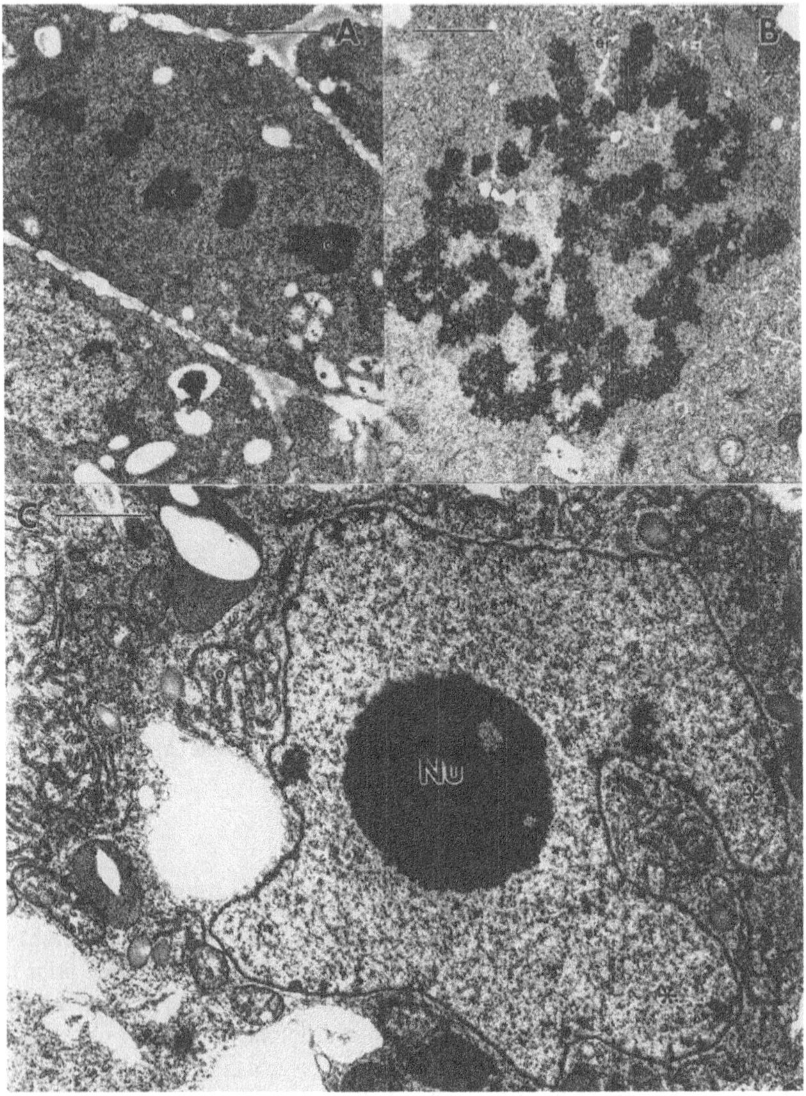

Figure 2. Electron micrographs of dinitroaniline-resistant (A) and dinitroaniline-sensitive (B and C) green foxtail roots after treatment with 1.0 mM oryzalin for 24 h.

A. The resistant biotype develops a normal metaphase plate on which the chromosomes are aligned, indicating no effect of the herbicide on microtubules.

B. The sensitive biotype forms a prometaphase figure with no microtubules present, a so-called arrested prometaphase or C-mitosis.

C. After a disrupted attempt at mitosis, the nuclear envelope reforms around the scattered mass of chromosomes, resulting in a lobe nucleus. The lobes are denoted with asterisks. c= chromosomes; er=endoplasmic reticulum; Nu= nucleolus. Bars= 1.0 mm

by mitotic index) to trifluralin was 1,000- to 10,000-fold compared to susceptible (S) goosegrass (Vaughn, 1986), although a lower level of resistance was determined at the whole plant level (Table 1). Vaughn *et al.* (1990) later reported a biotype of goosegrass [termed intermediate (I)] with only 50-fold resistance to trifluralin, based upon mitotic index figures. Resistant (R) biotypes of goosegrass have also been identified in Alabama, Georgia and Tennessee (Whitwell *et al.*, 1992), most of which have characteristics of the intermediate biotype. Relative levels of resistance to mitotic disrupters varied among biotypes as well, although cross-resistance to both dithiopyr and amiprophosmethyl was observed consistently.

Since this initial discovery, resistance to mitotic disrupters has been identified in rigid ryegrass (*Lolium rigidum* Gaudin) in Australia (McAlister *et al.*, 1995), green foxtail (*Setaria viridis* L. Beauv.) in Canada (Morrison *et al.*, 1989), as well as johnsongrass (*Sorghum halepense* (L.) Pers.) (Wills *et al.*, 1992) and Palmer amaranth (*Amaranthus palmeri* S. Wats) (Gossett *et al.*, 1992) in the cotton belt of the United States. The level of resistance and pattern of cross-resistance to mitotic disrupters varies greatly between these different resistant biotypes (Table 1). Palmer amaranth is the first dicot reported with resistance to mitotic disrupters. In England, blackgrass (*Alopecurus mysouroides* Huds.) (Moss, 1990) is resistant to pendimethalin, although this biotype was selected following repeated use of chlortoluron, which has an unrelated mechanism of action (Moss and Cussans, 1991). As was the case for goosegrass, selection of resistant weed biotypes followed years of repeated application with little or no crop rotation.

In addition to natural selection for resistance, mutants of the unicellular green alga, *Chlamydomonas reinhardtii*, have been selected *in vitro* for resistance to the microtubule depolymerizing drug colchicine (Bolduc *et al.*, 1988). Cross-resistance to the herbicides pronamide, amiprophos-methyl (APM), oryzalin, trifluralin, and profluralin was also detected.

The laboratory selection of higher plants resistant to mitotic disrupter herbicides has proven difficult. In our laboratory, a massive selection of M2 *Arabidopsis* on oryzalin-containing meidum resulted in no oryzalin mutants out of over 750,000 individuals. Only one marginally-resistant mutant selected on pronamide was selected utilizing a similar number of individuals (Vaughn, unpublished). Similar negative results were obtained by a Minnesota cytoskeletal group (Snustad, personal communication), utilizing an oryzalin screen. Generally, a mutant individual is recovered easily from such screening protocols, with herbicide resistant to sulfonylureas, isoxaben and other herbicides recovered and characterized. Using tissue culture selection protocols, Strashnyuk *et al.*, (1993) recovered a mutant cell line of tobacco with resistance to amiprophos-methyl .

Resistance Mechanisms

Variability in the level of resistance to mitotic disrupters both within (goosegrass) and between species (goosegrass versus green foxtail) indicates that one or more mechanisms

Table 1. Weed resistance to mitotic disrupter herbicides; compound used to select for resistance, relative level of resistance, and cross-resistance pattern.

Plant species	Selection Agent	Resistance Level (R/S)	Cross- resistance
Eleusine indica (R)	triflularin	1,000-10,000[a]	Other DNA´s, APM, dithiopyr
Eleusine indica (I)	triflularin	50	Other DNA´s, dithiopyr
Setaria viridis	triflularin	3.2[b]	Other DNA´s, APM, terbutol, DCPA
Sorghum halepense	triflularin	1.5-2[c]	Pendimethalin
Amaranthus palmeri	triflularin	5-6[d]	Other DNA´s
Lolium rigidum	triflularin	9.6-14[e]	Other DNA´s, APM, CIPC

[a]Vaughn, 1986; Vaughn *et al.*, 1987; Vaughn *et al.*, 1990; [b]Smeda *et al.*, 1992; [c]Smeda, unpublished data; [d]Gosset *et al.*, 1992; [e]McAlister *et al.*, 1995

underlies resistance. These potential mechanisms for resistance include, uptake/translocation, metabolism, compartmentalization/lipid content, and alterations in the site of action. Potential site of action changes include hyperstabilization of microtubules, alteration at the herbicide binding site, and post-translational modifications of microtubules. The remainder of this chapter will examine research regarding each mechanism.

Uptake/ Translocation

The first barrier for herbicides is uptake into plants and movement to the site of action. Restricted movement of herbicides has been proposed as a mechanism to explain paraquat resistance (Fuerst *et al.*, 1985), and studies have been conducted on some dinitroaniline-resistant biotypes to determine if uptake or translocation could be altered in the R biotypes. Although Chernicky (1985) found differences in uptake and translocation of ^{14}C-trifluralin and ^{14}C-oryzalin between R and S goosegrass, differences varied between herbicides and were concentration dependent. Therefore, uptake and translocation of dinitroanilines was not thought to explain goosegrass resistance. Similarly, McAlister *et al.* (1995) reported no differences in uptake and translocation between a trifluralin-susceptible and two resistant biotypes of rigid ryegrass.

Metabolism

There are numerous pathways for the degradation of dinitroaniline herbicides, but metabolism is often low in plants (Probst *et al.*, 1975). This lack of metabolism holds true for naturally resistant plants such as carrots, cotton, and soybean (Funderburk *et al.*, 1967; Golab *et al.*, 1967) and also resistant rigid ryegrass (McAlister *et al.*, 1995), a species in which accelerated metabolism does cause resistance to other herbicides. However, Moss (1990) showed chlortoluron resistant blackgrass was cross-resistant to pendimethalin but not other dinitroaniline herbicides. James *et al.* (1995) recently determined that the basis of pendimethalin resistance in one population of blackgrass was attributable to ring-methyl oxidation. Replacement of chemical substituents sensitive to oxidative degradation with insensitive groups, rendered blackgrass susceptible to that compound. Trifluralin, benfluralin, and ethafluralin do not have side chains sensitive to oxidative degradation, and pendimethalin-resistant blackgrass is susceptible to these herbicides. Cytochrome P-450 oxidase appears to be involved with blackgrass resistance based upon the kind of metabolites produced and their inhibition of production by P-450 inhibitors (Menéndez *et al.*, this volume). Herbicide metabolism does not appear operative in R goosegrass because one would expect an immediate effect on cellular processes following herbicide exposure, and then recovery. However, short-term experiments with R goosegrass show that microtubules are not affected following exposure to trifluralin.

Compartmentalization/Lipid Content

Little is known of the subcellular distribution of mitotic disrupter herbicides. Dinitroanilines are highly soluble in lipids, and Hilton and Christiansen (1972) established a positive correlation between seed lipid content and trifluralin tolerance. These data indicate that dinitroanilines are compartmentalized into high lipid regions of the seed. However, Chernicky (1985) reported that the seed lipid content in goosegrass is lower than many dinitroaniline-susceptible plants, and the R biotype has even less total lipid than the S biotype. This would argue that seed lipid content is not the mechanism of dinitroaniline resistance in goosegrass. Also, carrot and other members of the *Umbelliferae* are highly resistant to dinitroanilines, but do not have particularly high lipid contents nor particularly large seeds (Vaughan and Vaughn, 1988). Obviously, compartmentalization of the dinitroanilines into large lipid reserves cannot explain resistance in these cases.

Hyperstabilization of Microtubules

Normal cells contain a balanced level of tubulin in a stable, polymerized form (microtubules) and as a free pool (Cabral *et al.*, 1986). By blocking the polymerization process with microtubule disrupting herbicides, this balance is upset, and microtubules will disappear unless the depolymerization process is inhibited. One potential mechanism to increase stability is to physically cross-link the microtubules so that microtubule dynamics are slowed down. This process, termed microtubule bundling, would allow various microtubule arrays to continue their function in the presence of the herbicide. Certain elements or factors in the cell also modulate the assembly and function of microtubules [e.g. calcium, calmodulin, microtubule associated proteins (MAP's)]. Changes in these MAP's may alter herbicide affects on microtubules. To date, this mechanism has not been shown for any of the dinitroaniline-resistant weeds reported.

Other factors besides microtubule bundling can also stabilize microtubules and render them resistant to microtubule disrupting herbicides. The drug taxol stabilizes microtubules, S goosegrass treated with taxol prior to exposure to low concentrations of oryzalin resulted in microtubules which were not affected by the herbicide (Vaughn and Vaughan, 1991). Under higher concentrations of oryzalin, destabilization of S microtubules was observed. Taxol-treated S goosegrass cells exhibited abnormal cell plate formation. Interestingly, untreated R goosegrass appeared similar structurally to these taxol-treated S goosegrass, indicating that hyperstabilized microtubules in the R goosegrass cause the same herbicide resistance and morphological effects (Vaughn and Vaughan, 1991). Microtubules of R goosegrass are ultra-sensitive to taxol, and most cells exhibit abnormal cell wall formation and incomplete cell division after taxol treatment (Vaughn and Vaughan, 1991). In the presence of dinitroanilines, the hyperstabilized microtubules from R goosegrass are not affected in 24 h treatments. Taxol stabilizes microtubules rendering them resistant to most mitotic disrupters, although R goosegrass is primarily resistant to dinitroanilines and APM (Table 1). Thus, the mechanism of microtubule hyperstability is different between taxol treated microtubules and microtubules of R goosegrass. Hyperstabilization of R goosegrass microtubules apparently underlies the mechanism of dinitroaniline resistance in this weed biotype. This may also explain the differences in levels of resistance obtained by comparing short-term effects with mitotic indices with whole plant assays over days (Table 1). In the longer term experiments, effects would be observed on microtubules because the microtubules in the R biotype are dynamic, but with slower dynamics than in the S biotype. Thus, the prolonged effect of the herbicide could affect even microtubules that are stabilized, effects that would not be obvious in short-term experiments and increases in mitotic indices. Neither I goosegrass nor R green foxtail exhibits sensitivity to taxol, indicating that microtubule hyperstability is not the only mechanism of resistance to dinitroaniline herbicides in these weed biotypes (Smeda *et al.*, 1992; Vaughn and Vaughan, 1990).

Alteration at the Herbicide Binding Site

For certain classes of herbicides (e.g. s-triazines and sulfonlyureas), specific genetic mutations have resulted in altered polypeptides at the site of herbicide action, rendering plants highly resistant to the herbicide (Sherman *et al.*, 1996). Trifluralin and oryzalin both bind to free tubulin, and by incorporation of the tubulin-herbicide complex on the "+ end" of the microtubule, polymerization into microtubules is inhibited. Tubulin is a multigene family in most plants examined (Morejohn and Fosket, 1991). Theoretically, mutations in one or more tubulin genes could confer herbicide resistance. Hyperstabilization of microtubules from R goosegrass may be due to one or more alterations in a tubulin gene, however, definitive demonstration of a site of action change can only occur following sequencing of the tubulin genes in R and S plants. This has not been shown for any of the dinitroaniline-resistant plants reported, to date. A β-tubulin isoform unique to one biotype of R goosegrass was reported (Vaughn and Vaughan, 1990), but the importance of this isoform to dinitroaniline resistance has not yet been established. Changes in both α- and β-tubulin forms have been found in various goosegrass biotypes, but it is not known if these are simply biotypic differences or the cause of the resistance. In vitro polymerization of tubulin protein in microtubules reveals polymerization of tubulin into microtubules in the presence of oryzalin in the R but not the S biotype.

Alterations in tubulin encoding genes have been determined in two mutants of *Chlamydomonas reinhardtii* (Lee and Huang, 1990). Resistance was the result of two different single base substitutions in a β-tubulin gene. Strashnyuk *et al.* (1993) had preliminary evidence that their tobacco mutant selected for resistance to APM exhibited an altered β-tubulin. Undoubtedly, future genetic mutations in plant tubulin genes will give rise to herbicide resistant plants. The facts that two distinct mutations have already been characterized and that multiple genes control tubulin synthesis indicate that numerous viable mutations are possible.

Post-Translational Modifications of Microtubules

All microtubule arrays in plants are assembled with tubulin from highly conserved tubulin genes. The creation of different tubulin isoforms within a single plant can be accomplished in numerous ways, one of which is post-translational modification. Four reported modifications are acetylation, tyrosination, phosphorylation, and polyglutamylation (Kozminski *et al.*, 1993). Recent studies indicate that post-translational modifications can significantly affect the sensitivity of microtubules to depolymerization by mitotic disrupting herbicides (Hoffman and Vaughn, 1995, 1996; James, *et al.*, 1993). Acetylated tubulin is found in stable microtubule arrays, but is not thought to be the direct cause of increasing stability (Kozminski, *et al.*, 1993). James *et al.* (1993) generated a mutant of *Chlamydomonas* with higher levels of acetylated tubulin, and found the mutant more resistant to mitotic disrupters than the wild type *Chlamydomonas*. Similarly, Kozminski *et al.* (1993) determined that a transformant of *Chlamydomonas* deficient in acetylated tubulin was more sensitive to oryzalin than wild type algal cells. Recently, Hoffman and Vaughn (1996) showed acetylated microtubule arrays in spermatogenous cells of *Ceratropteris* were resistant to numerous mitotic disrupting herbicides. However, non-acetylated, tyrosinated tubulin in dynamic microtubule arrays were sensitive to these herbicides, and exhibited symptomology consistent with herbicide activity in sensitive plants. Therefore, post-translational acetylation of tubulin can give rise to organisms resistant to mitotic disrupters. To date, this mechanism has not been identified in any herbicide resistant higher plants.

Conclusions

Reports of herbicide resistance, once relegated primarily to photosynthetic inhibitors, has more frequently included herbicides which disrupt microtubules. Long term herbicide use with little or no rotation has selected for resistant biotypes. Unfortunately, differences among and between species in the level of resistance and pattern of cross-resistance gives evidence that multiple mechanisms of resistance have arisen, making agronomic control of resistant plants more difficult. Currently known mechanisms of resistance include high seed lipid content, herbicide metabolism, hyperstabilization of microtubules, and post-translational acetylation of microtubules. Although specific mutations in tubulin genes giving rise to altered tubulin has not yet been shown in higher plants, mutant strains of *Chlamydomonas* demonstrate that occurrence of this type of mutational event is merely a matter of time. However, mitotic disrupter herbicides have benefitted agriculture greatly, and precautions in the use of these compounds should ensure their continued use.

References

Bolduc, C., Lee, V.D. and Huang, B. (1988) β-Tubulin mutants of the unicellular green alga *Chlamydomonas reinhardtii. Proceedings National Academy Sciences USA* **85**: 131-135.

Cabral, F.R., Brady, R.C. and Schibler, M.J. (1986) A mechanism of cellular resistance to drugs that interfere with microtubule assembly. *Annals New York Academy Sciences* **466**: 745-756.

Chernicky, J.P. (1985) An investigation into the resistance of goosegrass (*Eleusine indica*) to dinitroaniline herbicides. *Ph.D. Thesis*, 82 pp. University of Illinois, Urbana, IL.

Fuerst, E.P., Nakatani, H.Y., Dodge, A.D., Penner, D. and Arntzen, C.J. (1985) Paraquat resistance in *Conyza. Plant Physiology* **77**: 984-989.

Funderburk, H.H., Jr., Schultz, D.P., Negi, N.S., Rodriguez-Kabana, R. and Curl, E.A. (1967) Metabolism of trifluralin by soil microorganisms and higher plants. *Proceedings Southern Weed Science Society* **20**: 389.

Golab, T., Herberg, R.J., Parka, S.J. and Tepe, J.B. (1967) Metabolism of carbon-14 trifluralin in carrots. *Journal Agricultural Food Chemistry* **15**: 638-641.

Gossett, B.J., Murdock, E.C. and Toler, J.E. (1992) Resistance of palmer amaranth (*Amaranthus palmeri*) to the dinitroaniline herbicides. *Weed Technology* **6**: 587-591.

Hess, F.D. (1987) Herbicide effects on the cell cycle of meristematic plant cells. *Review Weed Science* **3**: 183-203.

Hilton, J.L. and Christiansen, M.N. (1972) Lipid contribution to selective action of trifluralin. *Weed Science* **20**: 290-294.

Hoffman, J.C. and Vaughn, K.C. (1994) Mitotic disrupter herbicides act by a single mechanism but vary in efficacy. *Protoplasma* **179**: 16-25.

Hoffman, J.C. and Vaughn, K.C. (1995) Post-translational tubulin modifications in spermatogenous cells of the pteridophyte *Ceratopteris richardii. Protoplasma* **186**: 169-182.

Hoffman, J.C. and Vaughn, K.C. (1996) Acetylated microtubule arrays in *Ceratopteris* spermatogenous cells are resistant to mitotic disrupter herbicides. *Protoplasma.* In press.

James, E.H., Kemp, M.S. and Moss, S.R. (1995) Phytotoxicity of trifluoromethyl- and methyl-substituted dinitroaniline herbicides on resistant and susceptible populations of black-grass (*Alopecurus myosuroides*). *Pesticide Science* **43**: 273-277.

James, S.W., Silflow, C.D., Stroom, P. and Lefebvre, P.A. (1993) A mutation in the alfa-tubulin gene of *Chlamydomonas reinhardtii* confers resistance to anti-microtubule herbicides. *Journal Cellular Science.* **106**: 209-218.

Kozminski, K.G., Diener, D.R. and Rosenbaum, J.L. (1993) High level expression of nonacetylatable a-tubulin in *Chlamydomonas reinhardtii. Cell Motility Cytoskeleton* **25**: 158-170.

Lee, V.D. and Huang, B. (1990) Missense mutations at lysine 350 in β2-tubulin confer altered sensitivity to microtubule inhibitors in *Chlamydomonas. Plant Cell* **2**: 1051 1057.

McAlister, F.M., Holtum, J.A.M. and Powles, S.B. (1995) Dinitroaniline herbicide resistance in rigid ryegrass (*Lolium rigidum*). *Weed Science* **43**: 55-62.

Morejohn, L.C., Bureau, T.E., Mole-Bajer, J., Bajer, A.S. and Fosket, D.E. (1987) Oryzalin, a dinitroaniline herbicide, binds to plant tubulin and inhibits microtubule polymerization *in vitro. Planta* **172**: 252-264.

Morejohn, L.C. and Fosket, D.E. (1991) The biochemistry of compounds with antimicrotubule activity in plant cells. *Pharmacology Therapeutics* **51**: 217-230.

Morrison, I.N., Todd, B.G. and Nawolsky, K.M. (1989) Confirmation of trifluralin-resistant green foxtail (*Setaria viridis*) in Manitoba. *Weed Technology* **3**: 544-551.

Moss, S.R. (1990) Herbicide cross-resistance in slender foxtail (*Alopecurus myosuroides). Weed Science* **38**: 492-496.

Moss, S.R. and Cussans, G.W. (1991) The development of herbicide-resistant populations of *Alopecurus myosuroides* (black-grass) in England. *In: Herbicide Resistance in Weeds and Crops*, (J.C. Caseley, G.W. Cussans, and R.K. Atkin, eds.), 45-55. Butterworth-Heinemann Ltd., Oxford, U.K.

Mudge, L.C., Gossett, B.J. and Murphy, T.R. (1984) Resistance of goosegrass *(Eleusine indica)* to dinitroaniline herbicides. *Weed Science* **32**: 591-594.

Parka, S.J. and Soper, O.F. (1977) The physiology and mode of action of the dinitroaniline herbicides. *Weed Science* **25**: 79-87.

Probst, G.W., Golab, T. and Wright, W.L. (1975) Dinitroanilines. *In: Herbicide, Chemistry, Degradation, and Mode of Action* (P.C. Kearney and D.D. Kaufman, eds.), 453-500. Marcel Dekker, Inc., NY.

Sherman, T.D., Vaughn, K.C. and Duke, S.O. (1996) Mechanisms of action and resistance to herbicides. *In: Herbicide-Resistant Crops, Agricultural, Environmental, Economic, Regulatory, and Technical Aspects* (S.O. Duke, ed.), 13-35. CRC Lewis Publishers, Boca Raton, FL.

Smeda, R.J., Vaughn, K.C. and Morrison, I.N. (1992) A novel pattern of herbicide cross resistance in a trifluralin-resistant biotype of green foxtail [*Setaria viridis* (L.) Beauv.]. *Pesticide Biochemistry Physiology* **42**: 227-241.

Strashnyuk, N.M., Blume, Y.B., Smertenko, A.P., Sidorov, V.A. and Glebc, Y.Y. (1993) The obtaining of amiprophos methyl resistant lines of *Nicotiana plumbaginifolia* containing mutant tubulin. *Doklady Akademii Nauk* **332**: 240-243.

Vaughan, M.A. and Vaughn, K.C. (1988) Carrot microtubules are dinitroaniline resistant. I. Cytological and cross-resistance studies. *Weed Research* **28**: 73-83.

Vaughn, K.C. (1986) Cytological studies of dinitroaniline-resistant *Eleusine. Pesticide Biochemistry Physiology* **26**: 66-74.

Vaughn, K.C. and Lehnen Jr, L.P. (1991) Mitotic disrupter herbicides. *Weed Science* **39**: 450-457.

Vaughn, K.C., Marks, M.D. and Weeks, D.P. (1987) A dinitroaniline-resistant mutant of *Eleusine indica* exhibits cross-resistance and supersensitivity to anitmicrotubule herbicides and drugs. *Plant Physiology* **83**: 956-964.

Vaughn, K.C. and Vaughan, M.A. (1988) Mitotic disrupters from higher plants, effects on plant cells. *In: Biologically Active Natural Products: Potential Use in Agriculture* (H.G. Cutler, ed.), **380**: 273-293. The American Chemical Society, Washington D.C.

Vaughn, K.C. and Vaughan, M. A. (1990) Structural and biochemical characterization of dinitroaniline-resistant *Eleusine. In: Managing Resistance to Agrochemicals* (M.B. Green, H.M. LeBaron, and W.K. Moberg, eds.), **421**: 364-375. The American Chemical Society, Washington D.C.

Vaughn, K.C. and Vaughan, M.A. (1991) *Dinitroaniline* resistance in *Eleusine indica* may be due to hyperstabilized microtubules. *In: Herbicide Resistance in Weeds and Crops* (J.C. Caseley, G.W. Cussans and R.K. Atkin, eds.), 177-186. Butterworth-Heinemann, Ltd., Oxford, U.K.

Vaughn, K.C., Vaughan, M.A. and Gossett, B.J. (1990) A biotype of goosegrass *(Eleusine indica)* with an intermediate level of dinitroaniline herbicide resistance. *Weed Technology* **4**: 157-162.

Whitwell, T., Baird, V., Wells, J. and Tucker, K. (1992) Comparison of dinitroaniline resistant goosegrass biotypes from Alabama, Tennessee, Georgia and South Carolina. *Proceedings Southern Weed Science Society* **45**: 297.

Wills, G.D., Byrd Jr., J.D. and Hurst, H.R. (1992) Herbicide resistant and tolerant weeds. *Proceedings Southern Weed Science Society* **45**: 43.

CHAPTER 10
CYTOCHROME P450 ENDOWED HERBICIDE METABOLISM

F. DURST, J P. SALAÜN, D. WERCK-REICHHART and F. ZIMMERLIN
Laboratoire d' Enzymologie Cellulaire et Moléculaire, Institute de Biologie Moléculaire des Plantes-CNRS, 28 rue Goethe, F-67083 Strasbourg, Cedex

Introduction

Metabolism is a key component of herbicide efficiency, selectivity, and in the appearance of resistance. In many cases, oxygen transfer is the first, obligate step in the biotransformation process. These reactions are both slow and irreversible. Therefore, they determine to a large part the kinetics of elimination of the active compound (or its formation in some cases of conversion of a proherbicide to a phytotoxic compound). Although other enzymes like peroxidases may also be involved, it is now clear that cytochrome P450 plays a pivotal role in herbicide disposition. In this paper we will discuss some aspects of the biochemistry and genetics of plant P450 that are relevant to this topic, and show an example of an attempt for identifying and isolating P450 species involved in herbicide disposition.

P450s Catalyze the Oxidation of Herbicides

Cytochrome P450s are extremely versatile mixed function oxidases which carry out such diverse reactions as oxygenation, isomerization, dehydratation and even reduction. They are best known and most studied for their oxygenase activity, i.e. the transfer of one oxygen atom originating from atmospheric dioxygen to lipophilic physiological substrates and to xenobiotics. It is this last function, the capacity to attack a vast range of structurally unrelated and non reactive chemicals to transform them into more polar, soluble and disposable products, that has attracted the attention of pharmacologists, toxicologists, nutritionists... It is now well recognized that P450 plays a pivotal role in the biotransformation of drugs, pesticides, pollutants and in the activation of procarcinogens. Plant P450s catalyze all types of oxygenation displayed by their animal counterparts (aliphatic and aryl hydroxylations, heteroatom oxidation, N- and O-dealkylation, cleavage of carbon-carbon bonds, ring closure) and some more like ring migration and carbon-carbon coupling. All plant P450s characterized to date most resemble the enzymes found in animal and fungal microsomes, with the notable exception of the allene oxide synthase (Song *et al.*, 1993) which depends on fatty acid hydroperoxides rather than on NADPH and O_2, and may be localized in other membranes than the ER (Song *et al.*, 1993; Pan *et al.*, 1995).

Interestingly, among the first reactions hinting to the existence of P450 in plants was the N-demethylation of phenylurea herbicides described in the late sixties by Frear and his colleagues (Fear *et al.*, 1969) Since then, the existence of plant P450s, in the form of a microsomal multienzyme complex most resembling the mammalian hepatic oxygenases, has been firmly established. For reviews see: (Riviere and Cabanne, 1987; Durst, 1991; Durst and Benveniste, 1993; Bolwell *et al.*, 1994). The P450 mediated oxidation of more than 60 physiological (phenylpropanoids, fatty acids, terpenes alkaloids) and xenobiotic (pesticides, pollutants, drugs) substrates has been described (Durst and Benveniste, 1993; Bolwell *et al.*, 1994). Some of the P450-dependent oxidation reactions of herbicides which have been thoroughly established are listed in Table 1.

Table 1. Some P450-dependent oxidation of herbicides characterized in plant microsomes

Substrate	Reaction	Reference
Bentazon	Aryl-hydroxylation	1, 2, 3
Chlortoluron	N-dealkylation,hydroxylation	4, 5, 6
Clorsulforum	Aryl-hydroxylation	7, 8
Diclofop	Aryl-hydroxylation	7, 9
Flumetsulam	N-dealkylation,hydroxylation	10
Linuron	N-demethylation	8
Metolachlor	O-dealkylation	11
Monuron	N-dealkylation	12
Primisulfurom	Aryl-hydroxylation	13
Trialsulfuron	Aryl-hydroxylation	11, 14

1. McFadden *et al.*,1990; 2. Moreland and Corbin, 1991; 3. Haack and Balke, 1994; 4. Mougin *et al.*, 1990; 5. Fonné-Pfister and Klaus, 1990; 6. Pierrel *et al.*,1994; 7. Zimmmerlin and Durst, 1992; 8. Frear *et al.*, 1991. 9. McFadden *et al.*, 1989; 10. Frear *et al.*, 1993; 11. Moreland *et al.*, 1993; 12. Frear *et al.*, 1969; 13. Fonne-Pfister *et al.*, 1990; 14. Thalacker *et al.*, 1994.

Plant P450s Are Members of a Superfamily of Genes

The progress of molecular biology techniques has enabled the cloning of more than 50 P450 genes from a variety of plant species but only 5 physiological activities have been identified to date: the cinnamate hydroxylase (CYP73), the flavonoid 3',5'-hydroxylase (CYP75), a tyrosine hydroxylase (CYP79), berbamunine synthase (CYP80), and the allene oxide synthase (CYP74). In fact it appears now that cloning is much easier than assigning a function to the cloned genes and therefore the catalytic activity of most of the encoded proteins remains unknown. Sequence analysis, homology searches and phylogenetic studies show that the plant P450 belongs to the P450s superfamily of genes. They have been grouped in 16 new families (CYP71 to CYP86) and several sub-families. However, since several P450 have not yet been submitted to the nomenclature committee, and since the cloning effort is in an ascending curve, it is likely that the total number of P450 families in plants will be much higher. It is noteworthy that, since P450 families are defined as comprising those proteins which share more than 40% homology, this means that not a single of the plant P450s isolated until now shares more than 40% positional aminoacid identity with any other P450 gene from animals, fungi or bacteria. Phylogenetic trees of the plant P450s with members of the families from other organisms (Durst and Nelson, 1995) show that the plant oxygenases form two distinct groups:

i) one large group which stems from a single branch and comprises most families, subfamilies and individual P450s cloned till now: CYP71, CYP73, and CYP75-CYP85
ii) a second group of a small number of families (CYP72, and newly cloned P450s at Strasbourg,) which cluster near to P450s from fungi (CYP51, CYP52) and animals (CYP4).

It is our expectation that the P450s from the first group will be found to perform plant specific and species physiological reactions, while those from the second group do probably catalyze more general reactions like oxidation of fatty acids, of alkanes and of sterols. It is likely that the P450s which participate in the herbicide metabolism, and particularly those contributing to metabolism-based selectivity, belong to the first group.

Since plant microsomes have low stability and host a complex array of P450 isoforms, substrate specificity studies yield ambiguous results as to the number of P450 forms which oxidize a single herbicide, and, conversely, as to the number of herbicides that are oxidized by a single P450 form. The capacity of recombinant plant P450s to oxidize xenobiotics has only been reported with CYP73 (Pierrel *et al.*, 1994) and CYP71 (Bozak *et al.*, 1992). CYP71 performed p-chloro-N-methylaniline N-demethylation, confirming previous work using avocado microsomes. We have expressed CYP73 cloned from Jerusalem artichoke, integrated in a multicopy vector, in a dedicated yeast strain W(R) which overexpresses its own reductase. This system provides a suitable environment for the expressed protein which displays very high turn-over (about 400). Transformed yeast may be grown and induced in such a way that CYP73 represents the quasi totality of the P450 present.

Using this system, over 20 different xenobiotic molecules were tested as putative substrates. Only five, including chlortoluron, were metabolized to detectable amounts of products (Pierrel *et al.*, 1994; Schalk *et al.*, 1996; Table 2).

Chlortoluron was hydroxylated at the ring-methyl group but not dealkylated. Since we had previously shown that Jerusalem artichoke microsomes catalyze ring-methyl-hydroxylation and complete N-dealkylation, it is concluded that at least 2, probably 3, P450 isoforms metabolize this herbicide in this plant. Conversely, our results also demonstrate that a same P450 form may metabolize numerous (5 out of the 21 tested) xenobiotics It should be noted however that this enzyme was surprisingly selective for the physiological substrates: among the various compounds tested (known plant P450 substrates,

Table 2. Xenobiotic oxidation reactions catalyzed by yeast-expressed CYP73

Substrate	Product
Naphtoic acid	Hydroxynaphtoate
7-methoxycoumarin	Umbelliferone
7-ethoxycoumarin	Umbelliferone
p-chloro-N-methylaniline	Aniline
Chlortoluron	Hydroxychlortoluron
1-aminobenzotriole	Adduct
2,4-dichlorophenoxypropyne	Adduct

phenylpropanoids and flavonoids incorporating the cinnamoyl moiety), only cinnamic acid was turn-over (Pierrel *et al.,* 1994).

There has been a long debate about the number and identity of the P450 species involved in xenobiotic metabolism by plants. It is our view that P450s with relaxed substrate specificity such as those described in animal systems are probably not found in plants. Instead, the pesticide might be metabolized by one or more of the physiologically competent P450.

Herbicide Metabolism in Plants Transformed with P450

A promising avenue which is opened now by the availability ot P450 genes of various origins, is the transformation of crop plants with detoxifying enzymes. Recently tobacco plants have been transformed with a engineered gene coding for a fusion protein composed from CYP1A1 from the rat and the P450-reductase from yeast (Shiota *et al.,* 1994). The fused protein was expressed in the plant microsomes and conferred relative chlortoluron tolerance. Apparently, chlortoluron was the only tested herbicide to be detoxified by this system. This is somehow unexpected since CYP1A1 is generally considered as an archetype of the animal 'broad range' P450s. A very interesting work was recently described by O'Keefe *et al.,* (1994). Tobacco was transformed with CYP105A1 from *Streptomyces griseolus* which metabolizes sulfonylurea. Since this bacterial, soluble, P450 derives electrons from ferredoxin instead of from NADPH-P450 reductase the gene was transformed for chloroplast targeting of CYP105. The system was found to be active and capable of activating a sulfonylurea pro-herbicide. Heterologous plant-plant transformation has not yet been described.

In Vitro Versus *In Vivo* Evidence for P450 Involvement

In vitro and *in vivo* experiments (Table 3) do not provide the same information. There is ample evidence, from *in vitro* studies using microsomal preparations, that P450 is capable of oxidizing most classes of herbicides. However, these experiments do not prove that the reaction occurs within the plant at a rate susceptible to produce a tolerance or resistance.

Table 3: *In vitro* versus *in vivo* evidence for P450 involvement

In vitro	*In vivo*
Microsomal localization	Metabolites are `P450-like'
O_2 and NADPH dependent	P450 inhibitors decrease metabolism and tolerance
Inhibited by CO and reversed by light	P450 inducers (safeners) enhance metabolism and tolerance
Inhibited by anti reductase antibodies	

That these P450 reactions may endow resistance toward herbicides is mainly attested by *in vivo* studies showing that P450 inhibitors may act as herbicide synergists and sensitize tolerant plants while, conversely, inducers of P450 may protect susceptible plants against phytotoxic doses of herbicide (Burnet *et al.*, 1993a,b; Christopher *et al.*, 1994). Induction of P450 by safeners provides a rational basis for the understanding of the protective effect of these compounds. It is very likely that it is due to the selective induction of one or more P450 isozymes which catalyze a rate-limiting step of the herbicide detoxification. These inductions provide also much needed tools to increase the specific content of some P450s prior to purification. One surprising and disturbing fact is that in some cases, despite convergent evidence from both inhibitor and inducer studies, the reaction was not amenable to in vitro assay, despite the considerable progress that has been made in several laboratories. Whether this is due to the presence of powerful inhibitors or to the unstability of certain forms of P450 remains to be determined.

Identification of Herbicide Metabolizing P450s.

If P450 plays a role in herbicide resistance, some genetic mechanism such as gene amplification, alternative splicing, gene duplication, point mutation will be involved. Any understanding of these mechanisms will require access to the gene. This may be achieved either through genetic methods (use of mutants, degenerate PCR cloning), or via purification of the protein. Purification of plant P450 is an extremely strenuous task because of the low abundance of these proteins, their marked hydrophobic character and their lability. However, promising results have recently been published. A triasulfuron hydroxylase has been purified from wheat and hydroxylase activity was reconstituted (Thalacker *et al.*, 1994). Also, partial purification of a wheat flumetsulam hydroxylase has been recently obtained (Galitz and Coolbaugh, 1995).

Another route to the protein is through affinity or mechanism-based tagging. We have recently used the latter technique to obtain and micro-sequence P450s that might be involved in herbicide turn-over. Diclofop hydroxylation is the prime mechanism for tolerance of wheat to this herbicide (Zimmerlin and Durst 1992; McFadden *et al.*, 1989). Inhibitor and inducer studies have suggested that diclofop was in fact hydroxylated by a lauric acid (μ-1)-hydroxylase (Zimmerlin *et al.*, 1992; Figure 1).

We have devised mechanism-based inactivators which inhibit the laurate hydroxylase These compounds inhibit to a same extent diclofop and laurate oxidation. Furthermore, we have modified the inhibitors to render them stable in vivo and demonstrated that they are powerful and selective synergists for diclofop in wheat [to be published]. A radiolabeled inhibitor has been synthesized and shown to label proteins in the 51-53 Kd range in a reconstituted system comprising partially purified P450 from induced *Vicia sativa* seedlings, P450 reductase, cytochrome b5 and NADPH. The labeled proteins were isolated, digested with protease V8, and peptides were micro-sequenced. Two of these do map to the N-terminal membrane anchor and near the heme-binding domain. This opens a way for the cloning of this hydroxylase.

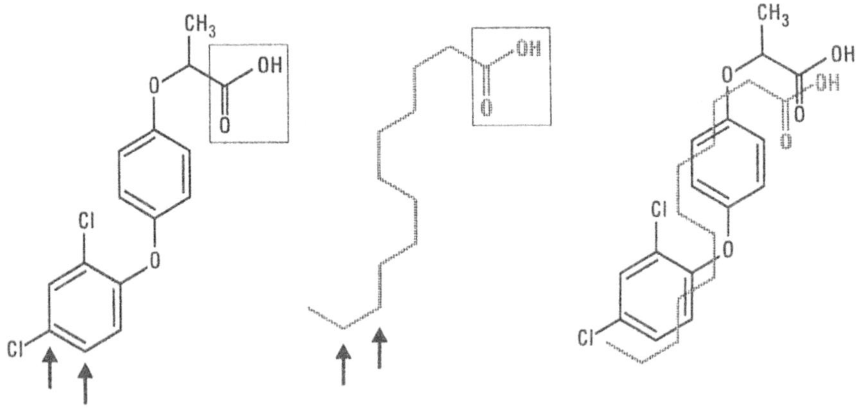

Figure 1. Diclofop and lauric acid may be substrates for the same P450. Free carboxilic groups are required for activity with both substrates. Major sites of oxidative attack are indicated by arrows.

Conclusion

There is now a considerable body of results concerning the involvement of P450 in the metabolism of and resistance to herbicides. Also, chemical means to induce and inhibit these oxygenases have been developed and progress is made in the purification of the proteins. The availability of molecular biology methods provides for a rapidly increasing number of DNA clones, which, upon heterologous expression, may be screened for herbicide metabolizing activity. These new developments should help to understand the mechanistic and molecular basis of P450 endowed herbicide resistance

References

Bolwell, G.P., Bozak, K. and Zimmerlin, A. (1994) Plant cytochrome P450. *Phytochemistry* **37:** 1491-1506.

Bozak, K.R., O'Keefe, D.P. and Christoffersen, R.E. (1992) Expression of a ripening-related avocado (*Persea americana*) Cytochrome P450 in yeast. *Plant Physiology* **100**: 1976-1981.

Burnet, M.W.N., Lovey, B.R., Holtum, J.A.M. and Powles, S.B. (1993a) Increased detoxification is a mechanism of simazine resistance in *Lolium rigidum. Pesticide Biochemistry Physiology* **46**: 207-218.

Burnet, M.W.M, Loveys, B.R., Holtum, J.A.M., and Powles, S.B. (1993b) A mechanism of chlorotolurom resistance in *Lolium rigidum. Planta* **190**: 182-189.

Christopher, J.T., Preston, C., and Powles, S.B. (1994) Malathion antagonizes metabolism-based chlorsulforun resistance in *Lolium rigidum. Pesticide Biochemistry Physiology* **49**: 172-182.

Durst F. (1991) Physiology and biochemistry of plant cytochrome P-450 enzymes. *In: Frontiers in Biotransformation, Vol. 4-Microbial and Plant Cytochrome P450: Biochemical Charcacteristics, Genetic Engineering and Practical Inplications,* (K. Ruckpaul and H. Rein, eds), 191-232. Akademie Verlag, Berlin.

Durst, F. and Benveniste, I. (1993) Plant P450. *In: Handbook of Experimental Pharmacology,* **105** (J.B. Schenkman and H. Greim, eds.), 293-310. Springer, Heidelberg.

Durst, F, and Nelson, D.R. (1995) Phylogeny and evolution of plant cytochrome P450 and P450 reductases. *Drug Metabolism. Drug Interactions* **12**: 189-206.

Fonnè-Pfister, R., Gaudin, J., Kreuz, K., Ramsteiner, K. and Ebert E. (1990) Hydroxylation of primisulfuron by an inducible cytochrome P-450-dependent monooxygenase system from maize. *Pesticide Biochemistry Physiology* **37**: 165-173.

Fonné-Pfister, R. and Klaus, K. (1990) Ring-methyl hydroxylation of chlorotoluron by an inducible cytochrome P-450-dependent enzyme from maize. *Phytochemistry* **224**: 835-844.

Frear, D.S., Swanson, H.R. and Tanaka, F.S. (1969) N-demethylation of substituted 3-(phenyl)-1-methylureas: isolation and characterisation of a microsomal mixed function oxidase from cotton. *Phytochemistry* **8**: 2157-2169.

Frear, D.S., Swanson, H.R. and Thalacker, F.W. (1991) Induced microsomal oxidation of diclofop, triasulfuron, chlorsulfuron, and linorun in wheat. *Pesticide Biochemistry Physiology* **41**: 274-287.

Frear, D.S., Swanson, H.R. and Tanaka, F.S. (1993) Metabolism of flutmetsulam (DE-498) in wheat, corn, and barley. *Pesticide Biochemistry Physiology* **45**: 178-192.

Galitz, D.S. and Coolbaugh, R.C. (1995) Partial purification of flumetsulam hydroxylase from wheat seedlings shoots. *In: Proceedings International Symposium Weed and Crop Resistance to Herbicides* (R. De Prado, J. Jorrín, L. García-Torres and G. Marshall, eds.), 75-77. University of Cordoba, Cordoba, Spain.

Haack, A.E. and Balke, N.E. (1994) Enhancement of microsomal bentazon 6-hydroxylase and cinnamic acid 4-hydroxylase activities from grain sorghum shoots. *Pesticide Biochemistry Physiology* **50**: 92-105.

McFadden, J.J., Frear, D.S. and Mansager, E.R. (1989) Aryl hydroxylation of diclofop by a cytochrome P450 dependent monooxygenase from wheat. *Pesticide Biochemistry Physiology* **34**: 92-100.

McFadden, J.J., Gronwald, J.W. and Eberlein, C.V. (1990) In vitro hydroxylation of bentazon by microsomes from naphthalic anhydride-treated corn shoots. *Biochemical Biophysical Research Communications* **168**: 206-213.

Moreland, D.E. and Corbin, F.T. (1991) Influence of safeners on the *in vivo* and *in vitro* metabolism of bentazon and metolachlor by grain sorghum shoots - A preliminary report. *Zeischrift Naturforschung* **46c**: 906-914.

Moreland, D.E., Corbin, F.T. and McFarland, J.E. (1993) Oxidation of multiple substrates by corn shoot microsomes. *Pesticide Biochemistry Physiology* **47**: 206-214.

Mougin, C., Cabanne, F., Canivenc, M.-C. and Scalla, R. (1990) Hydroxylation and N-demethylation of chlortoluron by wheat microsomal enzymes *Plant Science* **66**: 195-203.

O'Keefe, D.P., Tepperman, J.M., Dean, C., Leto, K.J., Erbes, D.L. and Odell, J.T. (1994) Plant expression of a bacterial cytochrome P450 that catalyzes activation of a sulfonylurea proherbicide. *Plant Physiology* **105**: 473-482.

Pan, Z., Durst, F., Werck-Reichhart, D., Camara, B., Cornish, K. and Backhaus R.A. (1995) The major protein of Guayule rubber particles is a cytochrome P450. Characterization based on cDNA cloning and spectroscopic and enzymatic analysis of the solubilized protein. *Journal Biological Chemistry* **270**: 8487-8494.

Pierrel, M.A., Batard, Y., Kazmaier, M., Mignottevieux, C., Durst, F. and Werck-Reichhart, D. (1994) Catalytic properties of the plant cytochrome-P450 Cyp73 expressed in yeast. Substrate-specificity of a cinnamate hydroxylase. *European Journal Biochemistry* **224**: 835-844.

Riviere, J.L. and Cabanne, F. (1987) Animal and plant cytochrome P-450 systems. *Biochimie.* **69**: 743-752.

Schalk, M., Pierrel, M.A, Zimmerlin, A., Batard, Y., Durst, F, and Werck-Reichhart, D. (1996) Interaction of xenobiotics and herbicides with CYP73, an isolated plant P450 expressed in yeast. *In: Proceedings International Symposium Weed and Crop Resistance to Herbicides* (R. De Prado, J. Jorrín, L. García-Torres and G. Marshall, eds.), 107-109. University of Cordoba, Cordoba, Spain.

Shiota, N., Nagasawa, A., Sakaki, T., Yabusaki, Y, and Ohkawa, H. (1994) Herbicide- resistant tobacco plants expressing the fused enzyme between rat cytochrome P4501A1 (CYP1A1) and yeast NADPH-cytochrome P450 oxydoreductase. *Plant Physiology* **106**: 17-23.

Song, W.C., Funk, C.D., Brash, A.R. (1993) Molecular cloning of an allene oxide synthase- A cytochrome-P450 specialized for the metabolism of fatty acid hydroperoxides. *Proceedings National Academy Sciences USA* **90**: 8519-8523.

Thalacker, F.W., Swanson, H.R. and Frear, D.S. (1994) Characterization, purification, and reconstitution of an inducible cytochrome P450-dependent trialsuforon hydroxylase from wheat. *Pesticide Biochemistry Physiology* **49**: 209-223.

Zimmerlin, A. and Durst, F. (1992) Aryl hydroxylation of the herbicide diclofop by a wheat cytochrome-P-450 monooxygenase - substrate specificity and physiological activity. *Plant Physiology* **100**: 874-881.

Zimmerlin, A., Salaün, J.-P, Durst, F. and Mioskowski, C. (1992) Cytochrome P-450 dependent hydroxylation of lauric acid at the subterminal position and oxidation of unsaturated analogs in wheat microsomes. *Plant Physiology* **100**: 863-873.

CHAPTER 11
GLUTATHIONE TRANSFERASES AND HERBICIDE METABOLISM AND SELECTIVITY

R. EDWARDS
University of Durham, Department of Biological Sciences, Durham, DH1 3LE, UK.

Introduction

In this short paper it is not intended to review glutathione conjugation of herbicides in plants (see Lamoureux and Rusness, 1993). Rather, some of the factors which regulate the role of glutathione transferases (GSTs) in herbicide tolerance and selectivity in plants will be critically evaluated using examples from both the literature and the author's research group. GSTs catalyse the conjugation of electrophilic substrates with glutathione (GSH) following either nucleophilic substitution or addition, with the resulting conjugates being generally non-toxic. Despite over twenty years of research, GSTs in plants remain poorly characterised even though it is clear that GSTs are present in many, if not all, plant species. With respect to herbicides, several factors appear to regulate their detoxification by plant GSTs and these are now considered in turn.

Structure and Reactivity of Herbicide Substrates

Several important herbicides including the chloro-s-triazine atrazine, the chloroacetanilides alachlor and metolachlor and the diphenylethers fluorodifen, formesafen and acifluorfen together with fenoxaprop ethyl and chlorimuron ethyl can serve as substrates for GSH conjugation directly without prior biotransformation. In contrast, the s-triazinone metribuzin and the thiocarbamate S-ethyl dipropylthiocarbamate (EPTC) will only undergo GSH conjugation following the formation of reactive sulphoxide intermediates (Lamoureux and Rusness, 1993) and 6-chloro-3-phenylpyridazine-4-ol derived from pyridate must be N-glycosylated to form the reactive keto derivative (Zohner, 1987). Some herbicides, or their derivatives, are so reactive that they will react with GSH at physiological pH in the absence of GST. This has led some researchers to suggest that the conjugation of herbicides in crops containing high concentrations of GSH does not require GSTs (Jablonkai and Hatzios, 1993). In some cases this is probably the case. For example, our studies have been unable to show a role for GSTs in the rapid conjugation of the N-glucoside metabolite of pyridate in maize or chickpea (Gimenez-Espinosa *et al.*, 1996). Similarly, EPTC and the more reactive chloroacetanilides undergo rapid spontaneous conjugation with GSH both *in vivo* and *in vitro* (Scarponi *et al.*, 1991). However, several lines of evidence support a role for GSTs in GSH conjugation of other chloroacetanilide and triazine herbicides *in vivo*, namely; i) susceptibility of maize to atrazine was observed in lines deficient in the corresponding GST activity (Timmerman, 1989); ii) atrazine-resistance in biotypes of *Abutilon threophrasti* was related to GST activity rather than GSH content (Anderson and Gronwald 1991); iii) tobacco plants transformed with maize GST I and GST II are resistant to alachlor even though GSH levels are similar to those in control plants (Jepson, personal communication); iv) selectivity toward triazine and chloroacetanilide herbicides in cereals and grass weeds correlates well with GST activities but not with GSH contents (Hatton *et al.*, 1996a). Similarly, it is difficult to envisage how the GSH-mediated cleavage of diphenylethers could occur without the intervention of a GST as such activities only occur non-enzymically at high pH.

GST Activities toward Herbicides in Various Species

Although many plant species contain genes encoding GSTs and show activities toward colorimetric substrates such as 1-chloro-2,4-dinitrobenzene (CDNB) only a relatively restricted range of plant GSTs show activities toward herbicides. Furthermore, plants which possess GSTs which are highly active in conjugating some classes of herbicides may be deficient in other activities. A classic example of this is seen in species-dependant variations in GST activities in peas and maize. Pea GSTs have little activity toward atrazine but efficiently catalyse the cleavage of the ether bond of fluorodifen (see Lamoureux and Rusness, 1993). Conversely maize contains GSTs with activities in the order atrazine=chloroacetanilides > fluorodifen (Dixon *et al.*, 1996). Maize contains at least four different isoenzymes with activities toward metolachlor and alachlor with the isoenzymes variously made up of subunits of 26 kD, 27 kD and 29 kD each encoded by distinct genes under separate control (Timmerman 1989, Fuerst *et al.*, 1993, Jepson *et al.*, 1994). However, the specificities of these isoenzymes toward alternative substrates is largely unknown and the distinct GST with activity toward atrazine still awaits final purification and molecular characterisation (Timmerman 1989). Once we consider the GSTs involved in herbicide metabolism in other plants there is far less information available. GSTs with activity toward chloroacetanilide herbicides have been described in sorghum (Dean *et al.*, 1990) and the GST involved in fluorodifen detoxification partially characterised in peas (see Lamoureux and Rusness, 1993) and purified from spruce (Schroder and Berkau 1993). In weed species the GST activities toward atrazine have been determined in *Abutilon threophrasti* (Anderson and Gronwald, 1991) and *Setaria faberi* (Boydston and Slife, 1986., Hatton *et al.*, 1995). Until the portfolio of all the GST activities in important crops and weeds have been determined it will remain difficult to predict their role in herbicide selectivity, even in crops whose detoxification capacities we consider relatively well characterised. The selectivity of fenoxaprop ethyl in barley and wheat illustrates this point. Thus, although herbicide metabolism in these cereals is more commonly associated with hydroxylations, the detoxification of fenoxaprop ethyl in tolerant cereals results from either bond cleavage and GSH conjugation (Tal *et al.*, 1993). Our recent studies have confirmed that wheat contains a GST activity highly active in cleaving such diaryl ethers, including fluorodifen (Pigg and Edwards, unpublished results). These findings suggest that the screening of other crops for GST activities toward candidate herbicides might be very useful in exploiting novel selectivity mechanisms.

Effect of Herbicide Safener Treatments

In the case of herbicides detoxified by GSH conjugation herbicide safeners appear to exert their activity, at least in part, by increasing GSH levels and inducing GSTs active in herbicide detoxification (Farago *et al.*, 1994). It has also recently been reported that safeners increase the activity of the vacuolar transporter involved in removing GSH conjugates from the cytosol (Gaillard *et al.*, 1994) and this may also have implications in enhancing GST mediated detoxification. Safening is usually associated with cereal crops (Farago *et al.*, 1994), but it is also clear that the induction of GSTs by safeners can also occur in legumes (Hunaiti and Ali, 1991). Safeners normally enhance GST activities which are constitutively present at lower levels. In maize, the best studied system, safeners such as dichlormid and benoxacor transcriptionally activate the 27 kD subunit form of GST which is a component of GST II and GST IV, both of which are highly active in the metabolism of chloroacetanilide herbicides (Timmerman, 1989, Fuerst *et al.*, 1993, Jepson *et al.*, 1994). We have recently extended these studies to examine the effect of safeners on other GST activities in maize (Dixon *et al.*, 1996). Treatment of maize seedlings with dichlormid increased the specific activities of GSTs active toward alachlor and metolachlor two fold and four-fold respectively in the roots and four-fold and two-fold respectively in the shoots. Safener treatment also induced the activities toward fluorodifen six-fold in the roots and three fold in the shoots, but had no effect in either tissue on the activity toward atrazine. These results clearly confirm earlier reports on the differential induction of GST activities in maize (Dean *et al.*, 1991) and indicate their tissue specific nature, which has also been neglected in many studies. Safeners also induce up to five novel GSTs with activities toward metolachlor in sorghum with untreated plants containing only a single minor isoenzyme with activity toward the herbicide (Dean *et al.*, 1990). We have recently determined the range of safener-inducible GST activities in wheat seedlings (Pigg and Edwards, unpublished results). Dichlormid and fenchlorazole ethyl, but not butylate or naphthalic anhydride, increased the specific activity of the GST with activity toward fluorodifen, but had no effect on the GSTs with activity towards atrazine and chloroacetanilides. Significantly, fenchlorazole ethyl has also been reported to elevate the GSH conjugation of fenoxaprop ethyl (Tal *et al.*, 1993) and the GST responsible for detoxifying fluorodifen appears to be identical to the enzyme with activity toward fenchlorazole ethyl. Safeners may increasingly prove to be a useful tool in identifying GST activities of interest and their effect on the GSTs of grass weeds and broad leaf crops warrants further study.

Relationship between GSH Availability and GST Activity

In order to catalyse herbicide detoxification GSTs require an available pool of reduced GSH as co-substrate. At a rudimentary level it would seem that the relatively high concentrations of GSH available in plants (10^{-4}M) would be easily capable of supporting the detoxification of herbicides. Few studies have addressed the relationship between GSH content and herbicide metabolism and selectivity though it has been reported that maize seedlings became more sensitive to metolachor injury following only a modest reduction in GSH content (Farago *et al.*, 1993). A factor which can often be overlooked is the role of compartmentation in controlling the availability of this GSH pool to GSTs. Thus, most of the GST activity is reported to be localised in the cytosol, while much of the cellular GSH is localised in the chloroplast (Rennenberg 1982). The observations that enhanced GSH conjugation of herbicides following safener treatment is invariably associated with both increased GST activities and elevated GSH content suggests that the increased availability of the co-substrate is required to support increased metabolism (Farago *et al.*, 1994). However simple calculations based on the concentrations of GSH required to conjugate the available herbicide and considerations of the catalytic characteristics of the GSTs would suggest that there is already enough GSH in the cell prior to safener treatment to support the increased metabolism. The co-dependence of GST induction and GSH accumulation following safener treatment is an area ripe for study and in particular the relative subcellular distributions of GSTs and GSH deserves attention. In addition numerous biotic and abiotic factors influence GSH levels in plants and it would be useful to understand the factors regulating GSH availability in more detail so as to be able to predict the likely effects of environmental stress on herbicide tolerance. Recent developments in the generation of transgenic plants engineered to contain modified levels of GSH may facilitate such studies.

Metabolism of GSH Conjugates and GST Activity

Many lipophilic GSH conjugates, including those of herbicides, are excellent competitive inhibitors of GSTs (Lamoureux and Rusness, 1993). A good example of this is seen with tridiphane which acts as a synergist of atrazine (Lamoureux and Rusness, 1986, Boydston and Slife, 1986). The synergistic activity is due, in part, to the formation of the GSH conjugate of tridiphane which is a more potent inhibitor of GSTs in weeds, such as *S. faberi,* than in maize (Lamoureux and Rusness, 1986). These results suggest that it is necessary to remove inhibitory GSH conjugates from the cytosol of plants in order to maintain effective GST-mediated detoxification. In the case of the GSH conjugate of tridiphane the reaction product was catabolised more effectively in maize than in the sensitive *S. faberi* (Lamoureux and Rusness, 1986). Plants are able to degrade GSH conjugates to a complex range of polar reaction products which are inactive as GST inhibitors (Lamoureux and Rusness, 1993). The relative abilities of plants to catabolise these conjugates may be just as important as their capacities to synthesize them and again this warrants evaluation. Our recent studies with [^{14}C] atrazine in maize and *S. faberi* have shown that significant quantities of these catabolites accumulate at the same time as the GSH-conjugates are being synthesized and the significance of these metabolites is under investigation (Hatton *et al.*, 1995). An alternative mechanism for the removal of GSH conjugates from the cytosol has recently been determined following the observation that plants contain an ATP-dependant transporter of herbicide conjugates, including GSH conjugates, in the tonoplast membrane

(Gaillard *et al.*, 1994). Significantly, treatment of barley with the safener cloquintocet-mexyl doubles the rate of metolachlor GSH import into the vacuole, apparently due to an increase in the expression of the transporter. This report demonstrates that enhanced GST mediated detoxification requires increased export capacity from the cytosol, presumably to prevent the build up of the conjugates to levels which would inhibit the GSTs. It will now be of interest to determine whether other routes of GSH-conjugate catabolism are enhanced by safeners.

Role of GSTs in Herbicide Selectivity

Having defined the factors likely to influence the activities of plant GSTs toward herbicides it should be possible to assess their role in herbicide selectivity. A good case already exists for the role of GSTs in herbicide selectivity within different cultivars or biotypes of a given crop (Timmerman, 1989) or weed (Anderson and Gronwald, 1991). However, although GSTs have often been cited as a major determinant of herbicide selectivity between crops and weeds, definitive data is rather scarce. This ambiguity has resulted from i) the difficulty in assigning any one factor a dominant role in something as complex as the selectivity of herbicides between completely different species and ii) the failure of many comparative studies to relate *in vivo* metabolism data to *in vitro* assays with herbicide substrates. We are addressing this latter point by determining the rates of metabolism of herbicides in graminaceous crops and weeds and relating this to herbicide selectivity, relative GSH content and GST activities toward the respective herbicides (Hatton *et al.*, 1995, 1996a, b). To date our results would suggest that GSTs, but not GSH content, accurately reflect relative herbicide tolerance in young plants, but that selectivity in older plants may be due to additional factors. For example, with atrazine the greatest selectivity between maize and *S. faberi* coincides exactly with the greatest difference in the respective GST activities (maize > > *S. faberi*) (Hatton *et al.*, 1996b). The specific activities of GSTs active toward herbicide substrates declined markedly in maize plants with increasing age while the GST activities in the weed remained largely unaffected. The net effect of these age-dependant changes was that the older the plants the more similar the detoxifying activities *in vitro*. Previous studies have reported that the GST activity toward atrazine in *S. faberi* remained similar up to the four leaf stage and then declined, but comparative studies with maize were not reported (Boydston and Slife, 1986). Lamoureux and Rusness (1986) determined the extractable GST activities toward CDNB in the two species at various growth stages and reported that GST activities declined in maize with age but increased in *S. faberi*.

Role of GSTs in Herbicide Resistance in Weeds.

As GSTs are a determinant of herbicide tolerance in crops it has also been tempting to speculate that the emergence of resistant weeds has resulted from changes in GST expression resulting in enhanced detoxification. Significantly, an atrazine-resistant biotype of *Abutilon threophrasti* contained four fold higher activities of the GST active toward the herbicide than did the normal susceptible plants (Anderson and Gronwald, 1991). Analysis of the GST isoenzymes suggested that the increased activity in the resistant plants was due to the enhanced expression of the GSTs normally observed. This in turn suggests that the resistance has evolved due to a regulatory mutation rather than a change in the catalytic characteristics of the GSTs. Interestingly, De Prado *et al.*, (1992) have reported that an atrazine-resistant population of *Setaria adherens* detoxifies the herbicide by GSH conjugation more rapidly than susceptible populations, in contrast to *S. glauca* populations which owe their tolerance to photosystem II site insensitivity.

References

Anderson, M.P. and Gronwald, J.W. (1991) Atrazine resistance in a velvetleaf (*Abutilon threophrastti* biotype due to enhanced glutathione S-transferase activity. *Plant Physiology* **96**: 104-109.

Boydston, R.A. and Slife, F.W. (1986) Alteration of atrazine uptake and metabolism by tridiphane in giant foxtail *(Setaria faberi)* and corn *(Zea mays)*. *Weed Science* **34**: 850-858.

Dean, J.V., Gronwald, J.W. and Eberlein, C.V. (1990) Induction of glutathione S-transferase isozymes in sorghum by herbicide antidotes. *Plant Physiology* **92**: 467-473.

Dean, J.V., Gronwald, J.W. and Anderson, M.P. (1991) Glutathione S-transferase activity in non-treated and CGA-154281- treated maize shoots. *Zeischrift Naturforschung* **46c**: 850-855.

De Prado, R., Romera, E., Menendez, J. and Tena, M. (1992) Differential mechanism of resistance to atrazine in *Setaria adherens* and *Setaria glauca*. *Mededelinge Faculteit Landbouwweten Schappen* **IV**: 1047-1052.

Dixon, D.P., Cole, D.J. and Edwards, R. (1996) Spectrum of glutathione transferase activities in maize. *In: Proceedings International Symposium Weed and Crop Resistance to Herbicides* (R. De Prado, J. Jorrín, L. García-Torres and G. Marshall, eds.), 72-74. University of Córdoba, Córdoba, Spain.

Farago, S., Kreuz, K. and Brunold, C. (1993) Decreased glutathione levels enhance the susceptibility of maize seedlings to metolachlor. *Pesticide Biochemistry Physiology* **47**: 199-205.

Farago, S., Brunold, C. and Kreuz, K. (1994) Herbicide safeners and glutathione metabolism. *Physiologia Plantarum* **91**: 537-542.

Fuerst, E.P., Irzyk, G.P. and Millar, K.D. (1993) Partial characterisation of glutathione S-transferase isozymes induced by the herbicide safener benoxacor in maize. *Plant Physiology* **102**: 795-802.

Gaillard, C., Dufaud, A., Tommasini, R., Kreuz, K., Amrhein, N. and Martinoia, E. (1994) A herbicide antidote (safener) induces the activity of both the herbicide detoxifying enzyme and of a vacuolar transporter for the detoxified herbicide. *FEBS Letters* **352**: 219-221.

Gimenez-Espinosa, R., De Prado, R. and Edwards, R. (1996) Role of glutathione transferases in the metabolism of pyridate in chickpea. *In: Proceedings International Symposium Weed and Crop Resistance to Herbicides* (R. De Prado, J. Jorrín, L. García-Torres and G. Marshall, eds.), 82-83. University of Córdoba, Córdoba, Spain.

Hatton, P.J, Edwards, R. and Cole, D.J. (1995) Glutathione transferases in major weed species. *Pesticide Science* **43**: 173-175.

Hatton, P J., Dixon, D.P., Cole, D.J. and Edwards, R. (1996a) Glutathione transferase activities and herbicide selectivity in maize and associated weed species. *Pesticide Science* **46**: 267-275.

Hatton, P.J., Cole, D.J. and Edwards, R. (1996b) Glutathione transferases and herbicide selectivity in maize and *Setaria faberi*. *In: Proceedings International Symposium Weed and Crop Resistance to Herbicides* (R. De Prado, J. Jorrín, L. García-Torres and G. Marshall, eds.), 87-88. University of Córdoba, Córdoba, Spain.

Hunaiti, A.A. and Ali, B.R. (1991) The induction of chickpea glutathione S-transferase by oxadiazon. *Phytochemistry* **30:** 2131 -2134.

Jablonkai, I. and Hatzios, K.K. (1993) In vitro conjugation of chloroacetanilide herbicides and atrazine with thiols and contribution of non enzymatic conjugation to their glutathione-mediated metabolism in corn. *Journal Agricultural Food Chem*istry **41**: 1736-1742.

Jepson, I., Lay, V.J., Holt, D.C., Bright, S.W.J. and Greenland, A.J. (1994) Cloning and characterization of maize herbicide safener-induced cDNAs encoding subunits of glutathione S-transferase isoforms I, II, and IV. *Plant Molecular Biology* **26**: 1855-1866.

Lamoureux, G.L and Rusness, D.G. (1986). Tridiphane (2-(3,5dichlorophenyl)-2-(2,2,2-trichloroethyl oxirane) an atrazine synergist: enzymatic conversion to a potent glutathione S-transferase inhibitor. *Pesticide Biochemistry Physiology* **26**: 323-342.

Lamoureux, G.L. and Rusness, D.G. (1993) Glutathione in the metabolism and detoxification of xenobiotics in plants. *In: Sulphur Nutrition and Assimilation in Higher Plants* (I.J. De Kok., I. Stulen, H. Rennenberg, C.Brunold and W. Rauser, eds.), 221-237. SPB Academic Publishers, The Hague.

Rennenberg, H. (1982) Glutathione metabolism and possible biological roles in higher plants. *Phytochemistry* **21**: 2771-2781.

Scarponi, L., Perucci, P. and Martinetti, L. (1991) Conjugation of 2-chloroacetanilide herbicides with glutathione: role of molecular structures and glutathione S-transferase enzymes. *Journal Agricultural Food Chemistry* **39**: 2010-2013.

Schroder, P. and Berkau, C. (1993) Characterization of cytosolic glutathione S-transferase in spruce needles. *Botanica Acta* **106**: 301-306.

Tal, A., Romano, M.L., Stephenson, G.R., Schwan, A.L. and Hall, J.C. (1993) Glutathione conjugation: a detoxification pathway for fenoxaprop-ethyl in barley, crabgrass, oat and wheat. *Pesticide Biochemistry Physiology* **46**: 190-199.

Timmerman, K.P. (1989) Molecular characterization of corn glutathione S-transferase isozymes involved in herbicide detoxication. *Physiologia Plantarum* **77**: 465-471.

Zohner, A. (1987) Mode of crop tolerance to pyridate in corn and peanuts. *Proceedings British Crop Protection Conference- Weeds*, 1083-1090.

CHAPTER 12
MECHANISMS OF HERBICIDE MULTIPLE RESISTANCE IN *LOLIUM RIGIDUM*

F.J. TARDIF, C. PRESTON and S.B. POWLES
Department of Crop Protection, The University of Adelaide, Waite Campus, PMB 1, Glen Osmond, 5064 Australia

Herbicide Resistance in *Lolium rigidum*

Lolium rigidum Gaud. (annual ryegrass) is an annual species that was introduced as a pasture plant in Australia at the end of the 19th century. Being well suited to the environmental conditions prevailing in southern Australia, *L. rigidum* became an integral component of the successful wheat/sheep rotation system. In this system *L. rigidum* is beneficial as the main component of the pasture phases but is detrimental as a weed in the cropping phase. The introduction in the late seventies of efficient selective post emergence herbicides such as diclofop-methyl and later chlorsulfuron has caused major changes in the rotations such that the importance of the pasture component was considerably reduced in favour of the more profitable cropping component. This has resulted in increasing herbicide usage (Powles and Matthews, 1992).

Cases of herbicide failure to control ryegrass populations previously well controlled were reported in the early eighties and resistance to diclofop-methyl was confirmed soon after (Heap and Knight, 1982). At first, resistance to diclofop-methyl did not present a large challenge in terms of management because alternative herbicides were available. However, it was soon realised that *L. rigidum* was different to previous cases of herbicide resistance (e.g. triazine resistant weeds) when it was observed that a biotype resistant due to selection with diclofop-methyl was also resistant to chlorsulfuron without ever having been exposed to the later (Heap and Knight, 1986). In the years following, many other cases of herbicide resistance were reported in ryegrass with different patterns of multiple resistance to herbicides of unrelated chemistry and mode of action. A survey conducted by the agrochemical industry in 1992 estimated that 2300 farms in Australia were infested with herbicide resistant ryegrass and this number is now thought to be over 3000. In a significant number of cases, ryegrass has multiple herbicide resistance which means that conventional control with selective herbicides has become impossible (Matthews, 1994).

To date, *L. rigidum* has developed resistance to 16 herbicide chemical groups affecting 11 target sites (Table 1). Not every population is resistant to each of these herbicides; indeed some are resistant to only one or two groups, but a majority are resistant to three or more. Extreme cases of herbicide multiple resistance are illustrated by two biotypes we have characterised, SLR 31 and VLR 69, which are resistant to 11 and eight herbicide classes, respectively.

Mechanisms Responsible for Multiple Herbicide Resistance.

In order to understand the phenomenon of herbicide multiple resistance, it is imperative to describe the mechanisms found in multiple resistant biotypes. These mechanisms can be divided into two main types depending on whether they involve a change at the herbicide target site or not.

Target Site Modifications

In multiple resistant *L. rigidum*, modification of the target site has been found for two herbicide groups: the acetyl-coenzyme A carboxylase (ACCase) inhibitors such as the aryloxyphenoxypropanoate (APP) and the cyclohexanediones (CHD), and the acetolactate synthase (ALS) inhibitors including the sulfonylureas and imidazolinones.

We have so far identified resistant ACCase in four *L. rigidum* biotypes (Table 2). These biotypes have forms of ACCase that are unaffected by high concentrations of APP and/or CHD herbicides (Tardif *et al.*, 1993, 1996; Tardif and Powles, 1994; Preston *et al.*, 1996). Resistant ACCase enzymes can vary in the level of resistance they confer to herbicides from the two classes. All the biotypes we have characterised so far have high resistance to the APP while the level of resistance to CHD varies. As was observed in other weed species, the response at the whole plant and enzyme levels observed in these *L. rigidum* biotypes suggest that different type of mutations in the gene coding for ACCase have been selected (see discussion by M. D. Devine, this volume). The important point here is that modified ACCase only provides resistance to the APP and CHD herbicides. This is well illustrated by biotypes such as WLR 96 and SLR 3 which only possess the modified ACCase as a resistance mechanism and are resistant only to APP and CHD (Tardif *et al.*, 1993, 1996).

Resistant ALS has been identified in two biotypes, VLR 69 and WLR 1 (Christopher *et al.*, 1992; Burnet *et al.*, 1994c). The ALS enzyme in these biotypes has been shown to confer high resistance to sulfonylureas and imidazolinones. A number of different phenotypes have been observed with plants having resistant ALS. The relative level of resistance to the sulfonylureas, imidazolinones and triazolopyrimidines varies quite extensively. This is due to the existence of a range of possible mutations and to the fact that the binding characteristics of the herbicide classes on the ALS enzyme vary (Saari *et al.*, 1994). As with the ACCase modification, ALS modification endows plants with resistance only to herbicides affecting ALS and not to any other herbicide group.

Table 1. Herbicide groups to which there is resistance in *Lolium rigidum* in Australia.

Mode of action	Herbicide group	Biotypes [3]	References [4]
ACCase inhibitors	APP	SLR 31, WLR 96	3, 5, 8
	CHD	SLR 3, VLR 69	
ALS inhibitors	Sulfonylurea Imidazolinone	SLR 31, WLR 1, VLR 69	3, 4, 5
PSII inhibitors	Substituted urea Triazine Triazinone	WLR 2, VLR 69	1, 3
Carotenoid biosynthesis inhibitors	Triazole	WLR 2	1
Mitosis disruptors	Chloroacetamide	SLR 31, VLR 69	2
Tubulin polimerisation inhibitors	Dinitroaniline	SLR 31, SLR 10	6
Phytoene desaturase inhibitors	Isoxazolidinone	SLR 31	7
Microtubule misorientation inducers	Carbamate[1]	SLR 31, SLR 10	6
Multipolar cell division inducers	Benzenedicarboxylate[1] Methylcarbamate[2]	SLR 31, SLR 10	6
Microtubule loss inducers	Phosphoric amide[2]	SLR 31, SLR 10	6

1 Resistance higher in germination assay than whole plant experiments. 2 Resistance assessed in germination assay only. 3 This list does not include all the biotypes studied in our laboratory. 4 References: 1, Burnet *et al.*, 1991; 2, Burnet *et al.*, 1994a; 3, Burnet *et al.*, 1994b; 4, Christopher *et al.*, 1992; 5, Heap and Knight, 1990; 6, McAlister *et al.*, 1995; 7, Tardif and Powles (unpublished results); 8. Tardif *et al.*, 1993.

Table 2. Biotypes of *Lolium rigidum* with resistant ACCase.

Biotypes	Resistance ratio [1]	
	Diclofop	Sethoxydim
SLR 3	>37	7,8
SLR 31 (subset)	6	26
VLR 69	31	1,2
WLR 96	>217	>1,6

1. Concentration of herbicide inhibiting 50% of the ACCase activity of the resistant biotypes compared to a susceptible biotype

Non Target Site Mechanisms

These refer to mechanisms of resistance not involving modification of the herbicide *get* site and include enhanced metabolism, reduced retention or absorption, reduced translocation and differential sequestration. In *L. rigidum*, we have documented a number of biotypes with enhanced metabolism. Recovery from herbicide induced membrane depolarization has also been observed in *L. rigidum* and may be indicative of some other mechanism of resistance.

Enhanced Herbicide Metabolism

In *L. rigidum*, enhanced degradation of herbicides from five chemical groups in four different biotypes has been observed (Table 3). Biotypes WLR 2 and VLR 69 are resistant to most substituted urea and triazine herbicides despite having a susceptible PSII target site. These two biotypes can metabolise the substituted urea herbicide chlorotoluron and the triazine herbicide simazine at least twice as fast as a susceptible biotype (Burnet *et al.*, 1993a, b). Moreover, application of the synergist ABT greatly increased herbicide toxicity while slowing the rate of herbicide metabolism in WLR 2 and VLR 69. These data together indicate that metabolism endows resistance to substituted ureas and triazines in these biotypes.

Similarly, enhanced metabolism of chlorsulfuron has been found in SLR 31, WLR 1 and VLR 69 (Christopher *et al.*, 1991, 1992; Burnet *et al.*, 1994c). The halflife of chlorsulfuron in SLR 31 was half that found in a susceptible biotype. In both SLR 31 and VLR 69, the organophosphate insecticide malathion can inhibit chlorsulfuron metabolism while reducing the level of resistance to this herbicide (Christopher *et al.*, 1994; Preston *et al*, 1996).

Enhanced diclofop-methyl metabolism has been found in SLR 31 and VLR 69. The half-life of diclofop acid, the herbicidally active form, was approximately 1.6 fold shorter in SLR 31 than in a susceptible biotype (Holtum *et al.*, 1991). A similar pattern of diclofop-methyl degradation has been found in VLR 69 (Preston *et al.*, 1996).

More information on multiple resistance can be obtained by further characterising

Table 3. Biotypes of *Lolium rigidum* with enhanced herbicide metabolism.

Biotypes	Herbicides
WLR 2	chlorotoluron, simazine
VLR 69	chlorotoluron, simazine, chlorsulfuron, diclofop-methyl, tralkoxydim
SLR 31	diclofop-methyl, chlorsulfuron
WLR 1	chlorsulfuron

the enhanced metabolism process and by identifying the enzymes involved. The question remains whether in these biotypes one or several enzymes are involved in the degradation on chemically unrelated herbicides. An answer to this can be provided from the use of specific synergists that act by inhibiting herbicide metabolism. Such a study has been done with biotype VLR 69, using the herbicides chlorotoluron, simazine, chlorsulfuron, diclofop-methyl and tralkoxydim. At the whole plant level, piperonyl butoxide (PBO) synergised only chlorotoluron while malathion synergised chlorsulfuron but not the other herbicides. These two compounds as well as aminobenzotriazole (ABT) and tetcyclasis were also used to examine the rate of metabolic degradation *in vivo*. ABT, PBO and tetcyclasis inhibited the metabolism of simazine and chloroloturon but not of the other herbicides. Malathion only inhibited chlorsulfuron degradation. None of the four chemicals inhibited tralkoxydim and diclofop metabolism. The main conclusion from these results is that there are at least three diferent herbicide degradation enzymes conferring resistance in VLR 69.

ABT, PBO and tetcyclasis are known broad range cytochrome P450 enzyme inhibitors while malathion has been shown to synergise sulfonylurea herbicides via competitive inhibition of cytochrome P450 (Kreuz and Fonne-Pfister, 1992). The fact that the synergists used in that study were efficient suggests that the enzymes involved are cytochrome P450 based. Ultimate proof of this will only come from the isolation and characterisation of the enzymes in microsomes isolated from herbicide resistant *L. rigidum* biotypes.

Recovery from Herbicide Induced Membrane Depolarization

It is now widely accepted that inhibition of ACCase is the mode of action of APP and CHD herbicides. However, the observation in a number of laboratories that these herbicides can depolarise the cell membrane potential has led to some speculation on this phenomenon as a mode of action. In theory, the dissipation of the electrogenic transmembrane potential will lead to cessation of cell function and eventually of plant growth (Devine and Shimabukuro, 1994). Some biotypes of *L. rigidum* have been found to have the capacity to recover from APP- and CHD-induced loss of membrane potential, while susceptible biotypes can not recover (Hausler *et al.*, 1991). The actual connection between that recovery of potential and a mechanism of resistance has never been clearly established, however it has been postulated that it may be indicative of a capacity to differentially sequester the herbicides from the site of action, therefore preventing phytotoxic effects (Holtum *et al.*, 1991, 1994). The presence of a mechanism of resistance involving differential herbicide sequestration is under investigation in our laboratory.

How Does *L. rigidum* Accumulate Multiple Resistance Mechanisms?

There is now no doubt that multiple herbicide resistance is due to the presence of multiple resistance mechanisms. Biotype SLR 31 has at least four different mechanisms of resistance: enhanced diclofop-methyl metabolism (Holtum *et al.*, 1991); enhanced chlorsulfuron metabolism (Christopher *et al.*, 1991); the membrane recovery response that is associated with resistance to APP and CHD (Hausler *et al.*, 1991); and a resistant ACCase present in 12% of the population (Tardif and Powles, 1994). Six different mechanisms of resistance have been documented in biotype VLR 69: enhanced diclofop-methyl metabolism (Preston *et al.*, 1996); enhanced metabolism of chlorsulfuron (Burnet *et al.*, 1994c); enhanced metabolism of chlorololuron and simazine (Burnet *et al.*, 1993a, b); a resistant ACCase enzyme (Preston *et al.*, 1996); a resistant ALS enzyme present in a 4% subset of the population (Burnet *et al.*, 1994c); and the membrane recovery response (Hausler *et al.*, 1991).

Multiple herbicide resistance is now very common in *L. rigidum* and to a lesser extent in black-grass (*Alopecurus myosuroides*) (Moss, 1992). This does not mean, however, that this phenomenon will not occur in other species. The speed of development of herbicide resistance is a function of many factors including the initial frequency of resistance genes in the population, the number of individuals being treated, the efficacy and duration of the selection pressure and the type of reproductive system (Maxwell and Mortimer, 1994). Some of these factors have favoured the development of herbicide multiple resistance in Australia. *L. rigidum* is widely distributed in the southern cropping area of Australia where it is often present at high densities. This ensures that at least a few individuals with resistance genes are likely to be present in a population even if the initial frequency is low. There has been, and there is still a high reliance on selective herbicides to control *L. rigidum,* which means that the selection pressure is high, eliminating all the individuals that do not carry resistance genes. Because *L. rigidum* is strictly cross-pollinated, the individuals that survive the herbicide treatment have no choice other than to exchange pollen (and therefore resistance genes) in order to produce seeds. If those survivors possess different resistance mechanisms, these will be mixed in subsequent generations. The presence of multiple mechanisms of resistance will increase the likelihood of rapidly selecting for resistance when a different herbicide is used.

A special set of circumstances have enabled the development of multiple herbicide resistance earlier in *L. rigidum* than in other species. Some are related to human intervention while others are due to the biology of the plant. The biological features that are found in *L. rigidum* are by no way unique. Indeed, although the problem has not been as acute as in *L. rigidum* the development of multiple herbicide resistance in *A. myosuroides* is worrisome. There are a number of weeds which possess biological features favourable to the selection of multiple herbicide resistance and, if over reliance on herbicides is continued, multiple herbicide resistance will inevitably become a common occurrence in world agriculture.

References

Burnet, M.W.M., Hildebrand, O.B., Holtum, J.A.M. and Powles, S.B. (1991) Amitrole, triazine, substituted urea, and metribuzin resistance in a biotype of rigid ryegrass (*Lolium rigidum*). *Weed Science* **39**: 317-323.

Burnet, M.W.M., Loveys, B.R., Holtum, J.A.M. and Powles., S.B. (1993a) A mechanism of chlorotoluron resistance in *Lolium rigidum. Planta* **190**:182-189.

Burnet, M.W.M., Loveys, B.R., Holtum, J.A.M. and Powles, S.B. (1993b) Increased detoxification is a mechanism of simazine resistance in *Lolium rigidum. Pesticide Biochemistry Physiology* **46**: 207-218.

Burnet, M.W.M., Barr, A.R. and Powles, S.B. (1994a) Chloroacetamide resistance in rigid ryegrass (*Lolium rigidum*). *Weed Science* **42**: 153-157.

Burnet, M.W.M., Hart, Q., Holtum, J.A.M. and Powles, S.B. (1994b) Resistance to 9 herbicide classes in a population of rigid ryegrass (*Lolium rigidum*). *Weed Science* **49**:369-377.

Burnet, M.W.M. Christopher, J.T., Holtum, J.A.M. and Powles, S.B. (1994c) Identification of 2 mechanisms of sulfonylurea resistance within one population of rigid ryegrass (*Lolium rigidum*) using a selective germination medium. *Weed Science* **42**: 468-473.

Christopher, J.T., Powles, S.B., Liljegren, D.R. and Holtum, J.A.M. (1991) Cross-resistance to herbicides in annual ryegrass (*Lolium rigidum*). II. Chlorsulfuron resistance involves a wheat-like detoxification system. *Plant Physiology* **95**: 1036-1043.

Christopher, J.T., Powles, S.B. and Holtum, J.A.M. (1992) Resistance to acetolactate synthase-inhibiting herbicides in annual ryegrass (*Lolium rigidum*) involves at least two mechanisms. *Plant Physio*logy **100**: 1909-1913.

Christopher, J.T., Preston, C. and Powles, S.B. (1994) Malathion antagonizes metabolism based chlorsulfuron resistance in *Lolium rigidum. Pesticide Biochemistry Physiology* **49**: 172-182.

Devine, M.D. and Shimabukuro, R.H. (1994) Resistance to acetyl coenzyme A carboxylase inhibiting herbicides. *In: Herbicide Resistance in Plants: Biology and Biochemistry* (S.B. Powles and J.A.M. Holtum, eds.), 141-169. Lewis Publishers, Boca Raton, FL, USA.

Häusler, R.E., Holtum, J.A.M. and Powles, S.B. (1991) Cross-resistance to herbicides in annual ryegrass (*Lolium rigidum*). IV. Correlation between membrane effects and resistance to graminicides. *Plant Physiology* **97**: 1035-1043.

Heap, J. and Knight, R. (1982) A population of ryegrass tolerant to the herbicide diclofop-methyl. *Australian Journal Agricultural Research* **48** :156-157.

Heap, I. and Knight, R. (1986) The occurrence of herbicide cross resistance in a population of annual ryegrass, *Lolium rigidum*, resistant to diclofop-methyl. *Australian Journal Agricultural Research* **37**:149-156.

Heap, I.M. and Knight, R. (1990) Variation in herbicide cross-resistance among populations of annual ryegrass (*Lolium rigidum*) resistant to diclofop-methyl. *Australian Journal Agricultural Research* **41**: 121-128.

Holtum, J.A.M., Hausler R.E., Devine, M.D. and Powles, S.B. (1994). Recovery of transmembrane potentials in plants resistant to aryloxyphenoxypropanoate herbicides: a phenomenon awaiting explanation. *Weed Science* **42**: 293-301.

Holtum, J.A.M., Matthews J.M., Häusler, R.E., Liljegren, D.R. and Powles, S.B. (1991). Cross-resistance to herbicides in annual ryegrass (*Lolium rigidum*). III. On the mechanism of resistance to diclofopmethyl. *Plant Physiology* **97**: 1026-1034.

Kreuz, K. and Fonne-Pfister, R. (1992) Herbicide-insecticide interaction in maize: malathion inhibits cytochrome P450-dependent primisulfuron metabolism. *Pesticide Biochemistry Physiology* **43**:232-240.

Matthews, J.M. (1994) Management of herbicide resistant weed populations. *In*: *Herbicide Resistance in Plants: Biology and Biochemistry* (S.B. Powles and J.A.M. Holtum, eds.), 317-335. Lewis Publishers, Boca Raton, FL, USA. 317-335.

Maxwell, B.D. and Mortimer, M. (1994) Selection for herbicide resistance. *In*: *Herbicide Resistance in Planls: Biology and Biochemistry* (S.B. Powles and J.A.M. Holtum, eds.), 1-25. Lewis Publishers, Boca Raton, FL, USA.

McAlister, F.M., Holtum, J.A.M. and Powles, S.B. (1995) Dinitroaniline herbicide resistance in rigid ryegrass (*Lolium rigidum*). *Weed Science* **43**: 55-62.

Moss, S. (1992) Herbicide resistance in the weed *Alopecurus myosuroides* (black-grass): The current situation. *In: Achievements and Development in Combating Pesticide Resistance* (I. Denholm, A. Devonshire and D. Holloman, eds.), 28-40. Elsevier, London, UK.

Powles, S.B. and Matthews, J.M. (1992) Multiple herbicide resistance in annual ryegrass (*Lolium rigidum*): a driving force for the adoption of integrated weed management. *In: Achievements and Development in Combating Pesticide Resistance* (I. Denholm, A. Devonshire and D. Holloman, eds.), 75-87. Elsevier, London, UK.

Preston, C., Tardif, F.J., Christopher, J.T. and Powles, S.B. (1996) Multiple resistance to dissimilar herbicide chemistries in a biotype of *Lolium rigidum* dute to enhanced activity of several herbicide degrading enzymes. *Pesticide Biochemistry Physiology* **54**: 123-134.

Saari, L.L., Cotterman, J.C. and Thill, D.C. (1994) Resistance to acetolactate synthase inhibiting herbicides. *In*: *Herbicide Resistance in Plants: Biology and Biochemistry* (S.B. Powles and J.A.M. Holtum, eds.), 83-119. Lewis Publishers, Boca Raton, FL, USA.

Tardif, F.J., Hollum, J.A.M. and Powles, S.B. (1993). Occurrence of a herbicide-resistanl acetyl-coenzyme A carboxylase mutant in annual ryegrass (*Lolium rigidum*) selected by sethoxydim. Planta **190**: 176-181.

Tardif, F.J. and Powles, S.B. (1994) Herbicide multiple resistance in a *Lolium rigidum* biotype is endowed by multiple mechanisms: isolation of a subset with resistant acetyl-CoA carboxylase. *Physiologia Plantarum* **91**: 488-494.

Tardif, F.J., Preston, C., Holtum, J.A.M. and Powles, S.B. (1996) Resistance to acetil-Coenzyme A carboxilase-inhibiting herbicides endowed by a single major gene encoding a resistant target site in a biotype of *Lolium rigidum*. *Australian Journal Plant Physiology* **23**: 15-23.

CHAPTER 13
IS THERE A MULTIPLE PESTICIDE METABOLIZING CYTOCHROME P450 IN MAIZE?

M. BARRETT, N. POLGE, R. BAERG and C. PONELEIT
Department of Agronomy, University of Kentucky, Lexington, Kentucky, USA 40546-0091

Introduction

Cytochrome P450 monooxygenases (P450) were long considered to be a potentially important enzyme system for the detoxification of herbicides in maize and other crop plants. Only recently has herbicide metabolism by P450 in maize microsomes been conclusively demonstrated. Fourteen herbicides in six chemical families have been shown to be metabolized by P450 in maize. Herbicides demonstrated as metabolized by P450 in maize include bentazon, chlortoluron, linuron, acetochlor, metolachlor, flumetsulam, CGA-152005, chlorimuron-ethyl, chlorsulfuron, nicosulfuron, primisulfuron, triasulfuron, imazethapyr, and AC 263,222. Reactions with the herbicides catalyzed by P450 include a number of ring hydroxylations, alkyl hydroxylations and N and O demethylations. These are all typical P450 reactions. In addition to the herbicides discussed, organophosphate insecticides (diazinon and malathion) are desulfurated by P450 in corn microsomes. The desulfurated metabolites are actually the active insect toxicant. This is the same reaction that can occur in insects and mammals. P450 is now considered of equal importance to glutathione-S-transferases (GST) for the metabolism of herbicides in maize.

P450 metabolism of herbicides in maize is characterized as an inducible enzyme activity which primarily catalyses hydroxylations and demethylations of the herbicide substrates. Induction of the herbicide metabolizing activity is specific, with only slight changes in the total P450 level accompanying several-fold induction of the herbicide metabolism. Despite the demonstration of P450 catalyzed metabolism of numerous pesticide substrates in maize, numerous questions concerning P450 in maize remain. These include determining how many P450's are involved in the herbicide metabolism and how their levels are regulated. Evidence is gathering that, for a number of the herbicides presently known to be metabolized by P450 in maize, there may be relatively few or even one P450 which metabolizes all the herbicide substrates.

We have been using a combination of genetic, *in-vivo* and *in-vitro* approaches to address the question of how many P450's in maize are involved in the metabolism of pesticide substrates.

Genetic Studies

Several years ago we discovered an inbred line of maize, GA209, which is susceptible to bentazon, a herbicide normally tolerated by maize. The susceptibility of GA209 to bentazon was shown to be due to the inability of the plant to detoxify the herbicide. Metabolism studies suggested that the limitation in bentazon metabolism in GA209 was the initial metabolic step, bentazon hydroxylation, and not the subsequent glucose conjugation with the hydroxylated bentazon metabolite (hydroxy-bentazon). It was also found that two recessive genes, tentatively named benl and ben2, were responsible for the susceptibility. Bentazon susceptibility is a recessive trait and required the homozygous expression of the recessive genes at both alleles (benlbenlben2ben2) in order for susceptibility to be present. Because the hydroxylation of bentazon is known to be catalyzed by P450, we were interested in whether the GA209 susceptibility to bentazon also expressed itself as increased sensitivity to other P450 metabolized herbicides. GA209 is known not to be sensitive to the GST metabolized herbicides atrazine and metolachlor.

Plants of the maize inbreds GA209 and B73, a bentazon tolerant inbred, were grown in greenhouse and treated postemergence with the herbicides bentazon, 2,4-D, dicamba, nicosulfuron, primisulfuron, and imazethapyr. All these herbicides have been shown to be, or are suspected of being, metabolized by P450 in maize. GA209 was more sensitive to all the herbicides, particularly bentazon, nicosulfuron, primisulfuron and imazethapyr, than B73. This indicates that the lack of P450 metabolism in GA209 is associated with increased injury from a number of P450 metabolized herbicides.

To test whether GA209 was also less efficient in the metabolism of some of these other P450 metabolized herbicides as well as bentazon, shoots of seedlings of GA209 and B73 were excised and placed into solutions containing ^{14}C-labeled bentazon, nicosulfuron or imazethapyr for a period of 4 hours. Shoots were harvested and herbicides plus metabolites extracted and quantified at the end of the 4 hour period. A second set of shoots were placed into water for 24 hours following the 4 hour ^{14}C-treatment period after which they were also harvested and extracted. Results of this experiment showed that GA209 metabolized all three herbicides less than B73.

Microsomes were isolated from shoots of etiolated seedlings of B73 and GA209 grown both from naphthalic anhydride (NA) treated and untreated seeds. Microsomes from NA treated B73 hydroxylated both nicosulfuron and bentazon while only bentazon hydroxylation was detected in microsomes from untreated B73. In contrast, no hydoxylation of either substrate was detected in either type of microsomes from GA209.

During this time, we began studying a third maize inbred, B90, which was discovered to be susceptible to nicosulfuron. Greenhouse experiments showed B90 was also more sensitive than B73 to the herbicides primisulfuron and imazethapyr but was tolerant of bentazon. *In vivo* metabolism studies, as described above, demonstrated that shoots of B90 did not metabolize nicosulfuron or imazethapyr but could metabolize bentazon. Similarly, microsomes from both NA treated and untreated B90 hydroxylated bentazon but did not hydroxylate nicosulfuron. Interestingly, while NA treatment induced bentazon hydroxylation in B73, NA did not alter the level of bentazon hydroxylation in B90.

Reviewing these data, we developed a working hypothesis that the two genes identified in GA209 both control the activity (or actually code for) two P450 isozymes. Expression of Benl results in the presence of a P450 which can metabolize several herbicides, including bentazon, nicosulfuron and imazethapyr, while expression of Ben2 results in the presence of a P450 which can only metabolize bentazon. Expression of the recessive forms of these two genes results in the loss of the respective P450 and increased herbicide sensitivity. In order for bentazon sensitivity to be observed, both alleles must be homozygous recessive. In addition, Benl is inducible by NA while Ben2 is not.

To further test this hypothesis, the repulsion phase homozygotes, genotypes BenlBenlben2ben2 (BENl) and benlbenlBen2Ben2 (BEN2), were developed from crosses between B73 (BenlBenlBen2Ben2) and Ga209 (benlbenlben2ben2). It was predicted that we would observe plants of these repulsion phase homozygotes which were either tolerant of both bentazon and nicosulfuron (BENl) or tolerant of bentazon but not nicosulfuron (BEN2). Greenhouse experiments confirmed this. Also, *in vivo* studies with excised shoots showed BENl could metabolize bentazon, nicosulfuron and imazethapyr more rapidly than BEN2. Finally, microsomes from NA treated and untreated BENl hydroxylated bentazon and nicosulfuron while microsomes from BEN2 only hydroxylated bentazon. In addition, bentazon hydroxylation was induced by NA in BENl but not BEN2.

Taken together, the inheritance data support the hypothesis of the presence of two P450's in maize which can hydroxylate bentazon. One of these, encoded by the gene Benl, can also hydroxylate nicosulfuron, imazethapyr, and, possibly, other herbicides. Benl is induced by NA while Ben2 is not.

Expression Studies

At the same time that the genetic studies were being conducted, we were also studying and characterizing the expression of several P450 activities in the maize hybrid Pioneer 3343IR (3343). This hybrid has been demonstrated to hydroxylate the herbicides bentazon, chlorimuron-ethyl, chlortoluron, imazethapyr, nicosulfuron and the normal plant component cinnamic acid. 3343 also desulfurates malathion. *In vitro* competition experiments were conducted with microsomes isolated from etiolated shoots of 3343 grown from NA treated seeds to determine whether these substrates could compete or interact at the same P450. Nicosulfuron, chlorimuron-ethyl, chlorsulfuron, linuron, chlortoluron, bentazon, 2,4-

D, dicamba, and cinnamic acid were tested for the ability to inhibit metabolism of chlorimuron-ethyl, nicosulfuron and imazethapyr. Chlorimuron-ethyl metabolism was strongly (>50%) inhibited by linuron and chlortoluron, moderately (~50%) inhibited by nicosulfuron, bentazon, 2,4-D and chlorsulfuron, and not inhibited by dicamba and cinnamic acid. Nicosulfuron metabolism was strongly inhibited by bentazon and chlorsulfuron, moderately inhibited by chlorimuron-ethyl, 2,4-D, and chlortoluron, and not inhibited by dicamba and cinnamic acid. Imazethapyr metabolism was strongly inhibited by chlorimuron-ethyl, bentazon, chlorsulfuron and chlortoluron, moderately inhibited by nicosulfuron, linuron, and 2,4-D, and slightly or not inhibited by dicamba and cinnamic acid. These data suggest that dicamba and cinnamic acid do not share a P450 in common with chlorimuron-ethyl, nicosulfuron and imazethapyr. However, the inhibitions by the other substrates imply the possibility that there is a common P450(s) among the herbicides. P450 isozymes which can metabolize multiple substrates are well established in the literature on mammalian P450's.

Although this data suggested that some of the substrates could be interacting at the same P450, it was not conclusive. To further examine this, more detailed competition experiments were conducted on the ability of bentazon to inhibit nicosulfuron metabolism, and the converse, plus chlortoluron inhibition of chlorimuron-ethyl hydroxylation and, again, the reverse. Kinetic analysis suggested that bentazon and chlortoluron were both noncompetitive inhibitors of nicosulfuron and chlorimuron-ethyl hydroxylation, respectively. Nicosulfuron was a much weaker inhibitor of bentazon hydroxylation than the reverse. However, the kinetic analysis of the data did imply that nicosulfuron was a noncompetitive inhibitor of bentazon. In contrast, chlorimuron-ethyl did not inhibit the metabolism of chlortoluron by the microsomes.

The competition data suggests that some of these herbicides may be interacting at the same (one or several) P450's but the results are not clear-cut. The situation is certainly more complex than the simple competitive interaction between substrates for a single enzyme. This is similar to what was found early in the history of P450 experimentation with mammalian microsomes.

Insecticide Inhibition

Although the inhibition of the metabolism of herbicide substrates by maize microsomes by classic P450 inhibitors such as tetcyclasis, l-aminobenzotriazole and piperonyl butoxide has been studied, an alternative to using these inhibitors is to examine the effects of other chemicals on the herbicide metabolism. Soon after the introduction of primisulfuron and nicosulfuron for weed management in maize, it was found that applications of organophosphate insecticides, particularly terbufos, resulted in injury to the maize when the normally tolerated herbicides were applied. Inhibition of the P450 mediated detoxification of the herbicides by the insecticide was suggested by the finding that another organophosphate insecticide, malathion, inhibited the in vitro metabolism of primisulfuron in maize microsomes. Malathion was proposed to cause the inhibition of primisulfuron metabolism through a mechanism-based or suicide substrate enzyme inactivation. We found that malathion was also an effective inhibitor of bentazon, imazethapyr and nicosulfuron metabolism in maize microsomes. Malathion is metabolized (desulfurated) by the maize microsomes. Malathion did not inhibit cinnamic acid hydroxylation. This implies that one P450 metabolizes primisulfuron, nicosulfuron, bentazon, imazethapyr and malathion or, if there is more than one P450 for the herbicides, all of the isozymes can metabolize malathion.

Interestingly, when terbufos was tested for the ability to inhibit nicosulfuron metabolism in intact maize plants and microsomes, it was only a weak inhibitor of the metabolism. It was found that the metabolites of terbufos, particularly the sulfone of terbufos, were strong inhibitors of nicosulfuron metabolism *in vivo* and *in vitro*. Further work found that the terbufos sulfone inhibited the metabolism of, not only nicosulfuron in maize microsomes, but also bentazon, chlorimuron-ethyl, imazethapyr and malathion, but not cinnamic acid. Again, this suggests that the P450(s?) for the pesticides share common features or that there is only one P450 that metabolizes all these substrates with the exception of cinnamic acid.

Summary

Several pieces of the data suggest that there is a P450 in maize which is capable of metabolizing a multitude of pesticide substrates, from a diversity of chemical families. These data include the inheritance studies, portions of the competition experiments, and the insecticide inhibition studies. However, conclusive demonstration of a single P450 which can metabolize multiple pesticide substrates is unlikely to be obtained in a microsomal system which contains multiple P450's. Rather, firm evidence for a multiple substrate P450 will require expression of a single P450 in a yeast, or other, P450 gene expression system. We are currently attempting to accomplish this task.

CHAPTER 14
D-1 PROTEIN TURNOVER AND SENSITIVITY OF HIGHER PLANTS TO PHOTOSYSTEM II-DIRECTED HERBICIDES

M. T. GIARDI, B. GEIKEN and A. CONA
IBEV-Area della Ricerca di Roma, CNR,. Via Salaria km. 29.3, 00016 Monterotondo Scalo, Italy

Introduction

Several mechanisms of resistance to photosystem II-directed herbicides have evolved in higher plants, such as slower translocation into the chloroplast, immobilization and detoxification by endogenous enzymes through conjugation and/or degradation and mutations in the target protein which preclude binding of herbicides (Gronwald, this volume). More than one mechanism often acts at the same time. The site of the PSII-directed herbicides is the Q-B site of the D1 protein (Draber *et al.*, 1991). This protein besides being the target site of triazine, diazine, urea and phenol-type herbicides is also the central component of PSII reaction center (Mattoo *et al.*, 1989). It is synthesized in the light on chloroplast ribosomes as a 33.5 kDa precursor, the precursor is integrated into the stroma-exposed lamellae, processed there to its mature size and subsequently translocated to the grana partition region.

Light-dependent rapid turnover is a characteristic of D1, at rates that are several times higher than any other PSII protein. A scission in the protein between the amino acid residues 248-247, yielding an N-terminal breakdown product of about 23 kDa, has been observed *in vivo* (Mattoo *et al.*, 1989). The D1 protein in grana membranes undergoes phosphorylation by a light-dependent kinase to a less electrophoretically mobile D1* form. The phosphorylation of D1 influences its association with the other phosphorylated PSII core proteins resulting in heterogeneous PSII core populations that have been isolated by isoelectrofocusing (Giardi *et al.*, 1992; Giardi, 1993b). The PSII core is constituted by the proteins CP47, CP43, D2, D1, cytb-559 and psbH gene product (Mattoo *et al.*, 1989; Giardi, 1993a). These heterogeneous PSII populations show differential properties in terms of electron transfer, sensitivity to strong light and affinity for the binding of photosynthetic herbicides (Giardi *et al.*, 1992; Giardi, 1993b). The aim of the present study was to test the implications of phosphorylation on D1 protein turnover and herbicide binding properties. Our studies indicate that it is possible to modulate D1 turnover rate, either at the level of its resynthesis or assembly or phosphorylation. It is observed that plant sensitivity to herbicide action directly responds to the D1 turnover modulation. The response is mainly seen in moderately resistant plants. Thus, the results suggest that D1 turnover, due also to its compartimentalization in the thylakoid membrane, could be important for the resistance of plants to herbicide action.

Materials and Methods

Phosphorylation profiles of the polypeptides and isolation of the differently phosphorylated PSII core populations were performed as reportd by Giardi *et al.* (Giardi *et al.*, 1992; Giardi, 1993b). D1 turnover was followed by pulse-chase experiments with ^{35}S-methionine incorporation *in vivo*, extraction of thylakoids, SDS-PAGE analysis and fluorography. For studies of combination effects of inhibitors with herbicides, seedlings were grown in a greenhouse at controlled conditions in vermiculate (30x30 cm pots). After germination, the seedlings (1 cm height) were watered with the inhibitors solutions (twice with 200 ml) and the measurements were performed after 4 days from treatment.

Results

Phosphorylation and D1 Turnover

In our previous work, it was observed that the Photosystem II exists as differently phosphorylated populations. The isolated phosphorylated PSII core populations were studied for their ability to bind photosynthetic herbicides (Giardi *et al.*, 1992; Giardi, 1993b) and D1 turnover. These populations showed different D1 turnover. In fact, pulse-chase experiments with ^{35}S-methionine indicated that, in the light, the least phosphorylated population was the first to incorporate the radioactivity. Moreover, immunoblot analysis revealed that mainly the phosphorylated populations generated the N-terminal breakdown fragment of D1*. When the protein synthesis was blocked with chloramphenicol (CAP), an inhibitor of protein synthesis at the level of translation on 70-S plastidic ribosomes, the D1 protein disappeared in the light without the formation of non-phosphorylated detectable fragments indicating that, before degradation, D1 was probably interconverted to D1* form (data not shown). It was concluded that, *in vivo*, while D1 synthesis occurs through the non-phosphorylated form, its degradation mainly proceeds through the phosphorylated form. Thus, these results suggested the possibility of modulating D1 turnover in higher plants. Conditions that have been previously shown to inhibit phosphorylation of PSII core, synthesis and assembly of protein or alternatively to increase protein phosphorylation were used for studying D1 turnover rate in peas and spinach plants as follows. i) Plants treated with low concentrations of $CdCl_2$ or N-ethylmaleimide (NEM) showed a slight dephosphorylation of the D1 protein. Protein synthesis seemed not affected in the stroma-exposed regions, however, the protein was not assembled into the grana region of the membrane (data not shown) perhaps mainly due to an immobilization of the -SH groups (Neumann *et al.*, 1994). A greatly decreased rate of methionine incorporation per unit of D1 protein was observed (Table 1), indicative of a reduced D1 turnover. ii) Plants treated with CAP showed a reduced incorporation of methionine, indicating again inhibition of D1 turnover rate (Table 1). iii) Water stress conditions were previously observed to increase the level of D1 phosphorylation without reducing the functionality of Photosystem II (Table 1); in these conditions, the incorporation of methionine was greatly enhanced (about 3 times), confirming that phosphorylation and D1 turnover are correlated.

Table 1. Incorporation of ^{35}S-met in a pulse-chase experiment (4 h in the light of 300 $\mu mol.s^{-1}.m^{-2}$) per unit of D1 protein (as measured by the ratio of the areas of the autoradiography scan to the corresponding Coomassie stained gel) relative to the control under different treatments with inhibitors or stress conditions (conditions as in Table 3).

Control	100
$CdCl_2$	20
CAP	40
water stress (RWC 70%)	300

Relationships between D1 Turnover Modulation and Herbicide Response

It was seen that the dissociation constant for the binding of herbicides increases with D1 phosphorylation indicating a reduced affinity at the Q-B pocket. The different heterogeneous phosphorylated populations show quite different affinity for the binding of herbicides (Table 2). Therefore, we studied herbicide action under different conditions of D1 turnover modulation. The effect of herbicide action under the different conditions reported above, was determined in sensitive, moderately sensitive and resistant plants. Table 3 reports the effect of conditions increasing or inhibiting D1 turnover, in combination with herbicide on the chlorophyll content and fresh weight. A synergistic effect between herbicide action and the tested inhibitors of D1 turnover rate (CAP, $CdCl_2$ and NEM) was observed. It is important to emphasise that no synergism was present with cycloheximide, an inhibitor of nuclear instead of chloroplast protein synthesis, indicating that the observed effect should be attributed to a primary action on D1. The synergism was weak in sensitive plants (*Avena sativa L., Senecio vulgaris*), greater in moderately resistant plants (*Pisum sativum, Triticum durum, Hordeum nigrum*), and absent in herbicide resistant plants (*Zea mays* and *Senecio vulgaris* mutant, in which it is known that the herbicide is immobilized by glutathione in the former or does not bind the Q-B site in the latter), again indicating that the synergism should be attributed to a primary action on D1.

This effect was mainly observed with triazine herbicides, while it was lesser using ioxinyl and phenolic-type herbicides (data not shown). Moreover, the different effects observed for the above mentioned plants can not be attributed to a differential translocation of these inhibitors, since their effect on the growth of the different plants was similar and even at their higher concentrations the relative synergistic effect with herbicide was maintained in the different species (data not shown).

The contrary herbicide action was observed in water stressed plants, where the D1 turnover and phosphorylation are enhanced (Table 1); in this case the herbicide (in the

Table 2. Dissociation constants for the binding of differently phosphorylated PSII cores. Herbicide binding experiments were performed according to Giardi *et al.*, (1992) and the dissociation constants (Kd) were calculated from double reciprocal plots of free versus bound herbicide. Each value given was obtained from three replicates.

	PSII Core populations			
D1 Phosphoryl. Index	1	9	16	41
Herbicide	Kd (nM)			
Chlorbromuron	70± 13	250± 23	180± 20	400±35
sites/cytb-559	0.9± 0.3	1.1±0.1	1.3±0.2	0.8±0.2
i-Dinoseb	94± 13	700± 61	400± 36	1000±77
sites/cytb-559	1.9± 0.1	0.8±0.3	0.9±0.3	0.9±0.3
Ioxynil	40± 6.5	180± 26	30±17	600±58
sites/cytb-559	1.1± 0.1	0.7±0.2	0.9±0.2	0.8±0.3

first 2 days of treatment) even accelerated the recovery from stress (Table 3). A similar effect was observed if atrazine was used in combination with 5 mM NaF, an inhibitor of the phosphatases responsible for dephosphorylation of proteins (data not shown) and presumably able to increase phosphorylation.

Discussion

The D1 protein undergoes at least five post-translational modifications during its life cycle: C-terminal processing, removal of the initiating methionine residue, N-acetylation of the resulting N-terminal threonine residue, covalent palmitoylation, and O-phosphorylation of the N-terminal threonine of the mature protein (Mattoo *et al.*, 1989). Processing of D1 occurs on stroma-exposed membranes while palmitoylation and phosphorylation occur in spatially distinct grana membranes (Mattoo *et al.*, 1989). The turnover of D1 protein is known to be very rapid, light-dependent and inhibited by herbicides such as atrazine and DCMU that displace the second electron-accepting plastoquinone of PSII (Mattoo *et al.*, 1989). A lower or no effect on D1 turnover has been observed with ioxynil herbicide. This rapid turnover has been correlated to the presence of a 14-amino acid helix in the protein, termed a destabilizing sequence, that is located adjacent to the putative cleavage domain (Mattoo *et al.*, 1989). Thus, it is consequential to think that the synergism occurs due to the additional effect of the inhibitor and the herbicide on D1 turnover. This observation seems to suggest that D1 turnover has a physiological importance for the plant life, since, if effectively blocked with more than one inhibitor, severe stress is caused to the plants. It was surprising to observe that the synergism was mainly present in moderately-resistant plants; in fact if the synergism were only due to the additional interaction of inhibitor and herbicide on the D1 turnover, a greater effect would have been seen in

Table 3. Effect of conditions changing D1 turnover on chlorophyll content and fresh weight.

Species Treatment	Chlorophyll/leaf area	Fresh weight (% reduction)
Pisum sativum		
+At	15	12
+CAP	4	5
+At+CAP	72	33
+Cycloheximide	5	4
+At+Ciclohex	6	9
+NEM	15	16
+At+NEM	50	48
+$CdCl_2$	5	30
At+$CdCl_2$	40	70
Avena sativa L.		
+At	38	40
+CAP	2	14
+At+CAP	50	45
Senecio vulgaris		
+At	31	27
+CAP	3	3
+At+CAP	43	41
S. vulgaris mutant		
+At	2	0
+CAP	0	10
+At+CAP	8	9
Zea mays		
+At	0	1
+CAP	6	11
+At+CAP	11	10
P. sativum		
+At	15	12
+NaF	0	0
+At+NaF	5	2
P.sativum		
Water Stressed	36	50
+At	20	40

Seedlings were treated after germination with inhibitors and herbicide (twice 200 ml for 30x30 cm vermiculate pots about 40 seedlings) four days before measurements. At, atrazine, 33 mg/l; CAP, chloramphenicol, 200 mg/l; Cycloheximide, 2mg/l; $CdCl_2$, 250 mg/l; NaF, 300 mg/l; NEM, N-ethylmaleimide, 100 mg/l. Stress conditions, 70% RWC and recovery time 4 days after watering with atrazine. SE was about 11%. The experiments were replicated seven times.

herbicide-sensitive plants compared to moderately-resistant plants. Thus, these experiments suggest that perhaps the D1 turnover is correlated to the resistance. To test this hypothesis, D1 turnover was forced to the other extreme, toward an increased rate. In this case a protection from herbicide action was observed, once again implying a function in the resistance. However, since the inhibitors as well as stress conditions could act at several levels in the membrane, at the present stage it is not possible to arrive at a definitive conclusion on the function of D1 turnover in the herbicide resistance. For instance, it is known that $CdCl_2$ treatment or water stress conditions also induce higher levels of glutathione (Neumann *et al.*, 1994) that is mainly responsible for atrazine conjugation and detoxification in higher plants. The present results indicate that it is possible to modulate D1 turnover through a modification of its synthesis or assembly or phosphorylation and that the sensitivity of a plant species to herbicide action directly responds to D1 turnover modulation. A differential D1 turnover is expected among different species with different morphological characteristics (such as leaf structure, thickness, chloroplast structure, organization of the thylakoid membrane etc.) that would influence the localization of D1 and of the kinases promoting phosphorylation. Thus, more experiments are needed to clarify the importance of D1 turnover in plant physiology and herbicide resistance.

References

Draber, W., Tietjen, K., Kluth, J.F. and Trebst, A. (1991) Herbicides in Photosynthesis Research. *Angewandte Chemistry* **30**: 1621-1633.

Mattoo, A.K., Marder, J.B. and Edelman, M. (1989) Dynamics of the photosystem II reaction center. *Cell* **56**: 241-246.

Giardi, M.T., Rigoni, F. and Barbato, R. (1992) PSII core phosphorylation: differential herbicide binding and electron transfer in PSII particle from spinach. *Plant Physiology* **100**: 1948-1954.

Giardi, M.T. (1993a) Phosphorylation and disassembly of photosystem II core as an early stage of photoinhibitions. *Planta* **190**: 107-113 .

Giardi, M.T. (1993b) Significance of photosystem II core phosphorylation heterogeneity for the herbicide binding domain. *Zeischrift Naturforschung* **48c**: 241-245.

Neumann, D., Lichlenberger, O., Günther, D., Tschiersch, K. and Noves, L. (1994) Heat-shock proteins induce heavy-metal tolerance in higher plants. *Planta* **194**: 360-367.

CHAPTER 15
PENETRATION, TRANSLOCATION AND METABOLISM OF PYRIDATE IN CHICKPEA

R. GIMENEZ-ESPINOSA and R. DE PRADO
Departamento Química Agrícola y Edafología, E.T.S. Ingenieros Agrónomos y Montes, Apdo 3048, 14080 Córdoba, Spain

Introduction

Chickpea is a major grain legume crop on the Indian subcontinent and West Asia, North Africa and the Americas with a world cultivated area of 10 million hectares, India being the main producer with 7 million has.

Spain is the main chickpea producer in Europe with a cultivated area fluctuating around 70 000 has. The crop is also grown in Turkey, Greece , Italy and Western Europe. The chickpea cultivated area in Spain has suffered a dramatic reduction since 1960 because of the introduction of more profitable crops, the rapid development of the oil seed industry, changes in diet where vegetable proteins have been substituted by animal proteins, lack of support for research, absence of selected material, cropping techniques, politics which stabilize markets, etc. In 1994 and 1995 the cultivated area in Spain has experienced a sudden increase due to changes in the EU agricultural policy.

There are several reasons to believe that chickpea, as one of the traditional crops of the Mediterranean region, has good prospects in the future.

- Firstly, legumes play an important role in farming rotations, improving the fertility of the soil.

- They have been recognized as a valuable source of dietary proteins in many developing countries (Williams and Singh, 1987).

- They are also important as animal food in developed countries.

- Moreover, chickpea is a crop with many possibities of improving its yield potential.

In the Mediterranean Basin, chickpea has traditionally been sown in spring time to reduce the incidence and severity of *Ascochyta* blight, the most damaging disease of the crop (Nene and Reddy, 1987). However, fall-winter chickpea sowings can increase the productivity of the crop if optimum materials and techniques are used. Research at ICARDA has provided cold tolerant, *Ascochyta* blight resistant chickpea germplasm and cultivars suited for extended cultivation in the Mediterranean Basin and elsewhere (Singh *et al.*, 1992). Nowadays, winter crop has been extended in Andalusia. However, yield loss due to weed competition is still a problem with winter chickpea in the Mediterranean Basin because of the lack of post-emergence herbicides.

There are more than 75 species which have been reported to infest chickpea fields in the Mediterranean region (Saxena, 1992). Grass weeds can be controlled by aryloxyphenoxypropanoate herbicides in post-emergence, but broadleaf weeds are more difficult to control at this stage.

Since1990, several trials carried out in our research group have shown that pyridate could be one of the most effective post-emergence herbicides for use with this crop. Pyridate applied after emergence controls several troublesome broadleaf weeds in corn and peanut.

This herbicide binds to the plastoquinone-binding D1 protein of photosystem II, inhibiting photosynthetic electron transport. After penetration, pyridate is rapidly hydrolyzed in the plant giving CL9673 (6-chloro-3-phenylpyridazine-4-ol). This compound has been established as the herbicidally active metabolite (Zonher, 1987). Crop tolerance has been explained in corn and peanut as the detoxification of CL9673 to conjugates that have no herbicidal activity in the plant, such as *N*-glycosides in peanut and more polar metabolites than *N*-Glycosides in corn (Gaillardon *et al.*, 1989).

Objectives

The following studies were carried out:

1. Effect of pyridate on growth in different chickpea varieties.

2. Effects of the herbicide on photosynthetic activity in chickpea and two common weeds (*Amaranthus blitoides* and *Lolium rigidum*).

3. Affinity of pyridate and its metabolites for the D1-protein in chickpea, *A. blitoides* and *L. rigidum.*

4. Penetration and translocation of pyridate in chickpea and *A. blitoides* and *L. rigidum.*

5. Metabolism of pyridate in chickpea, *A. blitoides*, and *L. rigidum.*

6. Role of glutathione transferases in the metabolism of pyridate in chickpea.

Results

Effect of Pyridate on Chickpea Growth

The effect of pyridate applied at 2 and 4 kg a.i. ha^{-1} on the growth of 15 different chickpea genotypes was studied. The fresh weight of shoots was measured 2 weeks after treatment. Results of fresh weights revealed a good tolerance to pyridate in all the varieties of chickpea studied. No phytotoxic effects were detected in any of the genotypes. However, two levels of tolerance were detected: High (13 genotypes) and Mid tolerance (2 genotypes).

Effect of Pyridate on Photosynthetic Activity

Illumination of plant leaves induced an increase in fluorescence. This is a specific property of photosystem II, in which the classic OIDP phases may be detected. O corresponds to a minimum fluorescence level produced before the excitation irradiance reaches the photosystem II centers by chlorophyll *a* molecules placed in the photosystem II antenna. The following phases (O-I, I-D, and D-P) reflect what is occurring in photosynthetic electron transport in photosystem II. Fluorescence yield is low when Q_A is oxidized (O), and high when Q_A is reduced to semiquinone form Q_A- (P). The I-D phase is explained by a disarrangement between Q_A (monoelectronic acceptor) and Q_B (bielectronic acceptor) (Ducruet *et al.*, 1991). The I fluorescence levels (F_I) are similar to P fluorescence levels (F_P) when electron transport is blocked between Q_A and Q_B. *I/P* values, defined as $(F_I - F_0) / (F_P - F_0)$, close to those of controls indicate only partial inhibition of photosynthetic transport in photosystem II, while values close to 1 indicate complete inhibition.

The effect of pyridate on photosynthetic activity was studied in two chickpea varieties (PV60 and JG62; Figure 1) and two common weeds in the Mediterranean region (*A. blitoides* and *L. rigidum*) treated at 2 Kg a.i. ha^{-1} pyridate. Fast fluorescence assays revealed that the degree of inhibition was higher in weeds than in the crop, although the maximum degree of inhibition was reached faster in chickpea and *A. blitoides* than in *L. rigidum*. Photosynthetic activity was gradually recovered in chickpea and *L. rigidum* but not in *A. blitoides*. Recovery of photosynthesis was faster in chickpea than in *L. rigidum*. Results show that tolerance of chickpea to pyridate is not caused by a change in the target-site but by mechanisms for the metabolic detoxification of pyridate into non-toxic compounds.

Affinity of the Herbicide for the Target Site

The concentration of pyridate and its metabolites that inhibits the Hill reaction by 50%, I_{50} values, was determined in the same chickpea genotypes and weeds. There were no differences among the I_{50} values obtained with chickpea, *L. rigidum* and *A. blitoides* to pyridate, CL9673 and CL9869, although the non-toxic metabolites (CL9673-*N*-Gly and CL9673-*O*-Gly) were more active in the weeds than in crops. This fact could explain the moderate tolerance in *L. rigidum* and the high tolerance in chickpea to pyridate. The I_{50} values for pyridate and CL9673, CL9673-N-Gly, CL9673-O-Gly and CL9869 metabolites showed that CL9673 has more herbicidal activity at chloroplastic level than the others. The order of phytoxicity was CL9673>>CL9673-N-Gly>CL9869>pyridate>CL9673-O-Gly (Table 1).

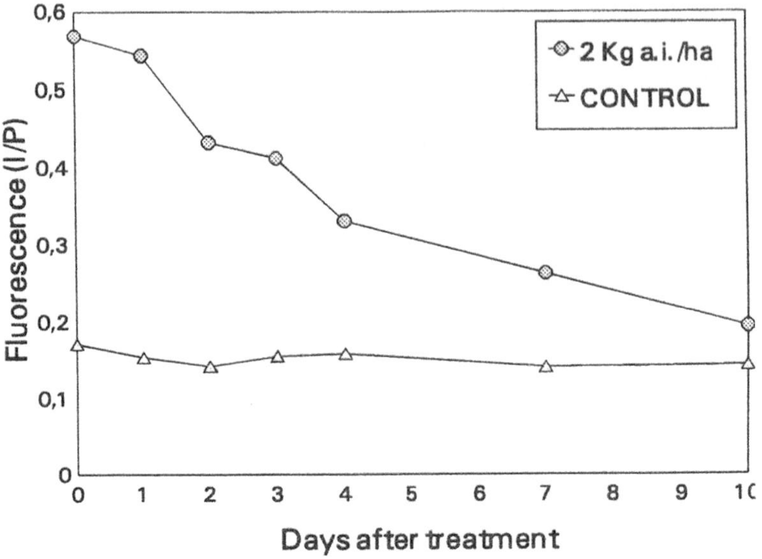

Figure1. Evolution of I/P ratios after pyridate treatment of chickpea (var. JG62).

Penetration and Translocation

Penetration and translocation of pyridate were studied in one chickpea genotype (cv. JG62) and in *A. blitoides* and *L. rigidum*. The penetration of pyridate was very low in *L. rigidum* in comparison with the broadleaf species studied. The lower absorption of the herbicide could be the reason for tolerance in this weed and in all the monocotyledonean species. Herbicide absorption was comparatively higher and quicker in chickpea plants than in *A. blitoides*. The study of translocation of pyridate in plants revealed that the herbicide hardly moves into the plant, with 60 to 90% of the applied herbicide remaining in the treated leaves.

Metabolism Assays

The metabolism assays, carried out on chickpea (cv. JG62, Figure 2), *A. blitoides* and *L. rigidum* showed that pyridate gives CL9673 when it penetrates into the plant, as reported by Zonher (1987). N-GlyCL9673, which is essential in the detoxification of pyridate, appeared in all three species. In chickpea, two conjugates were detected in the detoxification of pyridate and were named as C1 and C2. C1 appeared first and in greater amounts than C2. The formation of both conjugates was faster than in the weeds. The proportion of phytotoxic compounds decreased to 20% of total radioactivity. Metabolism was much slower in *L. rigidum*, C2 did not appear and the amount of C1 was lower, confirming that the tolerance observed in this weed cannot be solely explained by detoxification of the herbicide. In *A. blitoides*, C1 did not appear and the amount of phytotoxic compounds was considerably higher 5 days after treatment.

Table 1. Inhibition of the Hill reaction (I_{50}, μM) in isolated chloroplasts from leaves of chickpea and weeds by pyridate and its metabolites. Values are means ± S.E. of six replications.

Chickpea genotypes	Pyridate	CL9673	CL9673-N-Gly	CL9673-O-Gly	CL9869
PV60	46±2.9	0.9±0.1	14±2.5	120±6.8	15±3.7
ILC72	42±3.3	0.65±0.05	10±1.9	140±5.5	14±0.7
JG62	38±2.8	0.7±0.05	15±0.9	140±5.5	14±3.3
WR315	36±2.6	0.9±0.1	5±0.8	85±6.8	16±1.1
A. blitoides	48±1.4	0.7±0.4	9±0.2	65±5.9	15±0.8
L. rigidum	35±2.6	0.8±0.08	8±0.3	65±6.1	17±0.8

Role of GST on the Detoxification of Pyridate

The nature of the conjugates formed in chickpea in the detoxification of pyridate was identified using ^{14}C-pyridate. One of them was the conjugate formed by N-Gly-CL9673 and GSH (C1), and the second was the conjugate formed by N-Gly-CL9673 and cysteine (C2). The N-glucosylation of CL9673 was essential to the formation of these conjugates.

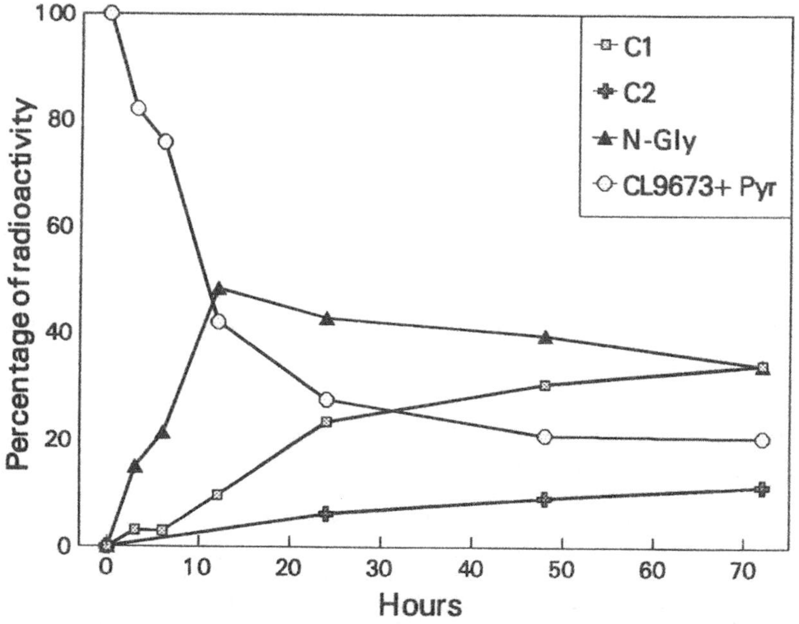

Figure 2. Evolution of metabolites in chickpea (var. JG62) after treatment with ^{14}C-pyridate.

Assays carried out *in vitro* showed that CL9673 does not react with GSH or cysteine at any pH, while its N-Gly-derivative is very reactive even at physiological pH.

Having identified the GSH conjugate, the role of the glutathione transferases (GSTs) in the metabolism of pyridate was studied. GST activity using CDNB (1-chloro-2-dinitrobenzene) as substrate was determined in crude extracts from four different varieties of chickpea. GST rates were very low in all cases. GST assays using ^{14}C-N-Gly-CL9673 in the presence and absence of enzyme and at different pHs demonstrated that the amount of GSH-conjugate formed was greatest at high pH. The rate of formation of the GSH-conjugate only increased in the presence of the enzyme at pH 6.

The results indicate that it is unlikely that GSTs are required in the conjugation of pyridate metabolites and in herbicide metabolism. Other mechanisms must be responsible for species-dependent variations such as the amount and bioavailability of GSH in different plants.

Conclusions

1. Tolerance of chickpea to pyridate is due to metabolic detoxification of the herbicide into non-toxic conjugates which have been identified as GSH and Cys-conjugates.

2. The phytotoxic CL9673 has to be N-glycosided before conjugation and chickpea GSTs seem to play no important role in detoxification.

3. Pyridate could be a good herbicide to control post-emergence broadleaf weeds in chickpea crop.

Our future research will focus on obtaining mixtures of herbicides to control post-emergence grass and broadleaf weeds in chickpea. Preliminary mixtures of pyridate and propaquizafop have shown excellent results in greenhouse assays.

Acknowledgements

The authors thank the CICYT for the financial support (Project No. AGF93-0332)

References

Ducruet, J.M. (1991) Les herbicides inhibiteurs due photosystème II. *In: Les Herbicides. Mode d'Action et Principes d'Utilisation* (R. Scalla, ed.), 79-114. INRA, Paris.

Gaillardon, P., Gasquez, J. and Scalla, R. (1989). Absorption, translocation and metabolism of pyridate in a tolerant crop (*Zea mays*) and two susceptible weeds (*Polygonum lapathifolium* and *Chenopodium album*). *Weed Research* **29**: 45-51.

Nene, Y. L. and Reddy, M. V. (1987). Chickpea diseases and their control. *In: The Chickpea* (M.C. Saxena and K. B. Singh, eds.), 233-270. C.A.B., Wallingford, Oxen, U.K.

Saxena, M.C. (1992). Current status and prospects of kabuli chickpea production. *In: Disease Resistant Breeding in Chickpea* (K. B. Singh and M. C. Saxena, eds.), 1-10. ICARDA, Aleppo.

Singh, K.B., Reddy, M.V. and Haware, M.P. (1992). Breeding to resistant for *Ascochyta* blight in chickpea. *In: Disease Resistant Breeding in Chickpea* (K. B. Singh and M. C. Saxena, eds.), 23-54. ICARDA, Aleppo.

Williams, P.C. and Singh, U. (1986). Nutritional quality and the evaluation of quality in breeding programmes. *In: The Chickpea,* (M.C. Saxena and K. B. Singh, eds.), 229. C.A.B., Wallingford, Oxen, U.K.

Zohner, A. (1987). Mode of crop tolerance to pyridate in corn and peanuts. *British Crop Protection Conference-Weeds* **10 B-2**: 1083-1090.

CHAPTER 16

DISTRIBUTION STUDIES OF PROPANIL RESISTANCE IN A BARNYARDGRASS BIOTYPE AND ELUCIDATION OF ITS RESISTANCE MECHANISM

R.E. HOAGLAND[1], V.F. CAREY III[2], S.O. DUKE[1] and R.E. TALBERT[3]
1. USDA-ARS, Southern Weed Science Laboratory, Stoneville, MS, USA; 2. Valent USA Corp., Seymour, IL, USA; 3. University of Arkansas Fayetteville, AR, USA

Abstract

Verification of barnyardgrass [*Echinochloa crus-galli* (L.) Beauv.] resistance to propanil has now been confirmed and its distribution found to be widespread in rice (*Oryza sativa* L.)-producing areas of Arkansas. We conducted laboratory and greenhouse experiments on propanil-resistant (R-BYG) and propanil-susceptible (S-BYG) barnyardgrass, and rice (cv. "Newbonnet") to determine the resistance mechanism(s). Propanil absorption and translocation were not different in the two barnyardgrass biotypes. *In vivo* chlorophyll fluorescence data from leaf disks exposed to propanil solutions at various concentrations indicated that the propanil binding site of photosynthetic electron transport in PS II was not altered. Propanil-R-BYG exhibited no cross-resistance to other PS II inhibitors such as atrazine, diuron, fluometuron, or linuron, since both biotypes had equal mortality and injury ratings when treated with these herbicides, and atrazine (50 μM) totally inhibited PS II as evidenced in chlorophyll fluorescence tests. Metabolic studies using ^{14}C-propanil applied to rice, R-BYG, and S-BYG indicated that detoxification of propanil to dichloroaniline (DCA) was more rapid in the R-BYG than the S-BYG biotype. Two additional polar metabolites were also detected in R-BYG, with TLC Rf values in two solvent systems equal to those of propanil metabolites found in rice. The detoxification appears to be via aryl acylamidase, since DCA and metabolites corresponding to DCA-saccharide conjugates were found and because the insecticide carbaryl, an inhibitor of this enzyme, increased injury or caused mortality of R-BYG when propanil and carbaryl were applied simultaneously. Metabolism of propanil appears to be the operative resistance mechanism in this propanil-R-BYG found in Arkansas.

Introduction

Propanil [*N*-(3,4-dichlorophenyl)propanamide] is widely used for post-emergence control of broadleaf weeds and barnyardgrass (*Echinochloa* spp.) in cultivated rice (*Oryza sativa* L.). This herbicide has been used in Arkansas (AR), USA, since 1962

(Smith, 1961), and is sometimes applied twice per season on a single crop. In 1989, rice producers in Poinsett County, AR, began to observe decreased control of barnyardgrass using standard propanil application concentrations that had previously provided control of this weed for about 30 years. Plants from seed collected in this area were first confirmed to exhibit propanil resistance in 1990 (Baltazar and Smith, 1994). Our objectives were to examine the distribution of propanil resistance in barnyardgrass in Arkansas rice-producing counties and to determine resistance mechanism(s). To accomplish the latter objective, we measured or examined uptake/translocation, site of action modification, cross-resistance, and metabolism. This paper comprises a summary of our results from other sources (Carey, 1994; Carey and Talbert, 1993; Carey, *et al.*, 1993; Carey *et al.*, 1994; Carey *et al.*, 1995a; Carey *et al.*, 1995b; Carey *et al.*, 1996) with additional discussion relevant to other herbicide resistance studies.

Materials and Methods

Plant Material, Growth and Resistance Tests

R-BYG and S-BYG seeds collected from Poinsett County, AR, in 1990 and Arkansas County, AR, in 1983, respectively, were planted and grown in the greenhouse. R-BYG (1-leaf stage) were treated with propanil (4.5 kg/ha) to insure resistance to propanil. Plants were then grown in the greenhouse (14 hr light, natural/artificial; 10 hr dark regime; day/night temperatures of 24° to 27°C/15° to 18°C) until further treatment.

Absorption/Translocation Studies

Greenhouse-grown plants (3- to 4-leaf stage) were treated with formulated propanil (Stam M4, Rohm and Haas, Philadelphia, PA) at 3.4 kg/ha in a carrier volume of 47 l/ha (250 mM) using a stationary spray chamber to simulate field application conditions. When dry, plants were moved to the lab and treated with uniformly-ring-labelled ^{14}C-propanil (specific activity = 21 µCi/mg) mixed with formulated propanil and distilled water at 100,000 dpm/µl (760 µM), pH 7.0. A 1-µl droplet was placed on the second leaf of R-BYG or S-BYG plants, and plants were placed in chambers (continuous light, PAR, 300 $\mu Em^{-2}s^{-1}$; 26 to 28°C). At harvest, (0, 4, 8, 12, 24, and 48 hr after treatment) plants were sectioned into 1) leaf tissue above treated leaf, 2) treated leaf, including collar, 3) tissue below treated leaf, and 4) root tissue. Each treated leaf was washed twice with 10 ml deionized water containing Triton X-100 (0.01%) to remove non-absorbed ^{14}C-propanil. Aliquots from washes were combined and ^{14}C quantified using liquid scintillation counting (LSC). Propanil translocation was determined by measuring ^{14}C in excised plant sections. Sections were oven-dried (90°C, 48 hr), oxidized using a biological oxidizer, and $^{14}CO_2$ was trapped in scintillation cocktail and quantified (LSC). ^{14}C from each plant section was calculated as percent of applied. Treatments were triplicated, and the experiment was repeated. Data were analyzed as a split-plot with barnyardgrass biotype as the main plot and harvest times as subplots.

Evaluation of Propanil Site of Action

The second leaf of six 6-wk-old R-BYG or S-BYG plants was cut into 4 mm diam. disks using a cork borer. Disks were placed in 10-cm petri dishes containing 50 μM propanil, 30 μM carbaryl (1-naphthyl *N*-methylcarbamate; inhibitor of propanil metabolism), 50 μM propanil plus 30 μM carbaryl, or deionized water (untreated). Pesticides were technical grade (>95% purity). Dishes were covered with foil to prevent light exposure during incubation (22°C). Chlorophyll fluorescence was measured using a plant productivity fluorometer after 1.5, 2.5, 5, and 10 hr incubation. Each leaf disk was removed from the treatment solution, blotted, and placed on the light electrode (adaxial side towards the light source), then covered with black tape to allow the photosystem to stabilize in the dark. Fluorescence was then measured for a 50-s illumination time; data were acquired with a data storage device and plotted. Three disks from each treatment solution were measured at each sampling time. The difference in peak and terminal fluorescences was used to calculate percent inhibition of photosynthesis compared to untreated disks. Data were analyzed as a split-plot with three replications and the experiment was repeated. Biotype was the main plot with incubation treatments as subplots.

Tests for Cross-Resistance to other PS II Inhibitors

R-BYG was tested for cross-resistance to other PS II inhibitors such as atrazine [6-chloro-*N*-ethyl-*N*'-(1-methylethyl)-1,3,5-triazine-2,4-diamine], diuron [*N*'-(3,4-dichlorophenyl)-*N*,*N*-dimethylurea], fluometuron {*N*,*N*-dimethyl-*N*'-[3-(trifluoromethyl)phenyl]-urea}, and linuron [*N*'-(3,4-dichlorophenyl)-*N*-methoxy-*N*-methylurea]. 'Newbonnet' rice, and R-BYG and S-BYG biotypes were seeded in silt loam soil in the greenhouse. Herbicide rates for atrazine, fluometuron, and linuron were 1.12 kg/ha, and 0.9 kg/ha for diuron, applied pre-emergence in a carrier volume of 187 l/ha. Plants were evaluated for injury 10 days after emergence. In addition to greenhouse whole-plant studies, chlorophyll fluorescence of R- and S-BYG barnyardgrass leaf disks was measured after incubation with 50 μM atrazine and the photosynthetic inhibition calculated.

Metabolism Studies

Greenhouse-grown rice and barnyardgrass plants (4-leaf-stage) were grown as above. Tests were performed on carbaryl alone and on the interaction of carbaryl and propanil applied to rice plants and the two barnyardgrass biotypes. The second leaf of each of six plants was excised under water and placed in tubes with 7.6 μM ^{14}C-propanil. The upper 2 mm leaf tips were clipped to increase transpiration and propanil uptake under continuous light (300 $\mu Em^{-2}s^{-1}$) for 16 hr, 20°C. Leaves were homogenized with methanol, homogenates filtered (0.4 μm), and filtrates concentrated by rotary evaporation. Plant residues were oxidized to quantitate non-extractable ^{14}C via LSC. Metabolites were spotted on silica gel TLC plates, developed 15 cm in either benzene:acetone (10:1, v:v) or pyridine:*n*-butanol:water (6:4:3, v:v:v), and radioactive zones quantified with a radiological chromatogram scanner.

Results and Discussion

Resistance Tests

Voluntary surveys of rice producers conducted over two years (1991 and 1992) resulted in a total of 154 barnyardgrass seed samples from 19 of 38 Arkansas counties (Carey, 1994; Carey *et al.*, 1995b). Of the samples collected, propanil resistance was confirmed in 115 samples from 16 counties, and the level of resistance could be grouped into several categories (Table 1). The average percent injury or control of S-BYG and R-BYG using propanil at 4.5 kg/ha was about 92% and 35%, respectively. Highly R-BYG samples were controlled 80% by propanil at 67 kg/ha.

Table 1. Propanil-susceptible/resistant categories of barnyardgrass biotypes from AR

Resistance Category	Resistance factor[a]	Calculated I_{50}[b]
S-BYG	-	6.7
Slightly R-BYG	6x	14
Moderately R-BYG	10x	20
Highly R-BYG	20x	39

[a] Based on propanil (3.4 kg/ha). [b] Propanil (kg/ha) required for 50% control, from regression analysis.

Propanil Absorption/Translocation Studies

Absorption of ^{14}C-propanil increased over the 48 hr test period in both R- and S-BYG biotypes, but statistical analysis indicated a lack of significant differences (95% confidence level) between biotype and subplots. Recovery efficiency of ^{14}C was 95% in the absorption studies. About 6% of the applied ^{14}C-propanil was absorbed by both biotypes after 48 hr. Although somewhat less than the 15% absorbed in an earlier study (Yih, *et al.*, 1968), our conditions were different (plant age, environment, etc.) and we avoided using acetone which could increase absorption by alteration of cuticular waxes and membranes.

Translocation studies of the absorbed ^{14}C-propanil also revealed no significant differences between biotypes (Table 2). The majority (91%) of absorbed radioactivity remained in the treated leaf with low radioactivity distributed among the other plant sections. Similar propanil translocation patterns have been reported for barnyardgrass and rice (Yih, *et al.*, 1968; Still, 1967). Lack of a BYG biotype effect in both the absorption and translocation studies indicates that they are not responsible for propanil resistance in barnyardgrass (Carey *et al.*, 1995b). Other absorption studies showed that barnyardgrass and rice foliage absorbed propanil equally; thus even diverse leaf morphological differences did not alter uptake and do not explain differential herbicide selectivity in the weed and crop (Yih *et al.*, 1968). Absorption/translocation differences do not explain propanil selectivity in green foxtail (*Setaria viridis*) and wheat (*Triticum aestivum*) (Eberlein and Behrens, 1984). A lack of significant differences in propanil uptake between propanil-susceptible and -resistant biotypes of junglerice [*Echinochloa colona* (L.) Link] at any growth stage has also been recently reported (Leah *et al.*, 1995).

Evaluation of Propanil Site of Action

Chlorophyll fluorescence analysis indicated a significant interaction between

Table 2. Translocation of ^{14}C-propanil in propanil-S- and -R-BYG[a].

Plant part	% ^{14}C Recovered
Treated leaf	91.0
Above treated leaf	3.2
Below treated leaf	0.8
Roots	5.0

[a] No significant differences between biotypes; data combined.

barnyardgrass biotype and incubation treatment (Figure 1A). Photosynthesis was inhibited in S-BYG when leaf disks were incubated in 50 µM propanil; i.e., >75% at 1.5 hr, and increased to 90% after 5 hr incubation. This high degree of inhibition indicates that propanil interferes with the photosynthetic electron transport chain in S-BYG (Carey *et al.*, 1995a). Propanil is a known PS II inhibitor (Mitsutake, *et al.*, 1986). Fluorescence transients of propanil-treated tissue were similar to those of other PS II inhibitors, for example, BAS 13-338 (Bose *et al.*, 1984) and MT-5950 (Gohbara *et al.*, 1988). Fluorescence analysis has been used in comparative studies of rice and S-BYG (Daniell *et al.*, 1981). Photosynthesis in R-BYG was inhibited only 50% by 50 µM propanil at 1.5 hr, and inhibition decreased as incubation time increased, indicating that R-BYG recovered from propanil exposure. If the resistance mechanism in propanil-R-BYG was due to a change in binding to the D-1 protein, as in triazine-resistance, there should be no initial effect on fluorescence (Gohbara *et al.*, 1988; Ahrens *et al.*, 1981).

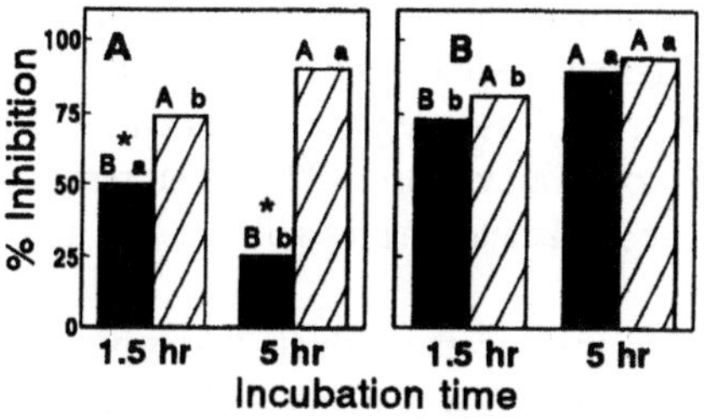

Figure1. Inhibition of photosynthesis determined by chlorophyll fluorescence analysis. A=propanil (50 µM), B=propanil (50 µM) + carbaryl (30 µM). Solid bars, R-BYG; hatched bars, S-BYG. Significance symbols (95% level): capital letters, comparison of biotypes at given time and treatment; lowercase letters, comparison of times for given biotype and treatment; *=treatment/time/biotype combination differs significantly from all others, both treatments.

Carbaryl inhibits the aryl acylamidase responsible for propanil detoxification in rice (Frear and Still, 1968), and was used with propanil in a preliminary test to block possible propanil metabolism in R-BYG. When leaf disks of both biotypes were incubated in a mixture of propanil and carbaryl, photosynthesis was decreased after 1.5 and 5 hr (Figure 1B). Thus, carbaryl not only blocked the recovery in R-BYG (propanil alone treatment), but also increased photosynthetic inhibition with time. Carbaryl alone (30 µM) did not affect photosynthesis. These combined effects suggested a lack of molecular binding site changes in the photosystem electron transport chain, and hinted that propanil was most likely being metabolized by the resistant biotype (Carey and Talbert, 1993; Carey, *et al.*, 1994; Carey *et al.*, 1995a).

Cross-Resistance to other PS II Inhibitors

Herbicide cross-resistance is most often caused by an altered binding site (Duke, 1993). Cases of cross-resistance have occurred in several triazine-resistant species (Duke, 1993; Feurst, *et al.*, 1986). Greenhouse-grown rice and both barnyardgrass biotypes were controlled 100% by atrazine, diuron, fluometuron, and linuron 10 days after emergence. Furthermore, other observations indicated no differential response between the BYG biotypes to these PS II-inhibiting herbicides at lower concentrations, indicating a lack of cross-resistance in the R-BYG. Likewise, chlorophyll fluorescence data indicated that photosynthesis was inhibited similarly by atrazine (50 µM) in R-BYG and S-BYG; i.e., inhibition of 40% after 1 hr and 100% after 2 hr incubation. These data indicate that R-BYG is not cross-resistant to atrazine or some other PS II-inhibiting herbicides (Carey, *et al*, 1995a). These findings are consistent with those of fellow researchers investigating possible cross- and multiple-resistance in R-BYG (pers. comm., M.J. Kitt and R.E. Talbert, Univ. Arkansas, 1994).

Metabolism Studies

To further examine possible *in vivo* metabolic mechanisms of resistance, carbaryl was applied alone and in combination with propanil to greenhouse-grown seedlings of R- and S-BYG and rice (Figure 2). Carbaryl alone did not injure plants, but in combination with propanil caused R-BYG to be controlled/injured at a level equal to that of S-BYG, and increased rice injury (although to a lesser degree) 7 days after treatment. Analogous trends were noted over a 20-day period (Carey, *et al.*, 1995a). Our results are similar to recent data showing that carbaryl decreased resistance in propanil-resistant junglerice (Leah, *et al.*, 1995).

In *in vivo* metabolism studies in which ^{14}C-propanil was fed to R- and S-BYG and rice leaves, methanol-extracted ^{14}C was 40, 49, and 62%, respectively, in R-BYG, rice, and S-BYG, with a total recovery in soluble and insoluble fractions of about 90% in all cases. TLC separation of extracts of ^{14}C-propanil-treated R-BYG, S-BYG, and rice using acetone:benzene showed that extracts of R-BYG and rice contained equal amounts of the propanil hydrolysis product DCA and relatively large amounts of polar material at the origin (Table 3). Development of this plate in a second solvent system (pyridine:butanol:H_2O) separated the polar metabolites, resulting in similar distribution patterns for rice and R-BYG.

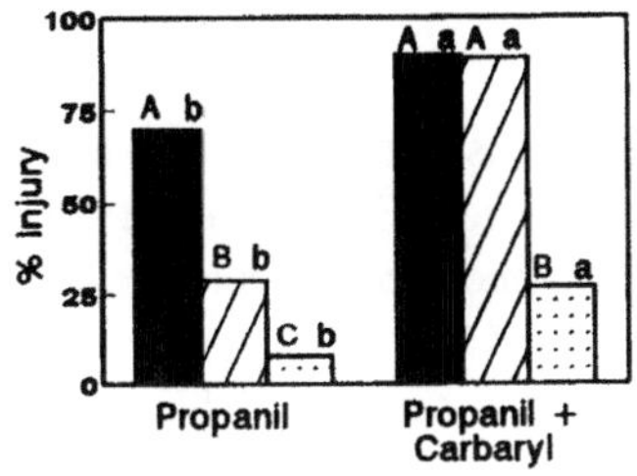

Figure 2. Response of R- and S-BYG and rice seedlings to foliar application of propanil (4.5 kg/ha) or propanil + carbaryl (1.1 kg/ha), 7 days after treatment. Solid bars, R-BYG; hatched bars, S-BYG; dotted bars, rice. Significance symbols (95% level): capital letters, comparison of biotypes/species within treatment; lowercase letters, comparison of treatments for given biotype/species.

Table 3. Summation of % ^{14}C and TLC Rf range of propanil or DCA standards and metabolic products extracted from leaves of ^{14}C-propanil-treated rice or BYG biotypes.

	Rf range					
	benzene:acetone			pyridine:*n*-butanol:water		
Spotted extract or standard	0 -.01	.33-.36	.49-.52	.60-.63	.78-.80	.87-.93
	% ^{14}C					
^{14}C-Propanil standard	-	100	-	-	-	100
^{14}C-DCA standard	-	-	100	-	-	100
S-BYG	5	95	-	-	-	100
R-BYG	51	45	4	57	23	20
Rice (tolerant)	43	54	3	44	28	28

No metabolic products were found in the S-BYG when treated under these conditions. The polar metabolites in rice and R-BYG are most likely the DCA conjugates 3',4'-dichlorophenyl-glucosylamine (Rf 0.78) and a 3',4'-dichlorophenyl-saccharide conjugate (Rf 0.60) previously found in rice (Yih, *et al.*, 1968; Still, 1967). Metabolism results show that propanil is metabolized at a higher rate in R-BYG than in S-BYG (Carey, *et al.*, 1996). S-BYG had previously been shown to have aryl acylamidase, but at a level 60-fold lower than in rice (Frear and Still, 1968). Generally, the mechanisms of herbicide resistance in weeds are different than selectivity mechanisms in the crops in which herbicides are used (LeBaron and McFarland, 1988), however, in both propanil-R-BYG (Carey, *et al.*, 1996)] and junglerice (Leah, *et al.*, 1994), the mechanism (metabolism) appears to be identical to that of the crop, rice.

Barnyardgrass is a serious weed problem in temperate and tropical zones and is considered to be the major weed in rice (Holm, *et al.*, 1977). The wide use of propanil for several decades has intensified some weed problems, since resistance to propanil now occurs in barnyardgrass (Baltazar and Smith, 1994; Carey, *et al.*, 1995b; Giannoppolitis and Vassiliou, 1989) and in junglerice (Fischer, *et al.*, 1993; Garro, *et al.*, 1992). The widespread distribution of resistant barnyardgrass in Arkansas suggests it may be an ever-increasing problem unless crops are rotated or herbicides with varying sites of action are used to control or reduce barnyardgrass populations.

References

Ahrens, W.H., Arntzen, C.J. and Stoller, E.W. (1981) Chlorophyll fluorescence assay for the determination of triazine resistance. *Weed Science* **29**: 316-322.

Baltazar, A.M. and Smith Jr, R.J. (1994) Propanil-resistant barnyardgrass (*Echinochloa crus-galli)* control in rice (*Oryza sativa*). *Weed Technology* **8**: 576-581.

Bose, S., Mannar, R.M. and Arntzen, C.J. (1984) Increased synthesis of photosystem II in *Triticum yulgare* when grown in the presence of BAS 13-338. *Zeischrift Naturforschung* **39c**: 510-513.

Carey, V.F. (1994) Propanil resistant barnyardgrass in Arkansas: Competitive ability, distribution, and mechanism of resistance. *PhD Thesis*. University of Arkansas.

Carey, V.F. and Talbert, R.E. (1993) Evaluation of propanil resistance with chlorophyll fluorescence. *Abstracts Arkansas Agricultural Pesticde Association* **32**: 8.

Carey, V.F., Talbert, R.E. and Smith Jr, R.J. (1993) Studies on the mechanism of propanil resistance in Arkansas barnyardgrass. *Proceedings Southern Weed Science Society* **46**: 266.

Carey, V.F., Hoagland, R.E. and Talbert, R.E. (1994) Determination of the resistance mechanism in propanil-resistant barnyardgrass. *Proceedings Rice Technical Working Group* **25**: 162.

Carey, V.F., Duke, S.O., Hoagland, R.E. and Talbert, R.E. (1995a) Resistance mechanism of propanil-resistant barnyardgrass: I. Absorption, translocation, and site of action studies. *Pesticide Biochemistry Physiology* **52**: 182-189.

Carey III, V.F., Hoagland, R.E. and Talbert, R.E. (1995b) Verification and distribution of propanil-resistant barnyardgrass in Arkansas. *Weed Technology* **9**: 366-372.

Carey, V.F., Hoagland, R.E. and Talbert, R.E. (1996) Resistance mechanism of propanil-resistant barnyardgrass: II. *In vivo* differential metabolism of propanil in susceptible and resistant barnyardgrass. *Pesticide Science*. In press.

Daniell, H., Sarojini, G., Kumarachinnayan, P. and Kulandaivelu, G. (1981) Action of propanil on *in vivo* chlorophyll *a* fluorescence in *Echinochloa crus-galli* and rice. *Weed Research* **21**: 171-177.

Duke, S.O. (1993) Mechanisms for resistance of weeds to herbicides. *Proceedings Beltwide Cotton Production Conference* **3**: 1509-1511.

Eberlein, C.V. and Behrens, R. (1984) Propanil selectivity for green foxtail (*Setaria viridis*) in wheat (*Triticum aesitivum*). *Weed Science* **32**: 13-6.

Feurst, E.P., Arntzen, C.J., Pfister, K. and Penner, D. (1986) Herbicide cross-resistance in triazine-resistant biotypes of four species. *Weed Science* **34**: 344-353.

Fischer, A.J., Granados, E. and Trujillo, D. (1993) Propanil resistance in populations of junglerice (*Echinochloa colona*) in Columbian rice fields. *Weed Science* **41**: 201-206.

Frear, D.S. and Still, G.G. (1968) The metabolism of 3,4-dichloropropionanilide in plants. Partial purification and properties of an aryl acylamidase from rice. *Phytochemistry* **7**: 913-920.

Garro, J.E., De la Cruz, R. and Merayo, A. (1992) The study of the growth of propanil susceptible and tolerant *Echinochloa colona* (L.) Link plants. *Manejo Integrado do Plagas* **26**:39-43.

Giannopolitis, C.N. and Vassiliou, G. (1989) Propanil tolerance in *Echinochloa crus-galli* (L.) Beauv. *Tropical Pest Management* **35**: 6-7.

Gohbara, M., Duke, S.O. and Takematsu, T. (1988) MT-5950, a new anilide herbicide, inhibits PS II at a site that slightly overlaps the triazine binding site. *Agricultural Biological Chemistry* **52**: 465-472.

Holm, L.G., Pluchnett, D.L., Pancho, J.V. and Herberger, J.P. (1977) *The World's Worst Weeds, Distribution and Biology* , 32-40. University Press, Hawaii, Honolulu.

Leah, J.M., Caseley, J.C., Riches, C.R. and Valverde, B. (1994) Association between elevated activity of aryl acylamidase and propanil resistance in jungle-rice *Echinochloa colona. Pesticide Science* **42**: 281-289.

Leah, J.M., Caseley, J.C. Riches, C.R. and Valverde, B. (1995) Age-related mechanisms of propanil tolerance in jungle-rice, *Echinochloa colona. Pesticide Science* **43**: 347-354.

LeBaron, H.M. and McFarland, J. (1988) Herbicide resistance in weeds and crops. *In: Managing Resistance to Agrochemicals*, 337-352. American Chemical Society, Washington, DC.

Mitsutake, K., Iwamua, H., Shimizu, R. and Fujita, T. (1986) Quantitative structure-activity relationship of photosystem II inhibitors in chloroplasts and its link to herbicidal action. *Journal Agricultural Food Chemistry* **34**: 725-732.

Smith Jr., R.J. (1961) 3,4-Dichloropropionanilide for control of barnyardgrass in rice. *Weeds* **9**: 318-322.

Still, G.G. (1967) Metabolism of 3,4-dichloropropionanilide in plants: the metabolic fate of the 3,4-dichloroaniline moiety. *Science* **11**: 992-993.

Yih, R.Y., McRae, D.H. and Wilson, H.F. (1968) Mechanism of selective action of 3',4'-dichloro-propionanilide. *Plant Physiology* **43**: 1291-1296.

CHAPTER 17
MECHANISMS OF RESISTANCE TO PROTOPORPHYRINOGEN OXIDASE-INHIBITING HERBICIDES

S.O. DUKE[1], H.J. LEE[1], M.V. DUKE[1], K.N. REDDY[1], T.D. SHERMAN[1], J.M. BECERRIL[2], U.B. NANDIHALLI[3], H. MATSUMOTO[4], N.J. JACOBS[5] and J.M. JACOBS[5]

1. USDA/ARS, Southern Weed Science Laboratory, Stoneville, MS 38776, USA. 2. Universidad del Pais Vasco, Facultad Ciencias, Departamento Biologia Vegetal y Ecologia, Apdo. 644, E-48080 Bilbao, Spain. 3. Hazleton Laboratories, P.O. Box 7545, Madison, Wl, 53707, USA. 4. Institute Applied Biochemistry, University Tsukuba, Ibaraki 305, Japan. 5. Department Microbiology, Dartmouth College, Hanover, NH 03755, USA

Abstract

Resistance to protoporphyrinogen oxidase-inhibiting herbicides can theoretically be generated by at least six different mechanisms. They include: 1) reduction of uptake or sequestration of the herbicide, 2) rapid metabolic destruction of the herbicide, 3) resistance at the site of action, 4) rapid metabolic destruction of protoporphyrinogen IX or protoporphyrin IX to non-photodynamic compounds, 5) inactivation of the herbicide-resistant enzyme that converts protoporphyrinogen IX to protoporphyrin IX, and 6) resistance to toxic oxygen species due to high levels of antioxidants and enzymes that destroy toxic oxygen species. Some of these mechanisms have been reported in cases of naturally-occurring resistance of crops and/or weeds, although resistance at the site of action is not well documented, and mechanism five has not been reported. Despite so many possible mechanisms of resistance, there are no documented cases of evolved resistance to these herbicides as a result of their use, and no crops have been genetically engineered to be resistant to these herbicides.

Introduction

Protoporphyrinogen oxidase (Protox)-inhibiting herbicides have a complicated mechanism of action (Duke, *et al.*, 1991, 1994; Nandihalli and Duke, 1993). After inhibition of plastid Protox, the substrate, protoporphyrinogen IX (Protogen) is lost from the plastid to the cytoplasm (Jacobs and Jacobs, 1993; Jacobs, *et al.*, 1994), where it is rapidly converted to protoporphyrin IX (Proto) by a plasma membrane bound, herbicide-insensitive, peroxidase-like enzyme (Lee and Duke, 1994; Lee, *et al.*, 1993). Proto accumulates to high enough levels in the plasma membrane and cytosolic components to cause photodynamic damage in the light (Duke, *et al.*, 1991). Protox inhibitors are very effective, in part, because they do not need to inhibit a high proportion of the Protox for Proto to accumulate, and they are competitive inhibitors of an enzyme that is not subjected to significant substrate accumulation when inhibited *in vivo*. The complicated mechanism of action of Protox inhibitors provides several sites at which resistance could evolve. Several of these sites are taken advantage of in cases of natural resistance. The mechanisms of natural resistance to Protox inhibitor herbicides have been reviewed previously (Kömives and Gullner, 1994; Matsumoto, *et al.*, 1994).

Multiple Mechanisms of Natural Resistance

There are normally thought to be three general mechanisms of resistance to herbicides: 1) resistance molecular site of action, 2) rapid metabolic degradation of the herbicide, and 3) prevention of movement of the intact herbicide from the site of application to the site action (Dyer, *et al.*, 1993). Because of the unique mechanism of action of Protox inhibitors, there are more potential mechanisms of resistance than these.

Some plants, such as soybean, metabolically degrade some of these herbicides rapidly to prevent herbicidal effects (Kömives and Gullner 1994; Frear, *et al.*, 1983). However, in most cases, natural resistance cannot be correlated with metabolic degradation (Kömives and Gullner 1994; Matsumoto, *et al.*, 1994; Ricotta and Masiunas 1992).

Rice appears to be largely resistant to the photodynamic effects of Proto through protective mechanisms against singlet oxygen (Matsumoto *et al.*, 1994). Cross resistance to acifluorfen of paraquat-resistant weeds with elevated levels of enzymes that protect against peroxidative damage suggests that this mechanism could evolve in response to selection pressure with Protox inhibitor herbicides (Gressel and Galun, 1994). However, claims that resistance of several crops is due to ratios of antioxidant (Finckh and Kunert, 1985) have been unsubstantiated (Matsumoto *et al.*, 1994; Sherman *et al.*, 1991).

Protox may have a low level of resistance in some species (Matsumoto *et al.*, 1994). However, other studies have suggested that this mechanism may not exist (Scalla and Matringe, 1994). Usually, resistance at the site of action is manifested at the molecular level by more than one hundred-fold resistance. This degree of difference in the I_{50} values of Protox inhibitors for Protox preparations from different sources has not been reported (Matsumoto *et al.*, 1994; Scalla and Matringe, 1994; Sherman *et al.*, 1991). Some plant species, such as rice and mustard, with quite low I_{50} values are some of the more naturally resistant plants (Matsumoto *et al.*, 1994; Sherman *et al.*, 1991). However, when comparing Protox preparations from different species, the relatively small variation in I_{50} values may be due to factors other than intrinsic differences in susceptibility to the herbicide. The only clear differences are between totally resistant bacterial Protox and the always susceptible forms from mitochondria and plastids of eucaryotic organisms.

In other species, such as mustards, Proto does not accumulate because Protogen is degraded to something less toxic than Proto (Jacobs *et al.*, 1994). The herbicide-insensitive plasma membrane peroxidase that converts Protogen to Proto can be manipulated in situ to reduce Proto accumulation and thereby safen the plant (Lee and Duke, unpublished), suggesting that it could be a site of a resistance mechanism. This mechanism deserves further study.

Reduced uptake of oxyfluorfen was determined to be one of the factors contributing to natural resistance of rice to oxyfluorfen (Lee *et al.*, 1991; Matsumoto *et al.*, 1994), and there is some evidence that it may contribute to natural resistance to other Protox inhibitors in other species (Kömives and Gullner, 1994). There are no cases of sequestration of Protox inhibitors accounting for any case of natural resistance. Figure 1 summarizes the known mechanisms of resistance to Protox inhibitors.

Prospects for Evolved Resistance in Weeds and Engineered Resistance in Crops

Although several crop and weed species have natural resistance to Protox-inhibiting herbicides, there are no verified cases of evolution of resistance to these herbicides due to selection pressure with Protox inhibitor herbicides. Some weeds that have evolved resistance to paraquat in the field have some cross resistance to Protox inhibitors (Gressel and Galun, 1994). In selection studies with mutagenized *Arabidopsis thaliana*, acifluorfen-methyl-resistant mutants were found in several hundred thousand viable seeds (Yamamoto and Duke, unpublished). Sato *et al.*, (1994) reported a mutant of the unicellular alga *Chlamydomonas reinhardtii* selected with a Protox inhibitor to be cross-resistant to a range of Protox inhibitors, but not to DCMU. Direct proof that the mutant had a resistant Protox was not produced. However, oxyfluorfen-resistant, non-chlorophyllous soybean cells were selected which contained Protox that was 12-fold less sensitive to the herbicide than the wild-type cells (Pomprom *et al.*, 1994). The resistant cells accumulated only about 25% less Proto in the light than the unselected cells. Whether this would translate into a significant increase in resistance in a plant regenerated from these cells is unknown.

The absence of evolved resistance to Protox inhibitors in the field is surprising for several reasons. Protox inhibitors appear to have a single site of action at which they are highly potent as inhibitors at the molecular level. These inhibitors mimic Protogen, competitively inhibiting the enzyme. Thousands of structurally diverse very herbicidally effective Protox inhibitors have been discovered, implying that the Protogen binding site is plastic, like the binding sites for other herbicides that have large numbers of effective herbicides (e.g., D-1, acetyl-CoA carboxylase and acetolactate synthase-binding herbicides). Resistance to these other herbicides has evolved relatively rapidly. There have been no reported cases of a higher plant Protox being highly resistant to Protox inhibitors.

There is, however, natural variation in weed populations susceptibility to Protox

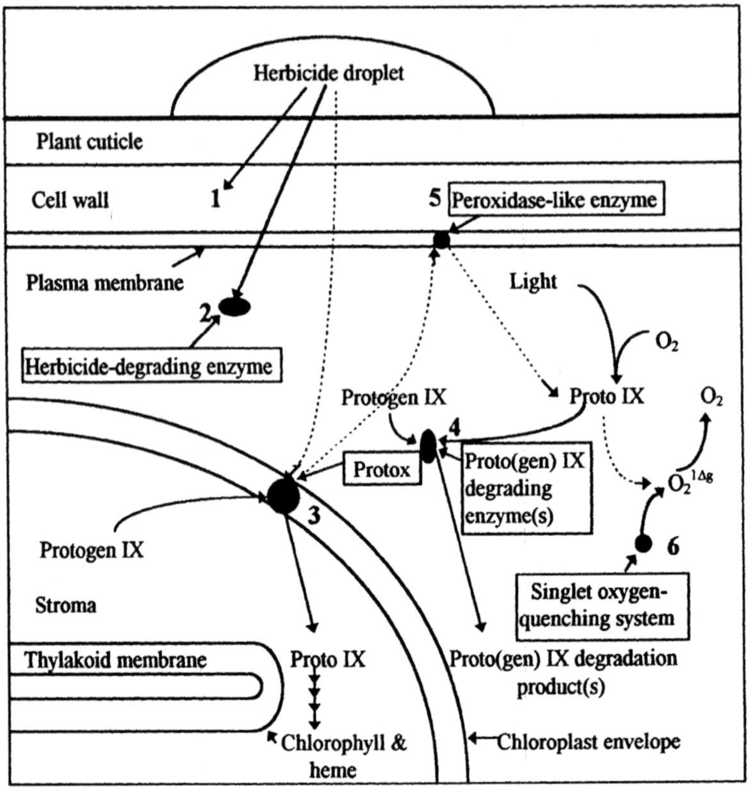

Figure 1. Schematic of possible mechanisms of resistance to Protox inhibitors. The normal herbicide mechanism is denoted by dotted lines. Enzymes are black dots or ovals with names in boxes. Potential sites of resistance are in bold numbers: 1 -inhibition of uptake or sequestration of the herbicide; 2 - rapid metabolic degradation of the herbicide; 3 - herbicide-resistant Protox; 4 - degradation of extraplastidic Protogen IX and/or Proto IX; 5 - inactivated herbicide-resistant, extraplastidic Protox; 6 - quenching of singlet oxygen and other toxic oxygen species.

inhibitors (e.g. Yu and Masiunas, 1992). Weed resistance to Protox inhibitors may not have evolved so far because of the usually short-lived selection pressure of most of the herbicides that have been used. Eventually, if the present methods and rates of use of these herbicides are continued, resistance should evolve, as it has for paraquat, another fast-acting herbicide.

Crop resistance to Protox inhibitors could be manipulated by alteration of any of at least four different enzymes; any enzyme that degrades the herbicide, Protox, the enzyme(s) that degrade Protogen and/or Proto, and the herbicide-resistant, peroxidase like enzyme that generates Proto in herbicide-treated plants. The herbicide-resistant Protox of procaryotic organisms such as *E. coli* (Jacobs *et al.*, 1994) could be moved to higher plants, but it is unlikely that this would produce a resistant plant unless the introduced enzyme could substitute for the plant's plastid and, perhaps, mitochondrial Protox. This would entail adding the proper molecular baggage for transport and processing for the proper organellar location, as well as inactivation of the plastid and mitochondrial versions of Protox, perhaps by antisense genetic engineering. Insertion of a gene into the crop that codes for a herbicide-degrading enzyme (e.g. Stalker *et al.*, 1996) might seem simpler. However, such an enzyme would have to work very fast, because Protox inhibitors work faster than any herbicides other than the bipyridyliums. So, to genetically engineer a Protox inhibitor-resistant crop is more problematic than for most other herbicides.

Acknowledgements

This work was supported in part by USDA National Reseach Initiative grant n°. 9000705.

References

Duke, S.O., Becerril, J.M., Lydon, J., Matsumoto, H. and T.D. Sherman (1991) Protoporphorinogen-oxidase inhibiting herbicides. *Weed Science* **39**: 465-473.

Duke, S.O., Lee, H.J., Nandihalli, U.B. and Duke, M.V. (1994) Protoporphyrinogen oxidase as the optimal herbicide in the porphyrin pathway. *American Chemical Society Symposium Series* **559**: 191-204.

Dyer, W.E, Hess, F.D., Holt, J.S. and Duke, S.O. (1993) Potential benefits and risks of herbicide-resistant crops produced by biotechnology. *Horticultural Reviews* **15**: 367-408.

Finckh, B.F. and Kunert, K.J. (1985) Vitamins C and E: an antioxidative system against herbicide-induced lipid peroxidation in higher plants. *Journal Agricultural Food Chemistry* **33**: 574-577.

Frear, D.S., Swanson, H.R. and Mansager, E.R. (1983) Acifluorfen metabolism in soybean: diphenylether bond cleavage and the formation of homoglutathione. *Pesticide Biochemistry Physiology* **20**: 299-310.

Gressel, J. and Galun, E. (1994) Genetic controls of photooxidant tolerance. *In: Causes of Photooxidative Stress and Amelioration of Defense Systems in Plants* (C.H. Foyerand and P.M. Mullineaux, eds.), 237-273. CRC Press, Boca Raton, FL.

Jacobs, J.M. and Jacobs, N.J. (1993) Porphyrin accumulation and export by isolated barley (*Hordeum vulgare* L.) plastids: Effect of diphenyl ether herbicides. *Plant Physiol*ogy **101**: 1181-1188.

Jacobs, J.M., Wehner, J.M. and Jacobs, N.J. (1994) Porphyrin stability in plant supernatant fractions: Implications for the action of prophyrinogenic herbicides. *Pesticide Biochemistry Physiology* **50**: 23-30.

Kömives, T. and Gullner, G. (1994) Mechanisms of plant tolerance to photodynamic herbicides. *American Chemical Society Symposium Series* **559**: 177-190.

Lee, H.J. and Duke, S.O. (1994) Protoporphyringen oxidizing activities involved in the mode of action of peroxidizing herbicides. *Journal Agricultural Food Chemistry* **42**: 2610-2618.

Lee, H.J., Duke, M.V. and Duke, S.O. (1993) Cellular localization of protoporphyrinogen-oxidizing activities of etiolated barley (*Hordeum vulgare* L.) leaves. Relationship to mechanism of action of protoporphyrinogen oxidase inhibiting herbicides. *Plant Physiology* **102**: 881-889.

Lee, J.J., Matsumoto, H., Pyon, J.Y. and Ishizuka, K. (1991) Mechanism of selectivity of diphenyl ether herbicides oxyfluorfen and chlomethoxynil in several plants. *Weed Research* (Japan) **36**: 162-170.

Matsumoto, H., Lee, J.J. and Ishizuka, K. (1994) Variation in crop response to protoporphyrinogen oxidase inhibitors. *American Chemical Society Symposium Series* **559**: 120-132.

Nandihalli, U.B. and Duke, S.O. (1993) The porphyrin pathway as a herbicide target site. *American Chemical Society Symposium Series* **524**: 62-78.

Pomprom, T., Matusumoto, H., Usui, K. and Ishizuka, K. (1994) Characterization and tolerance in selected soybean line. *Pesticide Biochemistry Physiology* **50**: 107-114.

Ricotta, J.A. and Masiunas, J.B. (1992) Acifluorfen tolerance in *Lycopersicon. Weed Science* **40**: 413-417.

Sato, R., Yamamoto, M., Shibata, H., Oshio, H., Harris, E.H., Gillaham, N.W. and Boynton, J.E. (1994) Characterization of a mutant of *Chlamydomonas reinharadtii* resistant to protoporphyrinogen oxidase inhibitors. *American Chemical Society Symposium Series* **559**: 91-104.

Scalla, R. and Matringe, M. (1994) Inhibitors of protoporphyrinogen oxidase as herbicides: diphenyl ethers and related photobleaching herbicides. *Reviews Weed Science* **6**: 103-132.

Sherman, T.D., Becceril, J.M., Matsumoto, H., Duke, M.V., Jacobs, J.M., Jacobs, N.J. and Duke, S.O. (1991) Physiological basis for differential sensitivities of plant species to protoporphyrinogen oxidase-inhibiting herbicides. *Plant Physiology* **97**: 280-287.

Stalker, D.M., Kiser, J.A., Baldwin, G. Coulumbe, B. and Houck, C.M. (1996) Cotton weed control using the BXN™ system. *In: Herbicide-Resistant Crops- Agricultural, Environmental, Economic, Regulatory, and Technical Aspects* (S.O. Duke, ed.). CRC Press, Boca Raton, FL.

Yu, C.-Y. and Masiunas, J.B. (1992) Characterization of acifluorfen tolerance in selected somaclones of eastern black nighthshade (*Solanum ptycanthum). Weed Science* **40**: 408-412.

CHAPTER 18
METABOLISM OF CHLOROTOLURON IN RESISTANT AND SUSCEPTIBLE *ALOPECURUS MYOSUROIDES*. BIOTYPES. A STUDY USING PLANT CELL SUSPENSIONS

J. MENENDEZ[1], J. JORRIN[2] and R. DE PRADO[1]
1.Departamento de Química Agrícola y Edafología; 2. Departamento de Bioquímica y Biología Molecula;. E.T.S.I.A.M., Universidad de Córdoba, Apdo 3048, 14080 Córdoba, Spain

Introduction

Among the phenylurea herbicides, chlorotoluron (*N'*-(3-chloro-4-methylphenyl)-*N,N*-dimethylurea) is one of the most widely used in winter wheat, due to its selectivity (Ryan *et al.*, 1981). This herbicide controls a range of broad-leaf weeds, and is also effective against slender foxtail (*Alopecurus myosuroides*) and wild oat (*Avena fatua*) (Ryan and Owen, 1982). The parent compound undergoes rapid metabolism in plants by oxidative *N*-demethylation and hydroxylation of the 4-methyl phenyl group, forming a number of derivatives with modified phytotoxicity (Ryan *et al.*, 1981). Chlorotoluron resistance has been commonly correlated to the presence of a detoxifying capability associated with Cyt P-450 monooxygenases. This superfamily encodes several enzyme isoforms which are involved in a number of biosynthetic pathways in plants, including those leading to phenolics, terpenoids, steroids, alkaloids, lignins and hydroxylated fatty acids. The relationship between plant Cyt P-450 and detoxification processes has been established for several crops and herbicides (Mouguin *et al.*, 1990; Fonne-Pfister *et al.*, 1990), this metabolic capability being the basis for the selectivity of chlorotoluron vs. weeds. However, the continuous use of chlorotoluron in monoculture systems has led to the appearance of several chlorotoluron-resistant biotypes of various weeds, including *Alopecurus myosuroides* (Niemans and Pestemer, 1984; Moss and Cussans, 1985; De Prado *et al.*, 1991) and *Lolium rigidum* (Burnet *et al.*, 1993).

An *Alopecurus myosuroides* biotype found in Spain has been characterized as chlorotoluron-resistant under laboratory conditions. This resistant (R) biotype was thirty times less sensitive to chlorotoluron than the susceptible (S) biotype and showed cross-resistance to several aryloxyphenoxypropionate and cyclohexanedione ACCase inhibitors (Menendez *et al.*, 1993; De Prado *et al.*, 1995). The chlorotoluron resistance was due to enhanced metabolism rather than differences in uptake and translocation of the herbicide or mutation at the target site level (Menendez *et al.*, 1994). Previous studies with whole plants showed that both the R and S biotypes metabolized chlorotoluron to *N*-dealkylated compounds and hydroxylated intermediary metabolites which gave inactive polar conjugates. This conjugate formation was greater and faster in R than in S plants, with the susceptible biotype accumulating twice the amount of phytotoxic metabolites (including chlorotoluron) than the resistant one after 24 h incubation in [^{14}C]-chlorotoluron solution (Jorrin *et al.*, 1992). This mechanism of resistance is similar to those observed in other chlorotoluron-resistant *A. myosuroides* populations (Kemp *et al.*, 1990).

Metabolism in Plant Cell Suspensions

Cell cultures of a variety of plant species have been used since the early 1980 to study the metabolism of xenobiotics (Davis and Shimabukuro, 1980). These systems have some advantages over whole plants when studying herbicide metabolism, such as (A) their relative homogeneity when they are used at an appropriate growth phase (Sterling and Balke, 1988); (B) a reduced content of pigments (which can mask results) that can facilitate analytical measurements (Matringe *et al.*, 1990) and (C) the avoidance of absorption and translocation barriers in species in which they complicate metabolism studies (Lamoureux *et al.*, 1991). Nevertheless, although no major qualitative differences have been found in pesticide metabolism between intact plants and cell cultures of the same species, quantitatively significant differences have been observed (Buhler *et al.*, 1985). In addition, tissue culture systems have been widely utilized during the last ten years for the selection of herbicide-resistant cells lines (Chaleff and Ray, 1984). These facts have led to the development of herbicide-resistant crops developed from genetically altered cell lines (Vasil *et al.*, 1992). However, tolerance to herbicides in tissue cultures must be carefully tested as the selection of herbicide-resistant cell lines may lead to the appearance of unexpected mechanisms of resistance, different from those observed in whole plants and which could be incompatible with plant regeneration (Shedletzky *et al.*, 1990).

We have studied the metabolism of [^{14}C]-chlorotoluron in plant cell suspension cultures of chlorotoluron-resistant and -susceptible biotypes of *A. myosuroides*, with specific regard to qualitative and quantitative differences in metabolite profiles between the R and S biotypes. As non-chlorophyllous cultures, they are quite insensitive to photosynthesis-inhibiting herbicides (Grossmann *et al.*, 1992). Therefore, herbicide metabolism could be evaluated in the absence of phytotoxicity.

Calli were established from resistant and susceptible *A. myosuroides* seedlings. Dehusked seeds were surface-sterilized and germinated on agar-solidified MS basal medium (Murashige and Skoog, 1962) supplemented with 30 g L^{-1} sucrose and 5 mg L^{-1} 2,4-D. After two weeks, the calli which developed in the hypocotile region were excised, transferred to fresh medium and subcultured monthly thereafter. The cultures were initiated and grown at 25 C in the dark. For suspension cultures, 6 month-old calli were placed in 50 mL of the medium (without agar), and grown in 250-mL Erlenmeyer flasks on a rotary shaker at 120 rpm. Cell suspensions were subcultured weekly by three-fold dilution into fresh medium. After four weeks of subculturing, both R and S cell suspensions appeared as fine thick lightyellow suspensions constituted of nonchlorophyllous free cells and clusters of cells. Only the clusters were able to grow.

Significant differences in [^{14}C]-chlorotoluron absorption were found between R and S cells (Table 1). Although the amount of radioactivity in the culture medium decreased continuously with both biotypes, resistant cells contained twice as much ^{14}C as susceptible cells after 12 and 24 h of incubation. A similar phenomenon was observed in bentazon resistant soybean cell cultures (Sterling and Balke, 1989). This was explained by compartmentalization of polar conjugates within cells in the R cell line. As these conjugates probably have lower partition coefficients and lipid solubilities than free metabolites, they are less able to diffuse across the lipophilic membranes of the cell. This is consistent with higher conjugate levels in R cells in the present work. The relatively low amount of conjugates present in the culture media (Table 2) may be due to the slow efflux of those metabolites from intact cells (Canivenc *et al.*, 1989) and/or the release of the conjugates from dead cells (Shimabukuro *et al.*, 1987).

Table 1.- Distribution of radioactivity between the aqueous medium and cell extract in chlorotoluron-resistant and -susceptible *Alopecurus myosuroides* cell suspension cultures.

Time of incubation (hours)	% recovered radioactivity			
	Resistant		Susceptible	
	Medium	Cells	Medium	Cells
3	78.0	22.0	81.3	18.7
12	58.4	41.6	78.1	21.9
24	37.9	62.1	70.0	30.0

Previous reports showed a time-related decline in the capability of cell cultures to metabolize chlorotoluron, as evidenced by the increase in the ratio of *N*-demethylation of chlorotoluron to 4-methylphenyl hydroxylation (Cole and Owen, 1988). This fact has been explained by the existence of distinct monooxigenase isozymes responsible for different hydroxylations which vary with the age of culture (Devine *et al.*, 1993). However, herbicide-metabolism patterns in both R and S *A. myosuroides* cell lines were essentially the same as those observed in whole plants (Jorrin *et al.*, 1992). Cell suspensions metabolized chlorotoluron into less phytotoxic (monomethyl-chlorotoluron) and non phytotoxic metabolites (didemethyl-chlorotoluron, hydroxy-derivatives and polar conjugates) (Table 2). Most of the ^{14}C recovered from the medium was in the form of free metabolites of chlorotoluron, whereas most of the ^{14}C recovered in the cells was in the form of polar conjugates (R biotype) or polar conjugates and chlorotoluron (S biotype). This metabolite partitioning between the medium and the cells appears to follow a mechanism of diffusion, and correlates with previous studies performed in wheat cell suspension cultures (Canivenc *et al.*, 1989). R cells formed conjugated chlorotoluron metabolites faster and to a greater extent than S ones, accumulating four times as many polar compounds after 24 h incubation. Phytotoxic compounds detected in S cells (namely, chlorotoluron and monomethyl-chlorotoluron) accounted for two to thirteen times the amount present in R cells, and remained in the cell extract as a constant value about 13% of the recovered radioactivity. However, R cells showed a maximum recovery of phytotoxic compounds within the first 3 hours of incubation, after which it declined gradually, as these compounds were substrates for further hydroxylation and conjugation reactions. In addition, R cells were able to rapidly hydroxylate chlorotoluron to hydroxy-chlorotoluron, whereas this metabolite was not found in the medium or cells of S cultures even after 24 h. This could explain the faster rise of hydroxymonomethyl-chlorotoluron in R cells as a result of either hydroxy-chlorotoluron demethylation or monomethyl-chlorotoluron ring-hydroxylation. However, the presence of hydroxy-chlorotoluron in S *A. myosuroides* whole plants after longer incubation periods (unpublished data) suggest an enhanced ability to hydroxylate the parent herbicide rather than the presence of a different detoxification pathway in the R biotypes (Table 2).

Concluding Remarks

Although the nature of the polar conjugates is still unknown, our data are consistent with other studies performed on *A. myosuroides* (Ryan *et al.*, 1981; Kemp *et al.*, 1990; Jorrin *et al.*, 1992). The enhanced ability of R *A. myosuroides* cells to oxidize the 4-methylphenyl group of both chlorotoluron and monomethyl-chlorotoluron justify their capability to detoxify this herbicide more efficiently than the S cells. In contrast, *N*-demethylation is the predominant chlorotoluron degradative route in susceptible cells. This agreement concerning metabolism of chlorotoluron in both whole plants and cell cultures validates the use of these *A. myosuroides* cell suspensions as an experimental system for herbicide resistance studies.

Acknowledgements

The authors thank the CICYT for financial support (Project No AGF 93-0332)

Table 2.- Metabolism of chlorotoluron in *A. myosuroides* resistant and susceptible cell suspensions. Data for a typical experiment is presented.

Time of incubation (hours)	Metabolite	% total recovered radioactivity			
		Resistant		Susceptible	
		Medium	Cells	Medium	Cells
3	Conjugate	9.40	10.21	0.02	3.43
	Hydroxydidemethyl-chlorotoluron	-	-	-	-
	Hydroxymonomethyl-chlorotoluron	1.95	3.05	-	-
	Hydroxy-chlorotoluron	25.18	1.88	-	-
	Didemethyl-chlorotoluron	5.74	-	-	-
	Monomethyl-chlorotoluron	19.69	3.49	2.98	1.26
	Chlorotoluron	16.05	3.22	78.26	13.25
12	Conjugate	27.88	28.11	2.11	7.00
	Hydroxydidemethyl-chlorotoluron	-	-	-	-
	Hydroxymonomethyl-chlorotoluron	2.28	6.05	7.85	1.19
	Hydroxy-chlorotoluron	18.83	2.80	-	-
	Didemethyl-chlrotoluron	3.34	-	-	-
	Monomethyl-chlrotoluron	2.23	2.04	7.58	2.37
	Chlorotoluron	3.86	1.69	60.59	11.20
24	Conjugate	21.76	52.36	5.80	13.72
	Hydroxydidemethyl-chlorotoluron	-	-	-	-
	Hydroxymonomethyl-chlrotoluron	2.67	-	15.57	1.54
	Hydroxy-chlorotoluron	11.74	5.08	-	-
	Didemethyl-chlorotoluron	1.61	3.86	1.15	-
	Monomethyl-chlorotoluron	-	-	7.47	3.99
	Chlorotoluron	-	-	40.02	10.66

References

Buhler, D.D., Swisher, B.A. and Burnside, O.C. (1985) Behaviour of ^{14}C-haloxyfop-methyl in intact plants and cell cultures. *Weed Science* **33**: 291-299.

Burnet, M.W.M., Loveys, B.R., Holtum, J.A.M. and Powles, S.B. (1993) A mechanism of chlorotoluron resistance in *Lolium rigidum. Planta* **190**: 182-189.

Canivenc, M.C., Cagnac, B., Cabanne, F. and Scalla, R. (1989) Induced changes of chlorotoluron metabolism in wheat cell suspension cultures. *Plant Physiology Biochemistry* **27**: 193-201.

Chaleff, R.S. and Ray, T.B. (1984) Herbicide-resitant mutants from tobacco cell cultures. *Science* **223**: 1148-1151.

Cole, D.J. and Owen, W.J. (1988) Metabolism of chlorotoluron in cell suspensions of *Latuca sativa*: a qualitative change with age of culture. *Phytochemistry* **27**: 1709-1711.

Davis, D.G. and Shimabukuro, R.H. (1980) Studies of herbicide toxicity and mode of action using mesophyll cells and callus-derived cell suspensions. *Canadian Journal Botany* **58**: 1482-1489.

De Prado R, Plaisance K, Menendez J, Gronwald J. (1995) Effect of graminicide herbicides on growth and ACCase activity in a chlorotoluron-resistant biotype of Alopecurus myosuroides Huds. *In: International Symposium Weed and Crop Resistance to Herbicides* (R. De Prado, J. Jorrín, L. garcía-Torres and G. Marshall, eds.), 59-61. University of Córdoba, Córdoba, Spain.

Devine, M.D., Duke, S.O. and Fedtke, C., eds. (1993) *Physiology of Herbicide Action*, Prentice Hall, Englewood Cliffs, NJ.

Grossman K, Berghaus R, Retzlaff G. (1992) Heterotrophic plant cell suspension cultures for monitoring biological activity in agrochemical research. Comparision with screens using algae, germinating seeds and whole plants. *Pesticide Science* **35**: 283-289.

Jorrín, J., Menendez, J., Romera, E., Tena, M and De Prado, R. (1992) Detoxification of chlorotoluron by *Alopecurus myosuroides* via conjugation. *Mededelingen van de Faculteit Landbouwwetenschappen. Universiteit Gent* **57/3B**: 1047-1052.

Kemp, M.S., Moss, S.R. and Thomas, T.H. (1990) Herbicide resistance in *Alopecurus myosuroides. In: Managing Resistance to Agrochemicals: From Fundamental Research to Practical Strategies* (M.B. Gree, H.M. LeBaron and W.K. Moberg, eds.), 376-393. American Chemical Society, Washington.

Lamoureux, G.L., Rusness, D.G., Schröder, P. and Rennenberg, H. (1991) Diphenyl ether herbicide metabolism in a Spruce cell suspension culture: the identification of two novel metabolites derived from a glutathione conjugate. *Pesticide Biochemistry Physiology* **39**: 291-301.

Matringe, M., Clair, D. and Scalla, R. (1990) Effects of peroxidizing herbicides on protoporphyrin IX levels in non-chlorophyllous soybean cell culture. *Pesticide Biochemistry Physiology* **36**: 300-307.

Menendez, J., Jorrín, J., Taberner, A. and De Prado, R. (1993) Penetration, translocation and metabolization of diclofop-methyl in chlorotoluron-resistant and -susceptible biotypes of *Alopecurus myosuroides. British Crop Protection Conference-Weeds* **1**: 213-220.

Menendez, J., Jorrín, J., Romera, E. and De Prado, R. (1994) Resistance to chlorotoluron of a slender foxtail (*Alopecurus myosuroides*) biotype. *Weed Science* **42**: 340-344.

Moss, S.R. and Cussans, G.W. (1985) Variability in the susceptibility of *Alopecurus myosuroides* (black-grass) to chlorotoluron and isoproturon. *Aspects Applied Biology* **9**: 91-98.

Murashige, T. and Skoog, F. (1962) A revised medium for rapid growth and bioassays with tobacco tissue cultures. *Physiologia Plantarum* **15**: 151-158.

Niemans, P. and Pestemer, W. (1984) Resistenz verschiedener Herkunfte vob Acker-Fuchsschwanz (*Alopecurus myosuroides*) gegenuber Herbizidbehandlungen. *Nachrich. Deuts. Pflanzen.* **36**: 113-118.

Ryan, P.J., Gross, D., Owen, W.J. and Loani, T.L. (1981) The metabolism of chlorotoluron , diuron and CGA 43057 in tolerant and susceptible plants. *Pesticide Biochemistry Physilogy* **16**: 213-221.

Ryan, P.J. and Owen, W.J. (1982) The mechanism of selectivity of chlorotoluron between cereals and grassweeds. *In: Proceedings 1982 British Crop Protection Conference-Weeds*, 317-324.

Shedletzky, E., Shmuel, M., Delmer, D.P. and Lamport, D.T.A. (1990) Adaptation and growth of tomato cells on the herbicide 2,6-dichlorobenzonitrile leads to production of unique cell walls virtually lacking a cellulose-xyloglucan network. *Plant Physiology* **94**: 980-987.

Shimabukuro, R.H., Walsh, W.C. and Jacobson, (1987) Aryl-*O*-glucoside of diclofop: A detoxification product in wheat shoots and wild oat cell suspension culture. *Journal Agricultural Food Chemistry* **35**: 393.

Sterling, T.M. and Balke, N.E. (1988) Use of soybean (*Glycine max*) and velvetleaf (*Abutilon theophrasti*) suspension-cultured cells to study bentazon metabolism. *Weed Science* **36**: 558-565

Sterling, T.M. and Balke, N.E. (1989) Differential bentazon metabolism and retention of bentazon metabolites by plant cell cultures. *Pesticide Biochemistry Physiology* **34**: 39-48.

Swisher, B.A. and Weimer, M.R. (1986) Comparative detoxification of chlorotoluron in leaf disks and cell cultures of two perennial weeds. *Weed Science* **34**: 507-512.

Vasil, V., Castillo, A.M., Fromm, M.E. and Vasil, I.K. (1992) Herbicide resistant transgenic wheat plants obtained by microprojectile bombardment of regenerable embryogenic callus. *Bio/technology* **10**: 667-674.

PART III

GENETICS AND BIOLOGY OF HERBICIDE RESISTANT WEEDS

CHAPTER 19
MOLECULAR ECOLOGY: ITS ROLE IN STUDYING HERBICIDE RESISTANCE

G. MARSHALL and R.P. FINCH
Scottish Agricultural College, Department of Plant Science, Auchincruive, Ayr, KA6 5HW, U.K.

Introduction

It is worth attempting a definition of molecular ecology. At the outset we may consider molecular ecology to be the application of molecular techniques to solve problems in ecology. Still, this definition is rather narrow and it may be more satisfactory to consider molecular ecology as a new discipline which supports the interests of plant ecologists, population biologists, geneticists, physiologists, biochemists and molecular biologists. It is via a multidisciplinary approach to problem solving in plant biology that molecular ecology is likely to continue to expand its activities and achievements. Presently, in molecular ecology, molecular approaches are undertaking studies on biodiversity, taxonomic identities, interspecific gene flow, assessment of kinship relationships, analysis of mixed genome samples and the production of specific diagnostic probes. The object of this review is to describe the latest molecular techniques with examples of their use for studies on herbicide resistance.

This review describes the principal approaches used in molecular biology for genetic fingerprinting based on restriction length polymorphism (RFLP) analysis or the polymerase chain reaction (PCR). The application of a range of molecular techniques are outlined in relation to some recent studies of the reproductive behaviour of weeds, assessments of intraspecific genetic variability, linking genotypes to physiological adaptations, molecular analysis of genes conferring herbicide resistance and introgression studies between weed and crop genotypes.

Techniques

DNA Hybridisation Techniques

Restriction fragment length polymorphism (RFLP) analysis has found wide utility in the study and characterisation of genetic variability. This approach utilises the specificity of restriction enzymes to recognise given DNA sequences and generate cuts thus giving a reproducible family of DNA fragments. Restriction of comparatively simple DNA molecules yields fragments that can be resolved by simple gel electrophoresis giving a profile specific to the sequence/restriction enzyme combination. Restriction of chromosomal DNA yields a vast array of fragments which cannot be clearly resolved by simple staining of electrophoresis gels. However, if a nitrocellulose/nylon print or 'Southern blot' of the gel is hybridised to a radioactively labelled DNA probe, complementarity between chromosomal and probe sequences gives rise to DNA:DNA hybrids which can be visualised by autoradiography. Length polymorphisms of these hybrid molecules provide distinct profiles which can clearly distinguish between genotypes and can provide heritability data. Probes may be based on homologous or heterologous chromosomal or cytoplasmic sequences. Often, probes may themselves be arbitrary restriction fragments or may be based on known length-polymorphic genomic regions.

For surveys of genetic variability, probes based on variable number tandem repeat (VNTR) or 'minisatellite' sequences which occur in length-polymorphic tandem repeat regions are often employed. The core units of a range of these regions have been isolated and sequenced and can be used for DNA fingerprinting in homologous and heterologous systems (Burke *et al.*, 1991). More recently, much simpler 'microsatellite' sequences have been utilised in a similar fashion (e.g. Sharon *et al.*, 1995). Variability in the spacer regions between genomic loci that encode ribosomal RNA is well documented in plants (Rogers and Bendich, 1987) and DNA fingerprinting assays using probes based on these sequences have been useful for assessing weed genetic variability (Richard *et al.*, 1995).

DNA Amplification Techniques

The polymerase chain reaction (PCR; Saiki *et al.,*1988) in its standard guise provides a powerful diagnostic tool capable of repeatedly copying, with great specificity, a given DNA fragment present in a mixed background of nucleic acids. The fragment, targeted by a pair of specific demarcating oligonucleotide primers, is amplified to such an extent that it can be stained and identified as a discrete band on an electrophoresis gel. The powerful technique of PCR is now being used in a number of ways to generate DNA profiles which are useful in molecular ecology.

Randomly Amplified Polymorphic DNA (RAPD) or 'Arbitrary Primer' PCR

RAPDs are manifested as multiple products, or profiles, following electrophoretic analysis of a completed PCR reaction mixture. They are produced by utilising a single short (10-base) oligonucleotide primer of arbitrary sequence which is induced to prime DNA polymerisation under comparatively low stringency conditions (Williams *et al.*, 1990; Welsh and McLelland, 1990). The primer is able to demarcate discrete fragments because it hybridises to inverted repeat sequences which occur in probably all genomes. Different arbitrary primer sequences can be used to generate different DNA profiles. The RAPD markers produced show Mendelian inheritance, but are dominant. Their inability, therefore, to discriminate between homozygotes and heterozygotes limits their usefulness in mapping and other inheritance studies. However, comparisons of multiple RAPD profiles generated from different collections/populations of organisms can yield useful information of relatedness and taxonomy (Hadrys *et al.*, 1992). Moreover, it may be possible in some cases to identify RAPD products associated with particular traits. Such fragments can be reamplified and sequenced to yield specific diagnostic PCR primers. There is some doubt as to whether DNA profiles generated by the RAPD technique are reproducible. Certainly, careful laboratory practice is essential. However, regardless of any controversy, there is no doubt that this extremely straightforward technique, which requires only small amounts of material and no DNA sequence information, has revolutionised studies of genetic diversity in a huge range of organisms in the last few years.

Directed Amplification of Polymorphic DNA (DAMD or 'Minisatellite-PCR ')

The simplicity of the RAPD approach for DNA fingerprinting has prompted the development of modified techniques which address the reproducibility problem. DAMD (Heath *et al.*, 1993) employs a similar approach to the RAPD technique by utilising a single primer in each reaction. However, the primer sequences are longer and are based on known variable number tandem repeat (VNTR) genomic regions which have been isolated and sequenced from a range of organisms. The PCR reaction can be carried out under comparatively high stringency conditions which leads to greater reproducibility. There is, however, a limited number of primer sequences available and in most cases these are necessarily used in heterologous systems. In our experience, only a few such primers give useful profiles in a given plant species.

Simple Sequence Repeat - PCR (SSR-PCR or 'Microsatellite PCR')

This technique is very similar to the DAMD approach but utilises primer sequences based on simple sequence repeats (SSRs) or 'microsatellites' which have been identified in a wide range of organisms. Like minisatellite repeat regions, SSR regions comprise tandem repeats of a core sequence but in the case of SSRs this may be just a single-base repeat, or repeats of 2 - 5 base sequences. Primers based on SSRs have been shown to generate reproducible DNA profiles in a range of organisms (Zietkiewicz *et al.*, 1994).

Combined Approaches - PCR-RFLP

PCR can be combined with RFLP analysis in at least three ways. One approach involves sequencing RFLP markers and then generating specific PCR primers which can then be used to amplify these sites without any need for DNA restriction of hybridisation. Secondly, restriction analysis of apparently similar PCR products provides a means of identifying sequence differences within the amplified region (Guttieri *et al.*, 1992). This combination of PCR-RFLP also avoids the need for DNA hybridisations. Thirdly, RAPD products can be used as specific hybridisation probes (Williams *et al.*, 1990).

The Reproductive Behaviour of Weeds

In the study of herbicide resistance it is clear that the balance of scientific discovery currently favours herbicide mechanism of action studies as compared to understanding the reproductive behaviour of weeds especially with respect to genetic analysis. Yet, forecasting herbicide resistance in weed populations requires gene flow processes to be incorporated in predictive models (Maxwell *et al.*, 1990).

One of the first requirements is to determine the mechanism of reproduction. Since in weeds this is generally not well-characterised, it follows that genetic diversity and the nature of genetic variation is poorly understood. As a starting point we may consider the reproductive behaviour of closely related crop species. This will be at best a guide only and it is necessary to assess gene flow by pollen movement. While morphological markers such as plant pigments (Griffiths, 1950) can help in the pursuit of such studies, specific interest in herbicide resistance has led researchers to consider the spacial pattern analysis of resistance in field situations. Assessments of herbicide resistance are normally made via *in vivo* assays or directly at the biochemical level using an *in vitro* enzyme assay. In a recent field study by Murray and Morrison (1995) out-crossing and pollen-mediated gene flow in wild oat (*Avena fatua* L.) was assessed using ACCase inhibitor resistance as a marker. The outcrossing frequencies were generally low (0.098-0.173 %) and influenced by the competitiveness of the companion crops, wheat and flax. Taking this approach one step further, Ghersa *et al.* (1994) studied the production of flowers and pollen dispersal patterns of diclofop-methyl-susceptible and -resistant Italian ryegrass to determine whether evolution of herbicide resistance could be controlled by cross-fertilization between the biotypes. These authors concluded that in a wheat field infested with resistant ryegrass, cessation of herbicide application and sowing of a susceptible ryegrass biotype could reduce the evolution of diclofop-resistance by at least 6 % per year.

Recently, Richard *et al.* (1995) utilised the known length polymorphic properties of the intergenic spacer (IGS) region of ribosomal DNA (Rogers and Bendich, 1987), to investigate the levels of genetic diversity in natural populations of *Holcus mollis* (creeping soft-grass), *H. lanatus* (Yorkshire fog) and their pentaploid hybrid. Using a DNA hybridisation approach these workers probed restricted *Holcus* DNA with several heterologous rDNA fragments from wheat. Surprisingly, some populations of *H. mollis*, a species thought to reproduce in a mainly vegetative manner due to its forest habitat, exhibited a high degree of intraspecific polymorphism. However, repeated analyses of the *H. mollis* populations using different subcultures in different years revealed identical DNA profiles thus confirming temporal stability of the rDNA in this species. Apparently, some populations of this species consist of single clones whilst others consist of individuals displaying polymorphic but stable rDNA profiles. Periodic changes in tree and shrub density in the vicinity of certain populations would favour flowering and lead to variability.

Intraspecific Variability of Weeds

Weeds which are predominantly self-pollinated produce fewer genotypes and reduced levels of heterozygosity compared to outbreeders (Warwick, 1991). By contrast, outcrossing species show less interpopulation differences but higher levels of genetic variation within populations (Hamrick and Godt, 1990). In weeds, allozyme variation and its application in the assessment of the levels of genetic variation in weeds is well established (Brown, 1979, Navas and Gasquez, 1991, Warwick and Black, 1993). These allozyme data are considered to provide an estimation of genetic diversity in the absence of selection pressure, and independent of morphological variation. Warwick, (1991) considered that allozyme variation appears to reflect with reasonable accuracy the genetic and adaptive diversity of predominantly outcrossing weeds, however, for self-pollinated species its value is much less.This is perhaps because allozyme variation reflects differences only in certain coding regions of the genome; it is assessed at the protein level by observing alterations in amino acid architecture. However, DNA techniques provide markers which may derive from coding or non-coding genomic regions and thus may be based on virtually any DNA-level polymorphism. Consequently, the latter provide many more opportunities to uncover genetic variability.

As aforementioned, Richard *et al.* (1995) demonstrated the use of a DNA hybridisation approach to investigate levels of genetic diversity in natural populations of *Holcus* spp. However, Cheng *et al.* (1995) employed a DAMD approach in an examination of genetic diversity in *Galium aparine* (cleavers), a species with no obvious pollinator which appears to establish clonal populations. The DAMD approach, utilising a heterologous VNTR sequence, was successful in the identification of genetic variation between ecotypes and gave rise to the generation of a 620bp amplification marker diagnostic of dark germination ability. López Martínez *et al.* (this volume) utilised RAPD, DAMD and SSR-PCR to investigate genetic variation in *Echinochloa crus-galli.* These assays could clearly distinguish between *E. crus-galli* individuals and between a range of species in the *Echinochloa* genus. Neither morphological or isozyme markers have proved reliable for this purpose in *Echinochloa.*

Introgression Studies

During the past decade considerable advances have been made in the production of transgenic crop plants. Transgenic crops will offer a wide range of novel traits including insect, disease and herbicide resistance, modified nutritional characteristics, especially novel fatty acids, and delayed ripening. Of these traits it is herbicide resistance which continues to be the most controversial in terms of its potential environmental impact. Within this context the introgression of herbicide resistance genes into wild or weedy relatives is of primary importance (Nickson and Fuchs, 1994). Clearly the indiscriminate or random movement of herbicide resistance genes into wild relatives of crops and perhaps back crossing into nontransformed crops is the worst fear scenario and one which has opened up a wide range of legislative, scientific, commercial and moral challenges.

General considerations of the impact of hybrids between genetically modified crop plants and their related species have recently been reviewed by Dale (1994) and

Darmency (this volume). They highlighted the importance of studying the consequences of hybridization between conventionally bred cultivars and their relatives and to develop test protocols to determine the likely effect of a transgene in a hybrid. There are already several important field-based studies reporting aspects of the ecological impact of the release of herbicide-resistant crops especially featuring rapeseed (*Brassica napus*) and its weedy relatives (LeFol *et al.*, 1991, Crawley *et al.*, 1993 and Darmency, 1994a). Typically, seedling populations of the crop, weeds and putative hybrids are screened to determine their response to the application of herbicide. However, Darmency (1994b) points out that such assessments can be of limited value in terms of making an accurate interpretation of the inheritance of herbicide resistance. Furthermore, there are often practical problems in obtaining sufficient seed production from crop-weed hybrids due to incompatibility mechanisms or even floral morphology. As outlined previously, the actual mechanics of pollination and fertilization in many weed species is poorly understood. Allozyme analysis has been used to determine paternity in a number of studies (Ellstrand and Marshall, 1986). However, simple molecular methods, such as RAPD analysis, can now be employed where allozyme variation is insufficient, and particularly when many individual offspring need to be examined.

Molecular markers will obviously be invaluable for investigations of transgene flow from genetically modified crops to their wild relatives. Specific DNA probes or PCR primers based on the transgene or linked marker genes, promoters, and vector sequences will provide robust diagnostic tests.

Physiological Adaptations

The genetic analysis of plant genotypes has played a crucial role in the genetic improvement of crop plants. Linkage or the association of genetic markers to traits of agronomic importance can substantially simplify their genetic analysis and perhaps ultimately gene extraction from genomic libraries. Isozyme loci have been used extensively for markers of quantitative genetic analysis. Indeed, predictive methods for identifying elite genotypes using isozyme alleles and ecological factors have been described by Nevo *et al.* (1992). These authors found several isozyme loci were highly and significantly correlated with herbicide resistance (phenylureas) in emmer wheat, the progenitor of cultivated wheats.

López-Martínez *et al.* (this volume) used RAPDs to study the apparent variability in acquired resistance to quinclorac in *Echinochloa crus-galli.* A collection of biotypes classified as *E. crus-galli* on the basis of morphological markers showed a range of phenotypic responses to quinclorac from susceptible through intermediate to resistant. However, the molecular data revealed that the susceptible biotype in fact clustered in a group encompassing *E. oryzicola* and *E. oryzoides* and was distinct from all other *E. crus-galli* biotypes analysed. This study clearly demonstrates the practical importance of a genetic fingerprinting approach for authenticating weed genotypes in studies of physiological adaptation such as the development of herbicide resistance.

Molecular Analysis of Herbicide Resistance

Understanding the biochemical and genetical basis of herbicide resistance in plants is the ultimate goal in this discipline. Molecular genetic analysis of the basis of herbicide resistance has been studied principally in relation to the production of transgenic crops (reviewed by Cole, 1994). Where resistance is conferred by one or two major genes it is likely that a molecular analysis will be necessary to provide effective characterisation. By contrast, where several genes are involved in determining resistance, uncovering the mechanism responsible for resistance is much less certain.

Acetolactate synthase (ALS), the enzyme conferred by the *als* gene, represents the site of action for sulfonylurea herbicides, but certain DNA sequence alterations at this locus can give rise to a sulfonylurea-resistance phenotype. Sequence conservation in *als* genes has allowed a yeast *als* gene to be used as a heterologous probe to screen genomic DNA libraries of plants including *Arabidopsis* and tobacco (Hartnett *et al.*, 1990) in what has almost become a general strategy for cloning target site genes (Holloman and Butters, 1994). DNA sequence analysis of cloned *als* genes from sulphonylurea-resistant tobacco cell lines has allowed point mutations responsible for the resistant phenotype to be identified (Hartnett *et al.*, 1990).

Guttieri *et al.* (1992) characterised the genetic mutations present in the *als* genes of kochia, prickly lettuce and Russian thistle in order to determine the basis of chlorsulfuron resistance in these species. These workers utilised sequence data from cloned *als* genes of rapeseed, tobacco and *Arabidopsis* to design PCR primers capable of amplifying a 234bp region of the *als* gene in their chosen weed species. DNA sequence analysis of the PCR products revealed that a chlorsulfuron-resistant prickly lettuce biotype differed from that of a susceptible biotype by a single point mutation in the coding region for domain A, a portion of the amino acid sequence known to play an important role in conferring resistance to herbicides that inhibit ALS. A similar result was observed in kochia but in a resistant Russian thistle biotype, no domain A sequence alterations were observed. On the basis of the sequence variation observed in kochia, it was possible to utilise a restriction analysis of the 234bp PCR product to give RFLP profiles diagnostic for phenotypic response to chlorsulfuron application in 6 kochia collections. This demonstrates the utility of a combined PCR-RFLP approach as a diagnostic tool for characterising herbicide resistance in weeds.

Conclusions

It is evident that there are a wide range of research issues in relation to the ecology of herbicide resistance in weeds and crops which could benefit from the selective adoption of sophisticated techniques in plant molecular biology. These techniques are become ever more user-friendly and affordable with the introduction of PCR technology. Molecular ecology is a rapidly emerging discipline which can make an important contribution to our understanding and management of herbicide resistance.

Acknowledgements

SAC receives financial support from the Scottish Office Agriculture and Fisheries Department to undertake research on plant molecular ecology (FF 647601).

References

Brown, A.H.D. (1979) Enzyme polymorphism in plant populations. *Theoretical Population Biology* **15**: 1-42.

Burke, T., Dolf, G., Jeffreys, A.J. and Wolff, R. (1991) *DNA Fingerprinting Approaches and Applications.* Birkhauser Verlag, Basel, Switzerland.

Cheng, J., Mitchelson, K.R., Ford, M.A., Wilson, F. and Atkinson, D. (1995) Genetic diversity study of *Galium aparine* by arbitrary primer amplification. *Molecular Ecology.* In press.

Cole, D.J. (1994) Molecular mechanisms to confer herbicide resistance. *In: Molecular Biology in Crop Protection* (G. Marshall and D.R. Walters, eds), 146-176. Chapman and Hall, London.

Crawley, M.J., Hails, R.S., Rees, M., Kohn, D. and Buxton, J. (1993) Ecology of transgenic oilseed rape in natural habitats. *Nature* **363**: 620-623.

Dale, P.J. (1994) The impact of hybrids between genetically modified crop plants and their related species: general considerations. *Molecular Ecology* **3**: 31-36.

Darmency, H. (1994a) The impact of hybrids between genetically modified crop plants and their related species: introgression and weediness. *Molecular Ecology* **3**: 37-40.

Darmency, H. (1994b) Genetics of herbicide resistance in weeds and crops. In: *Herbicide Resistance in Plants* (S.B. Powles and J.A.M. Holtum, eds.), 263-297. Lewis Publishers, Boca Raton.

Ellstrand, N.C. and Marshall, D.L. (1986) Patterns of multiple paternity in populations of *Raphanus sativus. Evolution* **40**: 837-842.

Ghersa, C.M., Martinez-Ghersa, M.A., Brewer, T.G. and Roush, M.L. (1994) Use of gene flow to control diclofop-methyl resistance in Italian ryegrass (*Lolium multiflorum*). *Weed Technology* **8**: 139-147.

Griffits, D.J. (1950). The liability of seed crops of perennial ryegrass (*Lolium perenne*) to contamination by wind-borne pollen. *Journal Agricultural Science* **40**:19-38.

Guttieri, M.J., Eberlein, C.V., Mallory-Smith, C.A., Thill, D.C. and Hoffman, D.L. (1992) DNA-sequence variation in domain A of the acetolactate synthase genes of herbicide-resistant and herbicide-susceptible weed biotypes. *Weed Science* **40**: 670-676.

Hadrys, H., Balick, M. and Schierwater, B. (1992) Applications of random amplified polymorphic DNA (RAPD) in molecular ecology. *Molecular Ecology* **1**: 55-63.

Hamrick, J.L. and Goldt, M.J. (1990). Allozyme diversity in plant species. *In: Plant Population Genetics, Breeding and Genetic Resources* (A.D.H. Brown, M.T. Clegg, A.L. Kahler and B.S. Weir, eds.), 43-63. Sinauer Associations, Sunderland.

Hartnett, M.E. Chui, C.F. Mauvais, C.J., McDevitt, R.E., Knowlton, S., Smith, J.K., Falco, S.C. and Mazur, B.J. (1990) Herbicide-resistant plants carrying mutated acetolactate synthase genes. *In: Managing Resistance to Agrochemicals. From Fundamental Research to Practical Strategies* (M.B. Green, H.M. Le Baron and W.K. Moberg, eds), 459-473. ACS Symposium Series 421, Washington DC.

Heath, D.D., Iwama, G.K. and Devlin, R.H. (1993) PCR primed with VNTR core sequences yields species specific patterns and hypervariable probes. *Nucleic Acids Research* **2**: 5782-5785.

Holloman, D.W. and Butters, J.A. (1994) Molecular determinants of resistance to crop protection chemicals. In: *Molecular Biology in Crop Protection* (G. Marshall and D.R. Walters, eds), 98-117. Chapman and Hall, London.

LeFol, E., Danielou, V., Darmency, H., Kerlan, M-C., Vallee, P., Chevre, A-M. and Renard, M. (1991) Escape of engineered genes from rapeseed to wild *Brassicaceae. Proceedings Brighton Crop Protection Conference- Weeds*, 1049-1054.

Maxwell, B.D., Roush, M.L. and Radosevich, S.R. (1990) Predicting the evolution and dynamics of herbicide resistance in plant populations. *Weed Technology* **4** :2-13.

Murray, B.G. and Morrison, I.N. (1995) Out-crossing and pollen-mediated gene flow in wild oat (*Avena fatua* L.). *Weed Science Society of America Abstracts,* **35,** 92.

Navas, M.L. and Gasquez, J. (1991) Genetic diversity and clonal structure of *Rubia peregrina* in Mediterranean vineyard and unmanaged habitats. *Weed Research* **31**: 247-256.

Nevo, E., Snape, J.W., Lavie, B. and Beiles, A. (1992) Herbicide response polymorphism in wild emmer wheat. Ecological and isozyme correlations. *Theoretical and Applied Genetics* **84**: 209-216.

Nickson, T.E. and Fuchs, R.L. (1994) Environmental and regulatory aspects of using genetically modified plants in the field. *In*: *Molecular Biology in Crop Protection* (G. Marshall and D.R. Walters, eds.), 246-262. Chapman and Hall, London.

Richard, M., Jubier, M.F., Bajon, R., Gouyon, P.H. and Lejeune, B. (1995) A new hypothesis for the origin of pentaploid *Holcus* from diploid *Holcus lanatus* L. and tetraploid *Holcus mollis* L. in France. *Molecular Ecology* **4**: 29-38.

Rogers, S.O. and Bendich, A.J. (1987) Ribosomal RNA genes in plants: variability in copy number and in the intergenic spacer. *Plant Molecular Biology* **9**: 509-520.

Saiki, R.K., Gelfand, D.H., Stoffel, S., Scharf, S.J., Higuchi, R., Horn, G.T., Mullis, K.B. and Erlich, H.A. (1988) Primer-directed enzymatic amplifications of DNA with a thermostable DNA polymerase. *Science* **239**: 487-91.

Sharon, D., Adato, A., Mohameed, S., Lavi, U., Hillel, J., Gomolka, M., Epplen, C. and Epplen, J.T. (1995) DNA fingerprints in plants using simple-sequence repeat and minisatellite probes. *HortScience* **3**: 109-112.

Warwick, S.I. (1991) The influence of intraspecific variation on the biology and control of agricultural weeds. *Proceedings Brighton Crop Protection Conference - Weeds,* 997-1005.

Warwick, S.I. and Black, L.D. (1993) Electrophoretic variation in triazine-resistant and susceptible populations of the allogamous weed *Brassica rapa*. *Weed Research* **33**: 104-114.

Welsh, J. and McClelland, M. (1990) Fingerprinting genomes using PCR with arbitrary primers. *Nucleic Acids Research* **18**: 7213-7218.

Williams, J.G.K., Kubelik, A.R., Livak, K.J., Rafalski, J.A. and Tingey S.V. (1990) DNA polymorphisms amplified by arbitrary primers are useful as genetic markers. *Nucleic Acids Research* **18**: 6531-6535.

Zietkiewicz, E., Rafalski, A. and Labuda, D (1994) Genome fingerprinting by simple sequence repeat (SSR)-anchored polymerase chain reaction amplification. *Genomics* **20**:176-183.

CHAPTER 20
GENETICS OF HERBICIDE RESISTANCE WITHIN WEEDS. FACTORS OF EVOLUTION, INHERITANCE AND FITNESS

J. GASQUEZ
I.N.R.A., Laboratoire de Malherbologie, B.V. 1540, 21034 Dijon, Cedex, France

Introduction

By definition, weeds are wild species adapted to agroecosystems. Moreover genetic variations are almost always present within wild populations and frequently at high rates. This leads to the fact that, given sufficient duration, evolutionary responses are inevitable within weed populations according to intensity of the selection. If there is some polymorphism for almost all the characteristics of the species we may expect polymorphism for herbicide susceptibility. As this is the highest selective pressure against weed species, this will lead to resistance i.e. the fact that some plants will survive and produce seeds despite a treatment which should have killed all the plants of the species in the field. At least a part of their progeny must survive the same treatment. Resistance can be defined either as being the inherited ability of a plant to not be controlled by (at least) the normal doses of a herbicide or as a peculiar physiological characteristic occurring in some plants.

When sprayed a herbicide must enter the plant and reach a susceptible target. During transport to its target it could be metabolized to a greater or lesser extent or, according to the health of the plant, set aside in order to prevent it from reaching the target. Thus resistance is any modification of one or several of these characteristics. From a genetic point of view, resistance may be due to one or several mutations on the genes encoding for these characteristics. But it could also be due to a recurrent selection which may act on quantitative variation and achieve a progressive increase in resistance from generation to generation with changes in gene frequency at several loci more or less correlated with resistance.

The mutation can occur by chance while the selection is being applied. In this case the probability of the observation of a resistance will depend on the relationships between mutation frequency and population size. But, generally, as many mutations preexist at many loci in a population one cannot reject the presence of at least one mutation on genes involved in herbicide efficiency. Thus the probability of resistance occurring in a weed population prior to herbicide selection will depend also on the mutation frequency and the size of the population, but overall on the relative fitness of the resistant genotypes apart from the herbicide selective pressure. This means that in some populations the frequency of resistant plants before the first treatment may be quite important leading quickly to huge populations of resistant plants after few treatments.

Factors of Evolution Linked to Herbicide Properties

According to their use, not all herbicides are suitable candidates for selecting resistant weeds. It follows that the duration and the constancy of the use of a product in the same field, the level the efficiency according to a given weed species and the persistence of the active ingredient in the soil will affect the intensity of the selective pressure and thus the probability of the selection of a resistance, provided the size of the treated area is enough important.

In intensive agriculture field doses are generally high enough to kill all plants of a susceptible species so, if these doses are far from the LD_{100} of a species, only mutations leading to target mutation resistances should be selected. If the field dose is around the LD_{100} of the species any type of non susceptible plant could survive and different resistances could be selected, as amplification of genes encoding for detoxification (Hall *et al.*, 1994). Thus, in France, one type of setaside is regularly treated with very low doses of glyphosate, this could be a way to select unknown resistance. Moreover, if the herbicide has several sites of action, these resistances would certainly never be selected, the only resistances which could appear are additive resistances like insecticide resistances within insects (Georghiou, 1990; Ferrari and Georghiou, 1991). Therefore, the selection and the spread of a resistant genotype becomes a reality in a field when a herbicide persistent enough to prevent any susceptible plant from producing seeds is used every year and when only one point mutation may shift the herbicide efficiency. This corresponds to herbicides having only one mode of action similar to all the recent registered molecules.

Thus, according to these conditions, in all the registered active ingredients only few molecules have led to the selection of resistances. A good example is atrazine. The study of the resistant genotypes has shown that only one point mutation is responsible for a target mutation resistance. Other target mutation resistances (to sulfonylureas and aryloxyphenoxypropanoates) have spread very fast. Any other type of resistance is less likely to appear and to spread.

Factors of Evolution Linked to Weeds

There are some interactions between genetic characteristics of the weeds and the mode of action of the herbicide. The probability of the selection for resistant genotypes depends on the number of mutations necessary to confer resistance and the cost of these mutations. As for weed species, the selection will obviously be correlated with the abundance of the species and the amount of seeds set. According to the mechanism of resistance, its spread will be highly correlated to its inheritance. The mating system of the species is also of major importance. If the mutation is correlated with a high cost, the frequency before the first treatment will be so low that the resistance may never be selected. If the inheritance is not simple the spread will be slow. As for cytoplasmic inheritance, there are only two genotypes; when the resistance is due to one, two or three genes there could be three, nine and 27 possible genotypes respectively, and as many levels of resistance. If the species is highly autogamous (within some weed species there is less than 1% allogamy) the spread of the resistance will be very reduced. As the emergence of seedlings is an important bottleneck, a field population is not always a representative sample of the seed production of an autogamous species as the mutation will never be distributed over the different genotypes.

Even if there is no reason for rejecting the hypothesis that any herbicide could lead to any kind of resistance within any weed species, according to these remarks we should expect to find in fields significantly more resistances due to a point mutation inherited by a single dominant or semidominant gene.

Inheritance of Resistances

As for broad spectrum herbicides not all the species are equally susceptible. Generally the field doses are several times higher than the LD_{100}, but for some species such a difference does not exist and sometimes the applied rate does not kill all the plants of the species.

However this could meet the needs of the farmer as the growth inhibition is sufficient to give an advantage to the crop. This situation could lead to the selection of additive resistances after repeated use of the same herbicide. Few examples have been reported. A positive linear relationship has been found between the percentage survival of a population of *Senecio vulgaris* treated at low rates of triazine and the number of herbicide applications (Holliday and Putwain, 1980). Thanks to genetic studies using selfing series procedures, broad sense heritability has been measured. Even although it is low (0.22) this indicates that a resistance inherited in a polygenic way has been selected. In the same way only four years of low rate simazine treatments have been suffficient to select larger plants with longer root systems within *Alopecurus myosuroides* (Darmency, 1981). Within *Avena fatua* several studies have dealt with variability in response to thiocarbamates after several years of treatments (Jana and Naylor, 1982) and barban within non treated populations (Price *et al.*, 1983). There is still no report for glyphosate resistance within weed populations. However, screening experiments have shown some variability to glyphosate treatments between several unexposed biotypes of *Convolvulus arvensis*. This means that within this species there is a significant variation in glyphosate susceptibility. After a diallel cross between five genotypes, analysis of variance components showed an additive inheritance (Duncan and Weller, 1987). This could lead to a possible resistance if environmental conditions are favourable, i.e. repeated sublethal doses. As this species is completely self-incompatible this may facilitate the accumulation of genes leading to the build up of resistant biotypes in a simple additive way.

Quantitatively inherited resistance mechanisms have still not been documented within weeds in fields. Some remarks could explain this situation. The herbicide doses have to be rather low. As several herbicides may be applied in crop rotations the resistance genes could be diluted in a susceptible genetic backgroud. The soil seed bank has a buffering effect. Such additive polygenic mechanisms are more prevalent within self-compatible or allogamous species whereas many weed species are highly autogamous. As quantitative mechanisms do not lead to more than two or three fold resistance this does not allow a clear-cut separation between resistant and susceptible plants. But the lack of such observations could also be due to the way in which we check resistances. The herbicide rates should allow discrimination between phenotypes. Several intermediate phenotypes should be estimated in an objective way and overall genetic experimental designs should be managed with enough progenies in order to be able to interpret more complex inheritance than single-gene traits.

Presently many resistances have been recognized, even if the mechanism is not known for each case, but there are less than 20 known cases of inheritance (Darmency, 1994). When the mechanism is well known and the resistance observed within many species, the inheritance observed in some species is inferred for all the others.

In some cases, cross-resistances have been reported and generally the inheritance for each product is not known. There are two major possibilities. First if resistance is due to only one mechanism which by itself gives the resistance to several herbicides, it is obviously the same inheritance. Alternatively the cross resistance could be a complex mechanism of detoxification due to several genes which each bring one the resistance to a specific herbicide. In that case the inheritance is not simple and it should be studied one by one in order to understand the relations between the different mechanisms.

Cytoplasmic Inheritance

To date, only one resistance has spread with a cytoplasmic inheritance: the triazine target mutation. However this resistance is very important as resistant genotypes have appeared within at least 50 species (almost all the species growing in triazine treated areas). Furthermore, as the target of only a few herbicide families is encoded by cytoplasmic genes, we may expect that the probability for the spreading of any other cytoplasmic resistance will be very low. Maybe because this resistance was the first huge spread of highly resistant biotypes, many weed scientists took part in the study of the mechanism which is now almost completely elucidated. But as for the state of the weed populations before the treatments, some points still remain in the dark. When drawing models, the first hypothesis which comes to mind is to use the expected frequency of random cytoplasmic mutation, but as the incidence of atrazine resistant biotypes in some populations is considerably higher than that expected in models, the only way is to ascribe a high cost to the mutation.

If the mutation is randomly distributed in a population, the more the field is sprayed with the same herbicide the more the frequency of resistant plants will increase. However in many cases of maize monoculture for more than twenty years no resistant genotype appeared. In fact several observations have shown that nucleo-cytoplasmic relations are involved in resistance inheritance. Within some species the first resistant populations were always nuclearly monomorphic (Warwick, 1991). The mutation did not appear in any genotype but could only be expressed through certain nuclear genotypes. Within *Chenopodium album* we have found the mutation at a high frequency in the progeny of specific genotypes (Darmency and Gasquez, 1990). Thus the spread of resistance depends on the presence of these specific susceptible genotypes. According to the fact that the polymorphism of a population is highly correlated with the intensity of cultivation and herbicide use (Al Mouemar and Gasquez, 1983), in regularly treated fields the rare specific genotype may have disappeared after several years of cultivation. Thus, when there is no resistance in fields treated during many years the probability of the spread of resistant plants of this very population is almost zero. On the other hand the resistant genotype will spread rapidly in new cultivated fields (i.e. after a meadow).

Nuclear Inheritance

From a statistical point of view there is no reason for finding more resistances to be due to one gene instead of several genes and no reason for more dominant rather than recessive genes. This is what actually appeared within 19 *in vitro* artificially selected resistances (Darmency, 1994): eight and eleven were encoded by several genes or only one gene respectively. Moreover, nine, six and five resistances are due to dominant, semi-dominant and recessive genes, respectively. This obviously points to a random distribution clearly due to the lack of selection.

On the other hand, in treated fields we may expect that the natural selection will allow only some resistances to appear. According to the target of many herbicides, one point mutation is sufficient to entail resistance, so the most frequent resistances will rather be due to monogenic inheritances. Thus in 19 confirmed cases, only two are due to more than one gene and within the 17 remaining, seven, nine and only one were dominant, semi-dominant and recessive respectively (Table 1). This could be due to the fact that the field doses generally only kill the homozygous susceptible genotypes or to the fact that, as demonstrated by computer simulations (Futuyma, 1986), whatever the original frequency of the gene the first fixed genotype is always the semidominant one.

Assuming that dominant and semi-dominant monogenic inheritances have really been proved, we could hypothesize that there could be two different mechanisms of resistance to paraquat as, within the six paraquat resistant species, a dominant gene has been observed for three of them, a semi-dominant for the others. These examples seem highly correlated with the characteristics we pointed out earlier. The only important case of resistance due to more than one gene is the resistance of *Alopecurus myosuroides* to chlorotoluron (Chauvel, 1991). As there are several genotypes with different levels of resistance, the spread will depend on the mating system (the species is allogamous with some self-incompatibility) many susceptible plants may be found within the progeny of a plant surviving a chlorotoluron treatment. Even if the resistance is due to several genes one may expect that herbicide selection would build up a resistant population with a reduced polymorphism. However all the populations of *A. myosuroides* have the same polymorphism with the same alleles whether the populations are resistant or not. Moreover according to the fact that resistances due to several genes need some allogamy in order to spread in a population, these resistances will hardly appear in autogamous species.

Furthermore, in *A. myosuroides* different resistances have been recognized characterized by different cross resistances which seem completely independent, but this could be only due to the fact that each gene is linked to a specific resistance and that the association of several genes in the same genotype could not only give cross resistances due to the addition of each gene but also produce new resistances emerging from this association like a pleiotropic effect.

Fitnesses of the Selected Resistances

As we have previously seen, the differential fitness of resistant and susceptible biotypes has been quickly recognized as an important factor which could influence the evolution of herbicide resistance even before the first treatment. When a resistant genotype is less fit than susceptible plants of the same population, stoppage of the selective herbicide allows natural selection to act for restoring relations between resistant and susceptible plants (i.e. the frequency of a resistant genotype decreases). But if there is no fitness difference or even if the resistant biotype is fitter, this genotype will remain dominant and never disappear.

Fitness is the capacity of a genotype to give fertile descendants in the normal competitive conditions of the environment from which it comes. In many cases fitness studies between very different genotypes conclude that differences in physiological performance exist due to the resistance mutation just because the growth or the productivity was higher for one genotype. If one wants to measure the effect of this mutation, only isogenic plants must be compared, but if one wants to know the relative fitness of the resistant, the measurements must take into account the natural conditions of the population and also compare the resistant genotype to the best fit genotypes of the population. This has generally never been done.

In some cases, resistance is accompanied by a pleiotropic effect due to the mutation. Thus the triazine resistant genotypes are less competitive because the mutation reduces the electron transfer in the PSII system. In experiments with triazine resistant *Chenopodium album* in pure stands and in mixture with maize, we observed clear cut differences in productivity between nearly isogenic resistant and susceptible genotypes: the susceptible genotype produced more seeds which have a better germination capacity. But

Table 1. Mendelian inheritance of herbicide resistances within weed species

Herbicide	Mechanism	Species	Inheritance
triazines	target mutation	almost 50	cytoplasmic
atrazine	detoxification	*Abutilon theophrasti*	1D nuclear
sulfonylureas	target mutation	*Lactuca serriola*	1 1/2D nuclear
	target mutation	*Kochia scoparia*	1 D nuclear
diclofop menthyl	target mutation	*Lolium multiflorum*	1 1/2 D nuclear
fluazifop methyl	target mutation	*Avena sterilis*	1 1/2 D nuclear
haloxyfop methyl	target mutation	*Lolium rigidum*	1 1/2 D nuclear
fenoxaprop ethyl	target mutation	*Avena sterilis*	1 1/2 D nuclear
	?	*Avena fatua*	1 1/2 D nuclear
paraquat	sequestration?	*Arototheca calendula*	1 1/2 D nuclear
	translocation?	*Conyza philadelphicus*	1D nuclear
	sequestration?	*Conyza bonariensis*	1D nuclear
	translocation?	*Erigeron canadensis*	1D nuclear
	sequestration?	*Hordeum glaucum*	1 1/2D nuclear
	translocation?	*Hordeum leporinum*	1 1/2D nuclear
chlorotoluron	detoxification	*Alopecurus myosuroides*	2 additive nuclear
siduron	?	*Hordeum jubatum*	3 complementary
trifluralin	target mutation	*Eleusine indica*	several linked D
trifluralin	?	*Setaria viridis*	1 recessive
sethoxidim	?	*Setaria viridis*	1D nuclear
2,4-D	diff.bind.protein	*Sinapis arvensis*	1D nuclear

the next year the same number of seedlings appeared for each genotype and also the year after, indicating that, instead of productivity differences, the fitnesses were not significantly different in natural conditions (data not published). As previously assumed for naturally occurring resistances, in many cases it is very difficult to observe any effect associated with resistance in the field certainly because the selective pressure favours resistances with no cost.

Conclusion

This short survey reveals that we are finding only the resistances that we seek. The doses applied in order to distinguish the possible resistant genotype and the variables used to test the reality of the resistance allow us to only observe mechanisms leading to high herbicide dose resistances with mendelian inheritances. These resistances are obviously the most spectacular because according to the genetic and biochemical characteristics which increase the spread of these mechanisms, a farmer will notice only huge infestations. Furthermore another important question is the lack of specific tests for almost all the herbicides which forces us to use only herbicide treatments for greenhouse tests. From a scientific point of view, it is impossible to confirm or reject a hypothesis for resistance by using as a test the same technique with the same sources of variation which allowed the farmer to suspect the presence of resistant genotypes. So, as long as we do not have this specific test, it will be difficult to determine exactly the characteristics of a resistance and especially the exact cross-resistances and their level.

In sustainable agriculture using reduced doses of total herbicides which are not very prone to select resistances, the repeated use could rather select progressive resistances with quantitative inheritances. Furthermore, nobody is presently aware of the build up of such populations and the farmers inform us too late. These resistances could be more serious and give rise to more difficult problems than we have had up to now.

According to these comments, predicting the spread of unknown genotypes through models drawn only with expected values is a difficult undertaking. Although it is quite easy to describe the herbicide characteristics (field rates, persistence, mode of action ...), the genetic characteristics of the species are generally unknown. Furthermore, they need to be accurately determined. As for inheritance, we need unambiguous genetic markers for the crosses, to be sure that the rates are high enough in order to distinguish a dominant from a semi-dominant gene. As for the mutation cost, the measurements of the relative fitnesses should be properly taken in field conditions. At the same time, especially for resistances with complex inheritances and cross-resistances, the study of the mechanism must be aware of the inheritance and use only real resistant progenies.

However it is clear that the repeated use of herbicides will select resistant genotypes; accordingly integrated weed management strategies must be introduced. The only outcome from the problems of controlling herbicide-resistant weeds is to realize that no single weed control method is likely to provide definitive control when used exclusively. Weed control should be a long term strategic process involving all the agronomic techniques.

References

Al Mouemar, A. and Gasquez, J. (1983) Environmental conditions and isozyme polymorphism in *Chenopodium album* L. *Weed Research* **23**: 141-149.

Chauvel, B. (1991) Polymorphisme genetique et selection de la resistance aux urees substituees chez *Alopecurus myosuroides* Huds. *Ph. D. Thesis*, University of Paris Orsay.

Darmency, H. (1981) Some effects of herbicide-selection on *Alopecurus myosuroides* Huds. *Plant Soil* **59**: 491-494.

Darmency, H. (1994) Genetics of herbicide resistance in weeds and crops. *In: Herbicide Resistance in Plants* (S.B. Powles and J.A. Holtum, eds.), 263-279. Lewis Publishers, Boca Raton.

Darmency, H. and Gasquez, J. (1990) Appearance and spread of triazine resistance in common lambsquarter (*Chenopodium album*). *Weed Technology* **4**: 173-177.

Duncan, C. N. and Weller, S.C. (1987) Heritability of glyphosate susceptibility among biotypes of field bindweed. *Journal Heredity* **78**: 257-260 .

Ferrari, J.A. and Georghiou, G.P. (1991) Quantitative genetic variation of esterase activity associated with a gene amplification in *Culex quinquefasciatus*. *Heredity* **66**: 265-272.

Futuyma, D.J. (1986) *Evolutionary Biology*. Sinauer Associates Inc. Sunderland, Massachusetts.

Georghiou, G.P. (1990) Overview of insecticide resistance. *In: Managing Resistance to Agrochemicals.*

From Fundamental Research to Practical Strategies (M.B. Green, H.M. LeBaron and W.K. Moberg, eds.). American Chemical Society Series **421**, 18-41. ACS, Washington.

Hall, L.M., Holtum, J.A. and Powles, S.B. (1994) Mechanisms responsible for cross resistance and multiple resistance. *In: Herbicide Resistance in Plants* (S.B. Powles and J.A. Holtum, eds.), 243-261. Lewis Publishers, Boca Raton.

Holliday, R.J. and Putwain, P.D. (1980) Evolution of herbicide resistance in *Senecio vulgaris:* Variation in susceptibility to simazine between and within populations. *Journal Applied Ecology* **17**: 779-791.

Jana, S. and Naylor, J.M. (1982) Adaptation for herbicide tolerance in populations of *Avena fatua. Canadian Journal Botany* **60**: 161 1-1617.

Price, S., Hill, J. and Allard, R. (1983) Genetic variability for herbicide reaction in plant populations. *Weed Science* **31**: 652-657.

Warwick, S.I. (1991) Herbicide resistance in weedy plants: physiology and population biology. *Annual Review Ecology Systematics* **22**: 95-114.

CHAPTER 21
EFFECTS OF MUTATION FOR ALS-INHIBITOR RESISTANCE ON ALS ACTIVITY IN RESISTANT AND SUSCEPTIBLE NEAR-ISONUCLEAR *LACTUCA* LINES

C.V. EBERLEIN, M.J. GUTTIERI, C.A. MALLORY-SMITH and D.C. THILL
University of Idaho, PO Box AA, Aberdeen, ID 83210

Abstract

The biochemical and physiological effects of target site mutation for ALS-inhibitor resistance were evaluated using chlorsulfuron-resistant (R) and chlorsulfuron-susceptible (S) near isonuclear 'Bibb' lettuce (*Lactuca sativa* L.) lines. The near isonuclear lines (isolines) were developed by backcrossing the resistance allele from a chlorsulfuron-resistant prickly lettuce (*Lactuca serriola* L.) biotype into Bibb lettuce. The specific activity of ALS from the R Bibb isoline was 46% of the specific activity of ALS isolated from the S Bibb isoline, suggesting that there are detrimental effects of the mutation for resistance on enzyme function, expression, or stability. The Km (pyruvate) values of ALS isolated from R Bibb and S Bibb were similar, 7.3 and 11.1 mM, respectively, suggesting that the mutation for resistance did not alter pyruvate binding on the ALS enzyme. ALS activity from both R and S Bibb isolines was inhibited by valine, leucine, and isoleucine. However, ALS from R plants was less sensitive to feedback inhibition by the individual amino acids than ALS from S plants. Concentrations of valine, leucine, and isoleucine were about 1.5 times higher in R seed than in S seed, and concentrations of valine and leucine in R Bibb lettuce leaves were 1.3 and 1.6 times higher, respectively, than in leaves of the S isoline. Therefore, mutation for resistance to ALS-inhibiting herbicides may alter regulation of branched chain amino acid synthesis.

Introduction

Weed biotypes resistant to herbicides that inhibit ALS (acetohydroxyacid synthase, EC 4.1.3.18), a key enzyme in the synthesis of branched chain amino acids, are an increasing agronomic problem in the United States and Canada. Resistance to ALS inhibitors was first documented in 1987, when a chlorsulfuron-resistant biotype of prickly lettuce was discovered near Lewiston, ID (Mallory-Smith *et al.*, 1990) in a winter wheat field that had been treated annually with sulfonylurea herbicides for the previous five years. Since then, ALS inhibitor resistance has been reported in 27 species.

The biochemical basis for resistance to ALS inhibitors has been characterized for some weed biotypes. For example, chlorsulfuron resistance in the Idaho prickly lettuce biotype was due to modified ALS rather than to differences in uptake, translocation, or metabolism of chlorsulfuron in the R vs the S biotype. The R biotype was cross resistant to eight sulfonlyurea herbicides. The R biotype also was cross resistant to the imidazolinone herbicides, imazapyr and imazethapyr, but was not cross resistant to imazaquin (Mallory-Smith *et al.*, 1990). Resistance was conferred by a point mutation that resulted in a proline to histidine substitution in the Domain A aminoacid sequence of the ALS enzyme (Guttieri *et al.*, 1995).

The resistance allele from the Idaho prickly lettuce biotype was backcrossed into a domestic lettuce background to obtain near isonuclear R and S Bibb lines (Mallory-Smith *et al.*, 1993). The near isonuclear lines provide an excellent set of materials for studying the specific effects of mutation for resistance on ALS activity, without confounding genetic background effects. Therefore, the objectives of our research were to characterize the biochemical and physiological effects of target site mutation for ALS-inhibitor resistance in the R Bibb isoline.

Results and Discussion

The apparent Km (pyruvate) values for ALS isolated from R and S Bibb were similar, 7.3 and 11.1 mM, respectively, which suggested that mutation for resistance did not alter pyruvate binding on the ALS enzyme. In previous studies, Km values reported for several field-selected ALS inhibitor resistant species also were similar for R and S biotypes (Saari *et al.*, 1992, 1994).

Specific activity of ALS from the R isoline was less than specific activity of ALS from the S isoline (Table 1), regardless of the ALS assay procedure used. For example, specific activity of ALS isolated from R Bibb with an *in vitro* procedure (Guttieri *et al.*, 1995) was 46% of the ALS activity isolated from S Bibb. Similar results were obtained when ALS activity in isolated chloroplasts was assayed, and when an in vivo leaf disk ALS assay (modified from Gollop *et al.*, 19833) was used. Lower specific activity of ALS from the R isoline suggests that there may be decreased expression of the ALS enzyme in R plants, or detrimental effects of the resistance mutation on enzyme function or stability.

Table 1 Specific activity of ALS from R Bibb lettuce isolines in three differed ALS assay procedures

Assay	Specific activity of R Bibb (% of S activity)
In vitro	40
Chloroplast	45
Leaf disk	53

ALS catalyzes two parallel reactions: condensation of two molecules of pyruvate to from acetolactate and condensation of one molecule of pyruvate with one molecule of 2-oxobutyrate to form acetohydroxybutyric acid. ALS activity may be regulated by several compounds, including 2-oxobutyrate and the branched chain amino acids, valine, leucine, and isoleucine. ALS activity from both R and S Bibb was strongly inhibited by 2-oxobutyrate and two values for both isolines were similar, suggesting that mutation for resistance did not affect 2-oxobutyrate binding.

ALS activity from both R and S Bibb was feedback inhibited by 1 mM valine, leucine, or isoleucine. However, ALS from R plants was significantly less sensitive ($p < 0.01$) to feedback inhibitor by the branched chain amino acids than ALS isolated from S plants (Figure 1). Leucine was a stronger feedback inhibitor of both the R and S isolines than valine or isoleucine. Decreased sensitivity of ALS to feedback inhibition by branched chain amino acids also has been reported for sulfonylurea resistant cell lines of *Datura innoxia* (Rathinasabapathi *et al.*, 1990) and for some lab-selected resistant mutants of *Nicotiana tabacum* and *Gossypium hirsutum* (Subramanian *et al.*, 1990). ALS from field-selected sulfonylurea resistant kochia and common chickweed (*Stellaria media*) also was less sensitive to feedback inhibition by leucine than ALS from herbicide-susceptible biotypes, but sensitivity to valine was not altered in R kochia and common chickweed biotypes (Subramanian *et al.*, 1991). In other studies, feedback regulation of ALS from corn mutants resistant to imidazolinone herbicides was similar to the susceptible wild-type (Newhouse *et al.*, 1991).

Previous studies have shown that *in vitro* extraction and assay conditions may mask effects of branched chain amino acids on ALS activity (Arfin *et al.*, 1973; Gollop *et al.*, 1983; Magee *et al.*, 1968; Miflin and cave, 1972; Relton *et al.*, 1986). Therefore, feedback inhibition studies were repeated using an ALS assay with isolated chloroplasts. Results were similar to those obtained with the *in vitro* assay procedure. ALS activity in chloroplasts from R Bibb was less sensitive to feedback inhibition by 1 mM concentrations of each branched chain amino acid than ALS activity from S Bibb chloroplasts (Figure 2) and leucine was the strongest inhibitor of ALS activity. Thus, mutation for ALS-inhibitor resistance in *Lactuca* alters regulation of ALS activity by branched chain amino acids.

To assess possible physiological consequences of reduced sensitivity of ALS to feedback inhibition, concentrations of free branched chain amino acids in leaves and seeds of R and S Bibb were measured. Concentrations of valine and leucine in R Bibb leaves were 1.3 and 1.6 times higher, respectively, than in S Bibb leaves; isoleucine concentrations were similar in R and S leaves (Figure 3). Concentrations of branched chain amino acids in leaves were much lower than concentrations of most other amino acids, suggesting that differences in valine and leucine concentrations in R vs S leaves might be of consequence. Concentrations of all three branched chain amino acids were about 1.5 times higher in R seeds than in S seeds (Figure 4). Free branched chain amino acids in seeds may be the result of *de novo* synthesis rather than remobilization of amino acids from the leaves

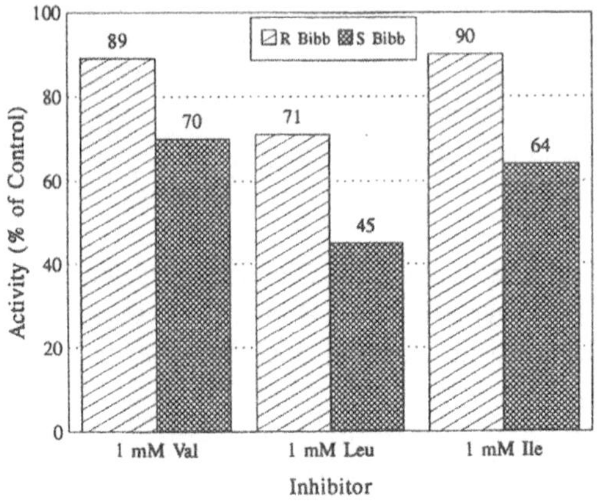

Figure 1. Feedback inhibition of in vitro ALS activity by valine, leucine and isoleucine. S.E. = 6.

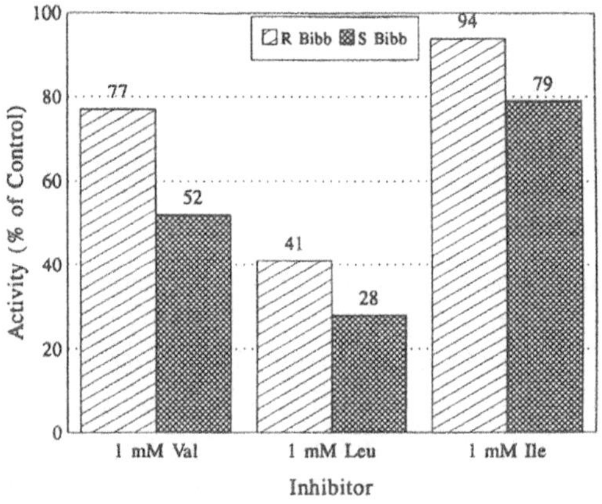

Figure 2. Feedback inhibition of ALS activity in isolated chloroplast by valine, leucine and isoleucine. S.E. =4.

(Muhitch, 1988). Muhitch (1988) reported that there was ALS activity in developing maize seeds and that activity was sensitive to feedback inhibition by branched chain amino acids. Similar studies are underway in our laboratory with developing seeds of R and S Bibb lettuce.

Elevated levels of branched chain amino acids have been reported in seeds of three field-selected sulfonylurea-resistant kochia accessions. These elevated levels of branched chain amino acids were correlated with faster germination of seeds from R vs S biotypes at cool temperatures (Dyer *et al.*, 1993). Interestingly, seeds of the R Bibb lettuce isoline also germinated faster at low temperatures than seeds of the S isoline (Saari *et al.*, 1994). Dyer *et al.* (1993) hypothesized that elevated levels of isoleucine and valine, which may be absolutely required for DNA synthesis, might allow cell division and growth to proceed more rapidly during low temperature germination in R vs S seed. Rapid early germination may allow R biotypes to capture space more effectively than S biotypes in certain environments (Dyer *et al.*, 1993).

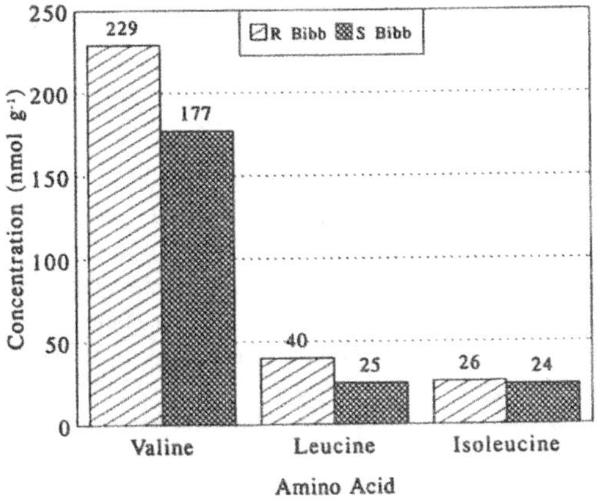

Figure 3. Amino acid concentrations in leaves of resistant and susceptible Bibb lettuce. S.E. (val) =9, S.E. (leu, ile) = 1.

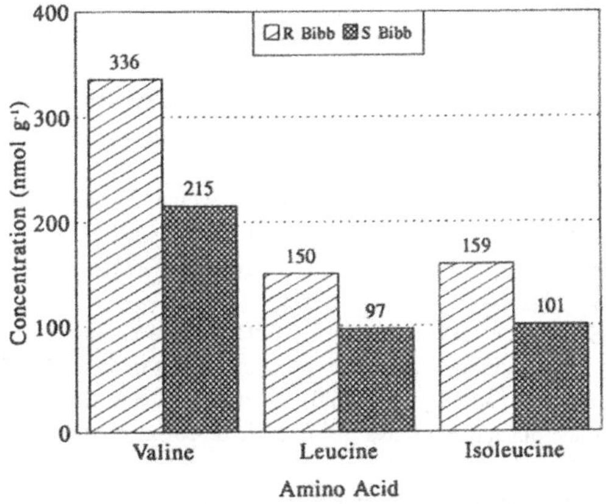

Figure 4. Amino acid concentrations in seeds of resistant and susceptible Bibb letuce. S.E. (val)=5, S.E. (leu,ile) =3.

Conclusions

Mutation for ALS-inhibitor resistance in *Lactuca* has biochemical and physiological consequences: specific activity of ALS is lower in the R vs S Bibb isoline;

ALS from the R isoline is less sensitive to feedback inhibition than ALS from the S isoline; concentrations of valine and leucine are higher in R Bibb leaves than in S leaves; and concentrations of all three branched chain amino acids are higher in R Bibb seeds than in S Bibb seeds. Lower specific activity of ALS from R vs S Bibb, but higher leaf and seed branched chain amino acid concentrations may seem contradictory. However, it is our hypothesis that greater accumulation of branched amino acids in the R isoline than in the S isoline is due to reduced downregulation of R Bibb ALS by branched chain amino acids, which more than compensates for differences in specific activity.

References

Arfin, S.M. and Koziell, D.A. (1973) Acetolactate synthase of *Pseudomonas aeruginosa*. I. Purification and allosteric properties. *Biochimica Biophysica Acta* **321**: 348-355.

Dyer, W.E., Chee, P.W. and Fay, P.K. (1993) Rapid germination of sulfonylurea-resistant *Kochia scoparia* is associated with elevated seed levels of branched chain amino acids. *Weed Science* **41**: 18-22.

Gerwick, B.C., Mireles, L.C. and Eilers, R.J. (1993) Rapid diagnosis of ALS/AHAS-resistant weeds. *Weed Technology* **7**: 519-524.

Gollop, N., Chipman, D.M. and Barak, Z. (1983) Inhibition of acetohydroxyacid synthase by leucine. *Biochimica Biophysica Acta* **748**: 34-39.

Guttieri, M.J., Eberlein, C.V., Mallory-Smith, C.A., Thill, D.C. and Hoffman, D.L. (1992) DNA sequence variation in Domain A of the acetolactate synthase genes of herbicide-resistant and -susceptible weed biotypes. *Weed Science* **40**: 670-678.

Guttieri, M.J., Eberlein, C.V. and Thill, D.C. (1995) Diverse mutations in the acetolactate synthase gene confer chlorsulfuron resistance in kochia (*Kochia scoparia*) biotypes. *Weed Science* **43**: 175-178.

Magee, P.T. and De Robichon-Szulmajster, H. (1968) The regulation of isoleucine-valine biosynthesis in *Saccharomyces cerevisiac*. 3. Properties and regulation of the activity of acetohydroxy-acid synthetase. *European Journal Biochemistry* **3**: 507-511.

Mallory-Smith, C.A., Thill, D.C. and Dial, M.J. (1990) Identification of sulfonyurea herbicide resistant prickly lettuce *(Lactuca serriola). Weed Technology* **4**:163-168.

Mallory-Smith, C.A., Thill, D.C. and Dial, M.J. (1993) ID-BR1: Sulfonylurea herbicide-resistant lettuce germplasm. *Horticultural Science* **28**: 63-64.

Miflin, B.J. and Cave, P.R. (1972) The control of leucine, isoleucine, and valine biosynthesis in a range of higher plants. *Journal Experimental Botany* **23**: 511-516.

Muhitch, M.J. (1988). Acetolactate synthase activity in developing maize (*Zea mays* L.) kernels. *Plant Physiology* **86**: 23-27.

Newhouse, K., Singh, B., Shaner, D. and Stidham, M. (1991) Mutations in com (*Zea mays* L.) conferring resistance to imidazolinone herbicides. *Theoretical Applied Genetics* **83**: 65-70.

Rathinasabapathi, B., Williams, D. and King, J. (1990). Altered feedback sensitivity to valine, leucine, and isoleucine of acetolactate synthase from herbicide-resistant variants of *Datura innoxia. Plant Science* **67**: 1-6.

Relton, J.M., Wallsgrove, R.M., Bourgin, J.P. and Bright, S.W.J. (1986) Altered feedback sensitivity of acetohydroxyacid synthase from valine-resistant mutants of tobacco (*Nicotiana tabacum* L.). *Planta* **169**: 46-50.

Saari, L.L., Cotterman, J.C., Smith, W.F. and Primiani, M.M. (1992) Sulfonylurea herbicide resistance in common chickweed, perennial ryegrass, and Russian thistle. *Pesticide Biochemistry Physiology* **42**: 110-118.

Saari, L.L., Cotterman, J.C. and Thill, D.C. (1994) Resistance to acetolactate synthase-inhibitor herbicides. *In: Herbicide Resistance in Plants: Biology and Biochemistry* (S.B. Powles and J.A.M. Holtum eds.), 83-139. CRC Press, Boca Raton.

Subramanian, M.V., Hung, H., Dias, J.M., Miner, V.W., Butler, J.H. and Jachetta, J.J. (1990) Properties of mutant acetolactate synthases resistant to triazolopyrimidine sulfonanilide. *Plant Physiology* **94**: 239-244.

Subramanian, M.V., Loney-Gallant, V., Dias, J.M. and Mireles, L.C. (1991) Acetolactate synthase inhibiting herbicides bind to the regulatory site. *Plant Physiology* **96**: 310-313.

CHAPTER 22
MOLECULAR CHARACTERIZATION OF HERBICIDE RESISTANCE IN *ECHINOCHLOA* SPP.

N. LOPEZ-MARTINEZ[1], R.P. FINCH[2], G. MARSHALL[2] and R. DE PRADO[1]
1. Departamento de Quimica Agricola y Edafologia, E.T.S.I. Agronomos y Montes, Apdo. 3048, 14080-Cordoba, Spain 2. Scottish Agricultural College, Auchincruive, Ayr, KA6 SHW, Scotland

Introduction

Echinochloa spp. are weeds of maize and rice that cause serious yield losses and require to be controlled. Two main herbicides used in these crops are atrazine and quinclorac. Triazine herbicides inhibit photosystem II and quinclorac enhances ethylene production in susceptible plants (Devine *et al.*, 1993; Grossmann and Kwiatkowski, 1993). Development of resistance to atrazine in weeds is well documented (Gronwald, 1994), but quinclorac resistance has not been reported in *Echinochloa* spp. We describe the characterisation of atrazine and quinclorac resistance in *Echinochloa* spp. Classification and identification of *Echinochloa* spp. is notoriously difficult but establishing the identity of herbicide susceptible/wild biotypes of *Echinochloa* is essential when undertaking comparative mode of action studies with putative herbicide-resistant biotypes of the same species. Accordingly, we employed molecular techniques to make genetic comparisons of test biotypes of *Echinochloa*, free from the vaguaries of morphological variation and subjective assessment (Carretero, 1981).

The aims of the study were first to classify the test biotypes/species of *Echinochloa* using molecular marker techniques. Second, to undertake mode of action studies which included herbicide dose-response experiments, physiological assays (Hill reaction and chlorophyll fluorescence) and uptake/translocation experiments using ^{14}C-labelled quinclorac.

Material and Methods

Plant Material

Seeds, thought to be of *Echinochloa crus-galli,* were collected from sites where atrazine or quinclorac had failed to provide control under normal practices. Biotypes which differed in their resistance to quinclorac were found in rice fields and designated resistant (R) and intermediate (Int.). A susceptible biotype (S), came from an area with no history of herbicide treatment. A cross-resistant biotype (X) was collected from a maize field subjected to the application of atrazine over several years. Seeds of other *Echinochloa* species: *oryzicola, oryzoides* and *colonum* were also used for molecular studies.

Molecular Assessments

For the molecular assessment, DNA was isolated from 9 plants of each of the four biotypes thought to be *crus-galli* (S, R, X and Int.) and from 9 plants of each of other *Echinochloa* species using a standard phenol:chloroform:isoamyl-alcohol procedure, except centrifugation of ethanol-precipitated DNA was avoided prior to washing since co-precipitants are known to inhibit RAPD reactions. In each case the DNA samples were mixed in equal quantities to give eight pooled samples for analysis of genetic variation.

For the statistical analysis of genome diversity, a similarity matrix was constructed from presence/absence data, scored form RAPD profiles, using Jaccard's coefficient. Hierarchical cluster analysis was performed using a furthest-neighbour clustering algorithm. A principal coordinated analysis was conducted to provide a graphical summary of similarities between samples. Approximately sixty primers were screened for their suitability for generation reproducible DNA profiles. An initial survey showed that 18 RAPD primers gave clear amplification products that were readily scoreable.

In order to assess inter-biotype variation, four pooled DNA samples prepared from *E. crus-galli* biotypes were analysed with the selected primers and resulting DNA profiles were scored. Presence/absence data were recorded for some 184 loci.

Physiological and Biochemical Assessments

To assess herbicide resistance, pots containing the test biotypes at the second leaf stage were treated with formulated herbicides: quinclorac (0.1-3.0 kg ha^{-1}), atrazine (0.2-8.0 kg ha^{-1}), propanil (0.8-2.0 kg ha^{-1}) and molinate (1.0-6.0 kg ha^{-1}). Plants were mantained for 15 days in a controlled-environment chamber. After this time plants were harvested and growth was evaluated by determining shoot fresh weight. The herbicide dose causing a 50% reduction in fresh shoot weight (ED_{50}) was calculated for each herbicide as previously described (Menendez *et al.*, 1994). Each treatment consisted of five plants replicated three times.

The fourth leaf from three replicate plants was removed for each biotype and incubated in a test tube with 10 ml of 20 ppm solution of technical grade quinclorac and atrazine. Leaf fluorescence intensity values were measured initially (F_0), 6 h after incubation in herbicide solution (F_6) and 24h after transferring from herbicide to water (F_{24}). The F_0/F_6 indicated the level of photosynthetic inhibition by herbicides. The F_{24}/F_6 indicated the level of recovery of photosynthetic activity (Menendez *et al.* 1994).

The uptake experiments were carried out with radioactively ring-labelled quinclorac in combination with an adjuvant. ^{14}C-quinclorac (1 μl, 30 000 dpm/μl) was applied by a microsyringe to the second leaf of the plants. Treated leaf samples were harvested at time intervals of 3, 6, 9, 24, 72 and 128 h and washed with 1ml of acetone to remove the surface residues. Samples of treated and untreated leaves and roots were dried, then combusted in an oxidizer and the $^{14}CO_2$ trapped and counted in 15 ml of Permfluor V:Carbo-sorb (2:1 by vol.). Radioactivity was quantified by liquid scintillation counting .

Results

Molecular Studies

Hierarchical cluster analysis of the RAPDs data revealed that the four "*E. crus-galli*" biotypes formed only two discrete groups: (I) The resistant biotypes and (II) the susceptible biotype. Interspecific variation was assessed in a similar manner using the pooled DNA samples from each *Echinochloa* species, including the four "*E. crus-galli*" biotypes. Presence/absence data were recorded for 238 loci.

The principal coordinates analysis of the RAPDs data revealed that the five *Echinochloa* species form only 3 discrete groups (I) *Echinochloa crus-galli+E. hispidula,* (II) *E. oryzoides+E. oryzicola* and (III) *E. colonum.* It was not possible to distinguish between *E. crus-galli* and *E. hispidula* or between *E. oryzoides* and *E. oryzicola* with the primers used. Interestingly, herbicide resistance was confined to group I. The susceptible biotypes, originally classified as *E. crus-galli,* clustered with group II biotypes and were thus reclassified as *E. oryzoides* (Figure 1 and 2).

All of our results to date indicate that primer OP-A04 (5'-AATCGGGCTG-3') is diagnostic of the three *Echinochloa* groups described above. Note in particular that a major 800 bp product appears to classify the samples as group I and a similar distinct 1600 bp product classifies samples as group II. *E. colonum* (group III), which is readily identifiable by its morphology, gives rise to profiles that lack both of these specific RAPD products (Figure 3).

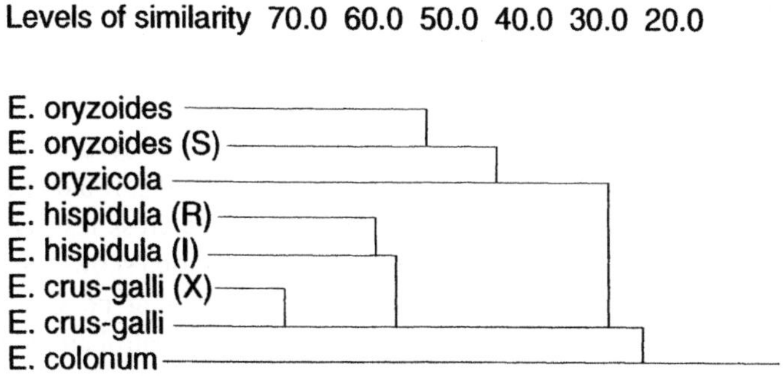

Figure 1. Dendogram constructed from furtherst-neighbour cluster analysis.

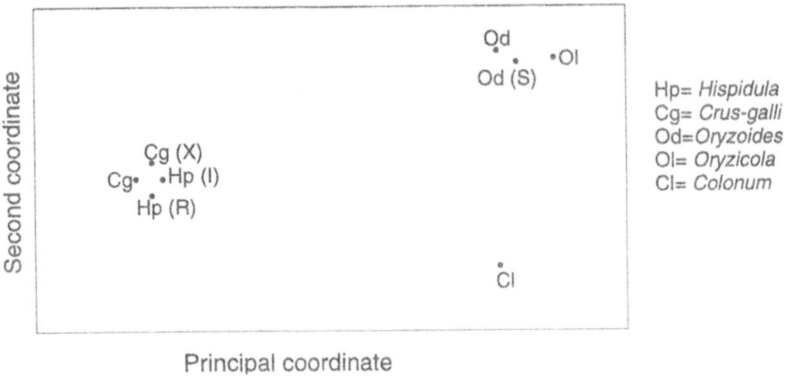

Figure 2. Two dimensional representation of cluster in eight *Echinochloa* accessions following principal coordinate analysis of RAPD data.

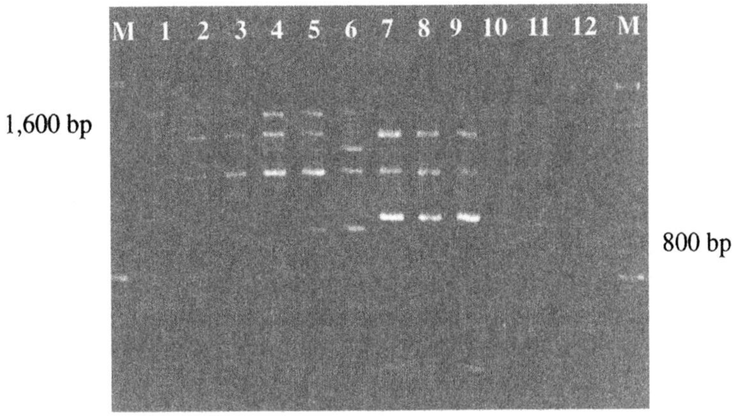

Figure 3. Amplification products from RAPD primer OP-A04. M: 100 bp marker; 1: *E. colonum*; 2-6: *E. oryzoides*; 7-12: *E. crus-galli.*

Biochemical and Physiological Studies

Resistance to quinclorac, atrazine, propanil and molinate differed significantly between the different biotypes studied (Table 1). The *Echinochloa* R biotype showed ED_{50} values 26 times higher than its susceptible counterpart. The cross resistant biotype (X) demonstrated considerable resistance to quinclorac but in addition showed an ED_{50} ratio for atrazine of an 80-fold difference compared to the susceptible biotype. Clearly, the cross-resistant biotype poses a considerable problem in classic weed control programs in Spain. It was apparent that propanil controlled all the biotypes successfully. By contrast, molinate was required at significantly higher rates of application to control the quinclorac-susceptible (S) biotype as compared to the other biotypes.

Accordingly, in an attempt to understand or characterise the biochemical and physiological basis of resistance, additional assessment was made of these biotypes. Fluorescence assays provide an *in vitro* assessment of photosynthetic capacity and therefore reveal changes in the photosynthetic performance of plant biotypes as influenced by herbicides (De Prado *et al.* 1993).

Quinclorac had no significant effect on the level of inhibition in all susceptible and resistant biotypes. However, atrazine treatment increased fluorescence emission levels in the quinclorac susceptible, intermediate and resistant plants, while the fluorescence emission values were close to control (F_0) in the cross-resistant biotype (Table 2). Results obtained for the Hill reaction assays were similar to those for fluorescence assays since they suggested that the cross-resistant biotype of *Echinochloa* may possess an altered target site (PS II) while this target site was not responsible for quinclorac resistance and was not affected by the herbicide (Table 3).

^{14}C-labelled quinclorac was absorbed rapidly in the treated leaf in all biotypes (Table 4). Subsequently there were no significant differences in the level of basipetal translocation of ^{14}C-label between biotypes within a 24 h period. Acropetal translocation was limited (less than 1.5% of applied) and again did not differ significantly between biotypes. The surface residue ^{14}C-label comprised a major proportion of the recovered ^{14}C-label at 24 h after treatment and declined in all biotypes. Therefore as ^{14}C-label increased, it was transported in acropetal and basipetal directions. Overall, there did not appear to be any notable differences in the pattern of uptake and translocation of ^{14}C-label in test biotypes.

Conclusions

Data present here support:

1. The importance of molecular characterization of *Echinochloa* spp. to verify the genetic material.

2. Resistance to quinclorac in two *Echinochloa* biotypes found in rice fields is not due to limited herbicide uptake and/or translocation.

3. Resistance to atrazine in a *Echinochloa* biotype found in maize field is due to target site mutation. This biotype also showed cross-resitance to quinclorac.

Table 1. Effect of different herbicides on ED_{50} of *Echinochloa* biotypes

	ED_{50} (kg ha^{-1})			
Biotype	Quinclorac	Atrazine	Propanil	Molinate
Resistant	2.6	0.1	0.4	1.1
X-Resistant	1.0	8.0	0.4	0.8
Intermediate	0.6	0.1	0.5	1.1
Susceptible	0.1	0.1	0.5	2.0

Table 2. Leaf fluorescence values of different populations of *Echinochloa* spp. measured at incubation time (F_0), after 6 hours in herbicide solution (F_6) and 24 h after transferring from herbicide to a nutrient solution (F_{24}).

	Atrazine (20 ppm)		Quinclorac (20 ppm)	
Biotype	F_6/F_0	F_{24}/F_0	F_6/F_0	F_{24}/F_0
Susceptible	2.9	3.2	1.0	1.0
Intermediate	3.0	4.1	1.4	1.4
X-Resistant	1.1	1.3	1.5	1.5
Resistant	2.8	3.9	0.8	0.8

Table 3. Values for 50% Hill reaction reduction (I_{50}) from linear plots of inhibition percentages versus logarithm of herbicide concentration

Biotype	Atrazine	Quinclorac
Susceptible	0.53±0.08	30,000±4,000
Intermediate	0.45±0,05	32,000±2,000
X-Resistant	310±16.6	23,000±7,000
Resistant	0.42±0.06	25,000±6,000

Table 4. Time-course distribution of radioactivity (expressed as percentage of the total radioactivity recovered) following the application of ^{14}C-Quinclorac to the second leaf of biotypes of *Echinochloa* spp.

Biotype	Susceptible		Intermediate		Resistant	
Time (h)	24	72	24	72	24	72
Organ						
Surface residue	55.1	35.6	55.0	34.8	54.8	30.8
Lower leaf	4.0	4.2	3.1	7.8	2.6	4.3
Treated leaf	39.3	57.4	39.7	46.3	41.1	56.2
Upper leaf	0.4	1.2	1.5	10.0	1.1	8.0
Root	1.2	1.6	0.7	1.1	0.5	0.7

Acknowledgements

The authors are thankful to BASF and C. Fabregas for supplying the seed and to Neil McRoberts for computational assitance. Financial support for this study was provided by CICYT (Project No. AGF93-0332). The Scottish Agricultural College receives financial support from the Scottish Office Agriculture and Fisheries Department.

References

Carretero, J.L. (1981) El Genero *Echinochloa* en el Suroeste de Europa. *Anales Jardín Botánico Madrid* **38:** 91-108.

De Prado, R., Domínguez, C. and Tena, M. (1993) Triazine-resistance in biotyes of *Solanum nigrum* and four *Amaranthus* species found in Spain. *Weed Research* **33:** 17-24.

De Prado, R., Menendez, J. and Romera, E. (1995) Atrazine detoxification in *Panicum dichotomiflorum* and target site *Poligonum lapathiflolium. Pesticide Biochemistry Physiology* **51:** 1-11.

Devine, M., Duke, S.O. and Fedtke, C. (1993) Herbicide inhibition of photosynthetic electron transport. *In: Physiology of Herbicide Action*, 113-140. PTR Prentice Hall, Englewood Cliffs, New Jersey.

Gronwald, J.W. (1994) Resistance to photosystem II inhibiting herbicides. *In: Herbicide Resistance in Plants* (S.B. Powles and J.A.M. Holtum, eds), 27-60. Lewis Publisher, Boca Raton.

Grossmann, K. and Kwiatkowski, J. (1993) Selective induction of ethylene and cyanide biosynthesis appears to be involved in the selectivity of the herbicide quinclorac between rice and barnyardgrass. *Journal Plant Physiology* **142:** 457-466.

Menendez, J., Jorrin, J., Romera, E. and De Prado, R. (1994) Resistance to chlorotoluron of a Slender foxtail (*Alopecurus myosuroides*) biotype. *Weed Science* **42:** 340-344.

CHAPTER 23
THE POTENTIAL FOR THE EVOLUTION OF HERBICIDE RESISTANCE: SELECTION, CHARACTERISATION AND POLYGENIC INHERITANCE OF RESISTANCE TO CHLORSULFURON IN PERENNIAL RYEGRASS

R. MACKENZIE[1], A.M. MORTIMER[1], P.D. PUTWAIN[1], I. B. BRYAN[2] and T. R. HAWKES[2]

1.University of Liverpool, Department of Environmental and Evolutionary Biology, PO Box 147, Liverpool, L69 3BX. U.K. 2. Zeneca Agrochemicals, Jealott's Hill Research Station, Bracknell, Berkshire RG12 6EY. U.K.

Introduction

The proportionately high number of cases of acetolactate synthase (ALS) inhibitor herbicide resistance in weed populations to date and the relatively short span of commercial availability of these herbicides suggest that genetic variation for resistance may be present in many weed species (Hawkes, 1993). If resistance genes are present in a population, their mode of inheritance is an important determinant of the rate of development of herbicide resistance (Maxwell and Mortimer, 1994). In those cases where ALS-inhibitor herbicide resistance has been carefully characterised, the trait has often been due to plants possessing an insensitive form of ALS (Devine *et al.*, 1991; Saari *et al.*, 1992). In addition, the mode of inheritance has been shown to involve a nuclear encoded single gene expressing partial dominance (e.g. Mallory-Smith *et al.*, 1990).

This paper reports the potential for evolution of resistance in a population of *Lolium perenne* previously unexposed to the sulfonylurea herbicide, chlorsulfuron.

Materials and Methods

Plant material

Seed of an unselected population of *Lolium perenne* L. cv Devon Eaver was obtained commercially.

Soil Culture

Seedlings were raised in trays containing a 1:1 mixture of sand and horticultural compost in a glasshouse under 14 h daylength for 21 days. Chlorsulfuron was applied as a postemergence spray using a Mardrive track sprayer. Herbicide was applied through a hydraulic pressure nozzle (Tjet n° 3, using one spray pass at a pressure of 2.2 x 10^5 Pa at a velocity of 1 m s^{-1}) 0.75 m above soil surface to give a spray volume of 175 l ha^{-1}. Six doses of chlorsulfuron (Glean®) in the range of 45-1125 g a.i. ha^{-1} were applied to trays containing on average 83 plants, each dose being replicated four fold. Twenty-one days after spraying the numbers of survivors were counted and the fresh weight of survivors measured.

Hydroponic Culture

Seeds were sown on to a raft of plastic beads floating on nutrient solution in 0.25 l plastic pots. Seedlings were thinned to 20 plants per pot and plants raised in a controlled environment chamber at 22°C with a 14 h daylength and 100 $\mu Em^{-2}s^{-1}$ light intensity. Eighteen days after sowing, the original nutrient solution was replaced with fresh solution containing chlorsulfuron in the dose range 30-30 000 µg a.i. l^{-1}. Five replicate pots received one of eight doses and survivorship and fresh weight of survivors were scored after 19 days.

Mass Selection

Glasshouse benches with a lipped edge were lined with polythene and covered with black plastic beads. Tap water was then added to float the beads and give a depth of ca 20 mm of water. Cotton muslin was attached to supporting metal frames (1.95m x 1m) that allowed the muslin to rest on the surface of the beads when the frame was immersed. Prior to immersion, seed was sown on to the dampened muslin at a density of 1 x 105 m^{-2}. Three weeks from sowing when seedlings had at least two leaves, chlorsulfuron was introduced in solution and thoroughly mixed. Three separate trials were conducted, chlorsulfuron concentrations (based on the dose response trials) being chosen to achieve 90% seedling mortality (trial 1) or 99.99% mortality (trials 2 and 3). Water volumes on the benches were kept constant by daily addition of tap water for the following three weeks to maintain constant chlorsulfuron concentration. Survivors selected on the basis of the presence of green biomass were transferred to nutrient culture for two weeks after which they were potted into soil and clonally propagated.

Screening of Resistant Phenotypes

Response to herbicide was assessed either by dose response trial (in phenotypes coded Rl...R3) or by comparative evaluation of sprayed and unsprayed plants at a single dose (in phenotypes coded rl...r95). Uniform tillers from clonally propagated plants were tested in soil to compare the response of R, r and susceptible phenotypes.

In dose response, ten doses and control were used in a range spanning 0.02 to 20 fold of the average recommended field rate of chlorsulfuron. Plants were grown for 14 days in a heated glasshouse (14 h daylength) and then sprayed by track sprayer as described above. Three weeks later, shoots were excised at the base and fresh weights measured. Plants which were assessed at a single dose were raised under similar conditions and sprayed with 230 g a.i. ha^{-1} chlorsulfuron after six weeks growth. Biomass was measured after a further 21 days growth.

Inhibition of ALS Activity

ALS enzyme extract was prepared by homogenisation of young leaf material from selected clones, protein being precipitated on ice in ammonium sulphate and purified by column chromatography (Mackenzie, 1994). Extracts were stored in liquid nitrogen prior to assay. To establish enzyme inhibition profiles for each clone, ALS activity was assayed by the method of LaRossa and Schloss (1984). Enzyme activity was measured in three replicate assays for concentrations in the range from 0.3mM to 1.3mM chlorsulfuron.

Inheritance of Resistance

Pair crosses

All cross combinations of three resistant (R1...R3) and two susceptible (S1 and S2) phenotypes were made in a glasshouse by placing paired inflorescences in pollination bags until seed set. In addition solitary infloresences were bagged from each plant to estimate self pollination. Mature seed was collected and stored on a per plant basis so as to identify the maternal parent. Seedling progeny from each cross were raised in soil and at the two/three leaf stage were sprayed with chlorsulfuron at a rate of 27 g a.i. ha^{-1}. Biomass was measured three weeks after spraying by harvesting excised shoots.

Narrow sense heritability

Fifty-six resistant phenotypes randomly chosen from rl...r95 were allowed to intercross by randomly placing flowering plants in close proximity to one another in a closed glasshouse. In addition inflorescences of 15 randomly chosen phenotypes were bagged to estimate selfing. Selfed and half-sib progeny from each cross were then collected and seedling populations raised from each family. Seedling and parental material were clonally propagated to produce tiller populations in the glasshouse. Tillers were then taken and planted into compost in 3 l pots. A factorial design was followed comprising two replicate blocks of three chlorsulfuron treatments (0, 12 and 60 g a.i. ha^{1}) applied to tillers of 15 parents and 20 offspring per parent. Each replicate block contained 135 pots into which tillers were randomly planted, 10 tillers per pot. Plants were sprayed with chlorsulfuron three weeks after transplantation and after a further three weeks, fresh weight per plant was measured by destructive harvest. The replicate blocks were placed in separate glasshouse bays giving differing growth conditions; in block 1 artificial lighting was constrained to a 12 h photoperiod and the mean ambient temperature was 22°C whilst in block 2 the photoperiod was 18h and the ambient temperature was 24°C. Narrow sense heritability was estimated by regression of mean offspring performance on mean parent performance, the data variable being the proportional yield reduction of sprayed to unsprayed plants for the two spray doses.

Results

Sensitivity to Chlorsulfuron

Mortality and reduction in fresh weight in survivors was observed in both soil and hydroponic culture over the dose ranges employed. Noticeable responsiveness to dose was evident in hydroponic culture as indicated by probit regression and reduction in fresh weight in comparison to plants grown in soil. Comparative LD_{50} values in soil and hydroponic culture were 15.18±1.04 g a.i. ha^{-1} and 532.76±1.11 μg a.i. l^{-1} respectively; ED_{50} values being respectively 40.13±27.38 g a.i. ha^{-1} and 59.89±1.22 mg a.i. l^{-1}.

Mass Selection

Over the three trials, a total of 9.9 x 10^5 seedlings were screened, but only two trials produced putatively resistant plants (Table 1). In the seedling population (4.1 x 10^5) screened with an intended selection intensity of 90% mortality, 95 survivors were recovered, whilst at the intended higher selection intensity (99.99% mortality) only 3 individuals survived.

Characterisation of Resistant Phenotypes

The ED_{50}'s of plants Rl...R3, calculated on the basis of fresh weight gain after spray, were in the range 13 - 25 g a.i. ha^{-1}, all three phenotypes being killed at doses in excess of 60 g a.i. ha^{-1}. A characteristic dose response curve (for R1) is given in Figure 1a. The relative resistances of phenotypes r1...r95 (calculated on the ratio of response of sprayed to unsprayed plants) to chlorsulfuron were continuously distributed (Figure lb), greatest resistance being observed in phenotype r52 (Table 2). The specific activity of acetolactate synthase in extracts varied amongst the three resistant phenotypes tested (Table 2) but concentrations of chlorsulfuron required to inhibit ALS activity (I_{50}) were not significantly elevated over the dose needed for the same inhibition of a susceptible standard.

Inheritance of Resistance

Pair crosses

All selfed progeny of susceptible phenotypes S1 and S2 were killed with a dose of 27 g a.i. chlorsulfuron ha^{-1}. Continuous variation in yield responses was present in Fl progeny of paired crosses amongst R1, R2 and R3 phenotypes and in progeny from crosses of susceptible and resistant parents. Mean yields of F1 progeny of crosses amongst resistant phenotypes exceeded both those exhibited by families from crosses of resistant and susceptible parents and by families from selfed resistant phenotypes.

Narrow sense heritability

Out-crossing predominated in the breeding population, very low levels of selfing being detected. Heritabilities of response to chlorsulfuron ranged from 0.4 to 0.9 and depended upon environmental conditions. Heritabilities did not differ according to dose for plants in block 1 in contrast to those measured in block 2 (Table 3).

Discussion

Mass selection of *L. perenne* seedlings was readily achieved in hydroponic

Table 1. Selection of putative resistant plants of *L. perenne* to chlorsulfuron.

Intended selection intensity (% mortality)	Number of plants screened x 10^5	Frequency of putative resistant plants Mean (±sem) x 10^{-3}	Number of survivors and code
90.00	4.1	0.4 (0.2)	95 (coded r1...r95)
99.99	4.8	0.0	-
99.99	1.0	0.03	3 (coded R1...R3)

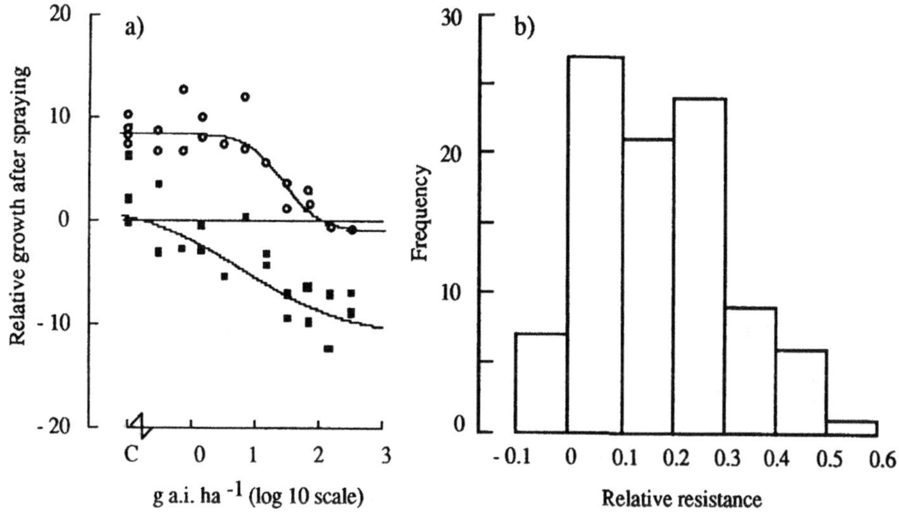

Figure 1. Characterisation of selected phenotypes. a) Response to chlorsulfuron in phenotype Rl (o) and a susceptible phenotype (n), C = unsprayed control; b) Distribution of relative resistance scores (see text) in phenotypes rl... r95.

culture and putative resistant phenotypes were isolated. The relative resistance of these phenotypes was however distributed across a response range varying from susceptible to highly resistant (Figure lb, Table 2). The inclusion of susceptible plants may have occurred as a result of uneven exposure to chlorsulfuron, choice being further favoured by prolonged retention of green biomass, a feature of chlorsulfuron phytotoxicity. Discounting these susceptible phenotypes, these data suggest that the frequency of phenotypes resistant to chlorsulfuron in the source population was greater than 1 x 10^{-6} under the selection regime imposed.

Resistance at the whole plant level (Table 2) was not achieved by altered sensitivity of the ALS enzyme even in the most resistant phenotype (r52). These *L. perenne* clones exhibited differing patterns of cross resistance to metsulfuron-methyl, isoproturon,

Table 2. Characterisation of selected phenotypes of *L. perenne*.

Phenotype	ED_{50}Chlorsulfuron (g a.i. ha $^{-1}$)	ID_{50}Chlorsulfuron (nM)	Specific activity (nmol acetoin min^{-1} mg protein^{-1})
R1	39.13*	7.55^{NS}	0.18
r26	255.50*	12.04^{NS}	0.25
r52	278.25*	6.04^{NS}	0.65
Susceptible	2.78	10.75	0.50

* = significant (P≤ 0.05), NS = non significant (P> 0.05) all comparisons versus susceptible.

tralkoxydim and imazamethabenz (Mackenzie, 1994). Since resistance to these herbicides may be mediated by the activity of mixed function oxidase enzymes (Sweetser *et al.*, 1982), herbicide degradation is postulated as a mechanism underlying the observed resistance in this study.

Pair crossing revealed that chlorsulfuron resistance was not recessive since the mean F1 performance of R x S crosses was intermediate between resistant and susceptible parental responses. Thus dominant chlorsulfuron resistance traits were evident in phenotypes R1, R2 and R3. It is unlikely that a single partially dominant allele was involved considering the variability in resistance observed within families from pair crosses (Mackenzie, 1994). Polygenic inheritance of resistance was inferred from these findings and the heritability values. The degree to which polygenic resistance is transmitted to offspring in a randomly mating population is determined by the additive genetic variance in a population (Falconer, 1981). Narrow sense heritability (range 0 to 1) measures the ratio of additive genetic variance to total phenotypic variance, but its estimation depends on the dose and environment chosen for study. Relatively high heritabilities were observed in three out of the four cases (Table 3) investigated, suggesting that the population of phenotypes (rl...r95) may strongly respond to selection, the speed being governed by the frequency and intensity of selection and the inheritance characteristics of the population (Mitchell-Olds and Rutledge, 1986).

Table 3. Narrow sense heritability of response to chlorsulfuron in *L. perenne*.

Environment	Chlorsulfuron (g a.i. ha-1)	Heritability
Block 1	12.0	0.588
	60.0	0.650
Block 2	12.0	0.366
	60.0	0.906

This study reports a successful deliberate attempt to select for herbicide resistance in a previously unexposed population of *L. perenne*. It is concluded that chlorsulfuron resistance was due likely to the additive effects of dominant minor genes, that herbicide degradation is a possible mechanism for resistance and that populations may exhibit transgressive segregation of further resistance to chlorsulfuron. Further work has since confirmed this.

Acknowledgements

The senior author gratefully acknowledges financial support from Zeneca Agrochemicals.

References

Devine, M.D., Marles, M.A.S. and Hall, L.M. (1991) Inhibition of acetolactate synthase in susceptible and resistant biotypes of *Stellaria media*. *Pesticide Science* **31**: 273-280.

Falconer, D.S. (1981) *Introduction to Quantitative Genetics*. Longman, London.

Hawkes, T.R. (1993) Acetolactate synthase: the perfect herbicide target? *Proceedings Brighton Crop Protection Conference-Weeds* **2**: 723-730.

LaRossa, R.A. and Schloss, J.V. (1984) The sulfonylurea herbicide sulfometuron methyl is an extremely potent inhibitor of acetolactate synthase in *Salmonella typhimurum*. *Journal Biological Chemistry* **259**: 8753-8756.

Mackenzie, R. (1994) The potential for evolution of herbicide resistance in weed species. *Ph.D. Thesis*. University of Liverpool.

Mallory-Smith, C.A., Thill, D.C., Dial, M.J. and Zemetra, R.S. (1990) Inheritance of sulfonylurea herbicide resistance in *Lactuca* spp. *Weed Technology* **4**: 787-790.

Maxwell, B.D. and Mortimer, A.M. (1994) Selection for herbicide resistance. *In: Herbicide Resistance in Plants: Biology and Biochemistry* (S.B. Powles and J.A.M. Holtum, eds.), 1-25. Lewis, Boca Raton.

Mitchell-Olds, T. and Rutledge, J.J. (1986) Quantitative genetics in natural plant populations: a review of the theory. *American Naturalist* **127**: 379-402.

Saari, L., Cotterman, J.C., Smith, W.F. and Primiani, M.M. (1992) Sulfonylurea herbicide resistance in common chickweed, perennial ryegrass, and Russian thistle. *Pesticide Biochemistry Physiology* **42**: 110-118.

Sweetser, P.B., Schow, G.S. and Hutchinson, J.M. (1982) Metabolism of chlorsulfuron by plants: biological basis for selectivity of a new herbicide for cereals. *Pesticide Biochemistry Physiology* **17**: 18-23.

PART IV

BIOTECHNOLOGICAL APPROACHES TO DEVELOP HERBICIDE RESISTANT CROPS. PROBLEMS AND POSSIBILITIES

CHAPTER 24
HERBICIDE RESISTANT CROPS GENERATED BY BIOTECHNOLOGY

G. DONN
Hoechst Schering AgrEvo GmbH, Biochemical Research, 65926 Frankfurt am Main, Germany

Introduction

Taking into account that it is a rare event to discover new herbicidal active compounds which give a sufficient broad spectrum of weed control, which are rapidly biodegradable and safe for non target organisms together with the fact that the costs to develop a new molecule for its introduction into agriculture as a safe herbicide are steadily increasing, it is a logical consequence to take efforts to broaden the use of already well characterised molecules that have been proven to possess the ecologically desirable properties as well as a good efficacy.

If crop selectivity is the only obstacle to use such a desirable compound in a broader scale, different methods to overcome the insufficient selectivity can be tested. Physical and chemical measures, for example the use of devices for directed spraying or the development of safeners are useful tools for some compounds and crops. Biological methods are based on the idea that an appropriate resistance gene as an inborn safener may fulfil the same task, in some cases perhaps even more perfect than the physical or chemical protectant.

In most cases where biologically active molecules were used in large scale over a longer time mankind had to recognise that evolution is a dynamic process that still goes on. A strong selection pressure together with the inherent, even small but detectable mutation frequency in all living organisms led within evolutionary extremely short periods of time to the appearance and dissemination of resistant biotypes in bacteria, fungi, insects and even in weeds. Whereas for weeds this process of micro-evolution needs several years or even decades, the selection of resistant crop plants can be accelerated using the methods of plant biotechnology, which emerged during the last twenty years. Using cells instead of whole plants mutant selection takes months and transferring appropriate herbicide resistance genes even further shortens the selection period to several weeks on the cellular level. Seventeen years ago it was demonstrated for the first time that plant cells can be used successfully for mutant selection leading to herbicide resistance (Chaleff and Parsons, 1978; Radin and

Carlson, 1978). Because plant cells are totipotent, from single cells whole fertile plants can be regenerated. At least in some cases from herbicide resistant mutant cells fertile mutant plants were obtained which transferred the mutant traits to their offsprings.

When methods for gene transfer into crop plants were developed during the last twelve years, genes conferring herbicide resistance were a very helpful tool to optimize the gene transfer technology (Rudelsheim, 1993). Plant cells expressing a herbicide resistance gene can be selected and identified very efficiently under culture conditions where a lethal dose of the corresponding herbicide suppresses the growth of unaltered cells completely. Here both methods to obtain herbicide resistant crops will be described and the advantages and drawbacks of both methods will be discussed.

In Vitro Selected Mutants

In principle all types of *in vitro* plant cultures are amenable for mutant selection. For practical reasons callus and cell suspensions from regenerable plant genotypes are the most appropriate source materials. Because millions of cells can be handled in a few Petri dishes or Erlenmeyer flasks, it is easy to handle sufficiently large cell populations to have the chance to find a mutant even without the use of mutagens. Therefore chemicals that are known to induce point mutations as EMS (Ethyl-Methyl-Sulfonic-Acid) are not a prerequisite. In Table 1 examples of selected herbicide tolerant plant cell mutants are listed.

It is obvious that a broad range of compounds and several crop species were used, leading at least to a pronounced level of resistance on the cellular level and in the cases listed also to regenerated plants.

The most successful and reliable mutant selections leading to resistant crop plants which inherit the mutant phenotype to their offsprings seem to be possible for classes of herbicides, for which also resistant weeds were discovered in fields on which these herbicides were used repeatedly over several years. Typical mutants that could be selected in vitro using sulfonylureas, imidazolinones, aryloxyphenoxy-proprionic acids and cyclohexanediones were in most cases mutants with an altered target enzyme. The acetolactate (acetohydroxy acid) synthase and acetyl-CoA carboxylase mutants were less sensitive to the corresponding herbicide used in the selection process and often showed cross resistance for compounds with the same mode of action (Marshall *et al.*, 1992; Parker *et al.*, 1990). The mutants showed no loss in their fitness. Therefore such mutants were already successfully introduced into breeding programmes and the first mutant varieties with agronomic useful levels of herbicide tolerance could be developed namely in maize. The first varieties with a mutated imidazolinone insensitive ALS were introduced into the market recently.

For other single target site herbicides like glyphosate or glufosinate for which up to now no resistant weeds were discovered in the field, the in vitro selection of mutants did not lead to resistant plants as well. Even though for both herbicides resistant cell

Table 1. Herbicide resistant mutants from *in vitro* selection.

Compound	Crop	Mode of Resistance	Regenerated Plants	Reference
Picloram	Tobacco	Unknown	yes	1,2
Amitrole	Tobacco	Unknown	yes	3
Bentazon	Tobacco	Unknown	yes	4
Paraquat	Tobacco	Unknown	yes	5
	Tomato	Unknown	yes	6
Atrazine	Soybean	Unknown	yes	7
Chlorsulfuron (and other SUs)	Tobacco	Insensitive ALS	yes	8
	Rapessed	Insensitive ALS	yes	9
	Rice	Insensitive ALS	yes	10
	Maize	Insensitive ALS	yes	
Imidazolinones	Maize	Insensitive ALS	yes	11, 12
Cyclohexanediones	Maize	Insensitive ACCase	yes	13, 14
Aryloxyphenoxypropionates	Maize	Insensitive ACCase	yes	13, 14

1. Chaleff and Parsons, 1978; 2. Chaleff, 1980; 3. Singer and McDaniel, 1984; 4. Radin and Carlson, 1978; 5. Hughes *et al.*, 1984; 6. Thomas and Pratt, 1982; 7. Wohlleben *et al.*, 1988; 8. Creason and Chaleff, 1988; 9. Swanson *et al.*, 1988; 10. Terekawa and Wakasa, 1992; 11. Anderson and Georgeson, 1989; 12. Newhouse *et al.*, 1991; 13. Marshall *et al.*, 1992; 14. Parker *et al.*, 1990.

variants could be isolated (Donn *et al.*, 1984; Steinrücken *et al.*, 1986), these variants failed to regenerate into plants. Even the protoplast fusion with regenerable genotypes could not overcome this failure (Deak *et al.*, 1988). In both cases the cell variants overproduced the target enzyme up to 40 fold, but the target enzymes were as sensitive as the wildtype enzymes (Donn *et al.*, 1984; Steinrücken *et al.*, 1986). In both cases the plant genes of the target enzymes EPSP-synthase as well as glutamine synthase could be expressed in *E.coli.* The genes that were altered by site directed mutagenesis coded for mutated enzymes that showed a reduced binding affinity for the respective inhibitor as well as the natural substrate.

Because both herbicides are transition state analogues which bind at the active site of the target enzymes is explainable that the alteration of the active sites of the two enzymes inevitably cause considerable loss of binding affinity to the natural substrate as well, if a mutated enzyme shows a reduced binding affinity of the analogue. If a mutation towards an insensitive enzyme occurs, it is most likely lethal or at least reduces the fitness of the mutant considerably. This may explain why neither in weed populations nor in cell populations even after a repeated exposure to the two herbicides structural mutants could be detected up to now.

Transfer of Herbicide Resistance Genes into Crops

For most crop plants nowadays methods are developed to introduce isolated genes. For dicots namely *Agrobacterium tumefaciens* is routinely used as a vector. For cereals direct gene transfer methods were developed. The most widespread used method is the biolistic approach (Spencer *et al.*, 1990; Vasil 1994) If a regenerable protoplast system is available, protoplasts are a convenient and reliable method (Donn *et al.*, 1990).

As Table 2 shows, herbicide resistance genes were isolated for several herbicides of different modes of action. Except for the mutant ALS genes from Arabidopsis and tobacco the most effective resistant genes for the other herbicides were isolated from bacteria.

Except for the glyphosate tolerance genes from different sources including an *Agrobacterium tumefaciens* glyphosate insensitive EPSP synthase gene, all these genes code for enzymes that detoxify the different herbicides in a sufficient rate. If these genes were transferred into crop plants using standard methods like *Agrobacterium* mediated transformation or direct gene transfer methods, regenerants that were selected in presence of toxic levels of the herbicides in the culture media were resistant to rates of the active ingredients sufficient for weed control.

For glufosinate, the transfer and overexpression of a plant glutamine synthetase gene (Eckes *et al.*, 1989) did not lead to transgenic tobacco plants, which showed an agronomic sufficient level of resistance, even though the overexpressed alfalfa glutamine synthetase is besides the rubisco the most prominent protein in these transgenic plants.

Not all of the resistance genes listed in Table 2 are currently used for the development of commercial herbicide resistant crop varieties. Only if the properties of the herbicide namely of its ecotoxicological profile as well as its potential for broad spectrum and cost efficient weed control justify the investment into the further development as a selective herbicide, further steps towards commercialisation eventually pay off. Glyphosate tolerant transgenic soybeans, bromoxynil tolerant cotton and glufosinate tolerant canola are the most advanced examples for this development. These three crops got the clearance for their commercial release and they are ready for market introduction in the near future.

A prerequisite for comercialisation is the registration of the herbicide for its selective use as well as a detailed characterisation of the inserted genes of the genetically modified plants and of the new gene products. Before the commercialisation phase the transgenic plants are tested in the field. In a risk assessment dossier the safety of the gene and its gene product, the properties of the modified plants, the expression stability of the introduced gene and the safety of the modified plant must be considered. Potential interactions of the modified plants with wild relatives and the even remote probability of undesired interaction with other organisms must be taken into account and must be evaluated. If a herbicide detoxification gene from a microbial source is introduced into crops, the substrate specificity of the new enzyme must be evaluated as well as the toxicological properties of the herbicide metabolite(s).

Glufosinate is unique because the active ingredient (L-phosphinothricin) as well as the detoxifying enzyme are the result of an evolutionary metabolic fine tuning. L-phosphinothricin is synthesised as a tripeptide by two *Streptomyces* strains in a way that the precursor molecule desmethyl-phosphinothricin is acetylated at its amino group. The resulting N-acetyl-desmethyl-phosphinothricin then is converted and linked with two alanines and methylated. The acetylating enzymes from *Streptomyces hygroscopicus* and *Streptomyces viridochromogenes* were analysed in detail. They show a high degree of sequence homology. Only 28 of 184 amino acids are different and both enzymes show a very high substrate specificity. The substrates are desmethyl-L-phosphinothricin, phosphinothricin and acetyl CoA. Even though phosphinothricin is an analogue of glutamic acid, neither the latter amino acid is acetylated nor are the other proteinogenic amino acids. This high substrate specificity is the result of an evolutionary process that may have taken millions of years. It is the reason for the perfect selectivity observed in transgenic plants expressing either the *S. hygroscopicus* or the *S. viridochromogenes* variant of the phosphinothricin-acetyl-transferase (PAT). Because already traces of the PAT enzyme in a transgenic cell are sufficient for inactivation of the herbicidal active L-phosphinothricin, the corresponding genes from *Streptomyces hygroscopicus,* first described as Bialaphos-resistance (BAR) (DeBlock *et al.*, 1987) gene and the related gene from *S. viridochromogenes*, first described as PAT (Wrather and Freytag, 1991) gene, are powerful selectable marker genes. They enabled researchers to develop gene transfer methods for cereals where the widely used NPTII gene conferring kanamycin resistance was not efficient enough (Nehra *et al.*, 1994; Spencer *et al.*, 1990; Vasil 1994).

Table 2. Transfer of isolated herbicide resistance genes into plants.

Herbicide	Source of Resistance Gene	Mode of Resistance	Transfer in Plants	Evaluation in Breeding Programmes	References
Glyphosate	*Salmonella typhymurium*	Altered EPSPS overexpression	Tobacco and others	yes	1
	Escherichia coli	Overexpression			2
	Petunia hybrida	Overexpression	Tobacco and others	yes	2, 3, 4, 5
	Arabidopsis thaliana	Overexpressed natural and mutated gene	Rapeseed	yes	
		Overexpression of EPSPS			6,7
	Agrobacterium tumefaciens		Soybean	yes	6,7
			Rapeseed		6,7
			Rapeseed		6,7
Glufosinate	*Streptomyces hygroscopicus*	Expression of PAT	Tobacco		10
			Tomato		11
			Potato		11
			Rapeseed	yes	11
			Sugarbeet	yes	12
			Maize	yes	13
			Soybean	yes	
			Wheat		14, 15
	Streptomyces viridochromogenes	Expression of PAT	Tobacco		16
			Tomato		
			Potato		
			Rapeseed	yes	17
			Sugarbeet	yes	
			Maize	yes	17, 18
			Soybean	yes	
			Wheat		

Table 2. . Transfer of isolated herbicide resistance genes into plants (Cont.).

Chlorsulfuron	*Nicotiana tabacum* *Arabidopsis thaliana*	ALS mutated gene ALS mutated gene	Tobacco Rice Rapeseed		 8 9
Bromoxynil	*Klebsiella ozenae*	Nitrilase	Cotton Clover Rapeseed	yes	19, 20
Norflurazon	*Erwinia uredovora*	Enhanced carotenoid biosynthesis	Tobacco		21
Dalapon	*Pseudomonas putida*	Dehalogenasa	*N. plumbaginifolia*		22
2,4-D	*Alcaligenes eutrophus*	Monooxygenase	Tobacco Cotton		23 24
Phenmedipham	*Arthrobacter oxidans*	Carbamate hydroxylase	Tobacco		25
Cyanamide	*Myrothecium verrucosa*	Cyanamide hydratase	Tobacco		

1. Stalker *et al.*, 1985; 2. Gasser and Fraley, 1989; 3. Kishore *et al.*, 1992; 4. Klee *et al.*, 1987; 5. Padgette *et al.*, 1991; 6. Barry *et al.*, 1992; 7. Wrather and Freytag, 1991; 8. Li *et al.*, 1992; 9. Miki *et al.*, 1991; 10. De Block *et al.*, 1987; 11. Rudelsheim, 1993; 12. D'Halluin *et al.*, 1992; 13. Spencer *et al.*, 1990; 14. Nehra *et al.*, 1994; 15. Vasil *et al.*, 1993; 16. White *et al.*, 1995; 17. Donn and Eckes, 1992; 18. Donn *et al.*, 1990; 19. Stalker *et al.*, 1988; 20. Ward *et al.*, 1993; 21. Misawa *et al.*, 1992; 22. Buchanan-Wollaston *et al.*, 1992; 23. Streber and Willmitzer, 1989; 24. Bayley *et al.*, 1992; 25. Streber *et al.*, 1994.

Precautions

Even though the PAT and BAR gene as well as different EPSP synthetase as ALS genes are already introduced in a range of crop species it is not advisable to develop for all crops resistant varieties bearing the same resistance gene. This also holds true for the other resistance genes as well as for herbicide resistant crop mutants. For crops as oats which freely interbreed with their wild relatives the release of a herbicide resistant crop inevitably would lead to the introgression of the resistance gene into the weed species (Gressel, 1992). For rapeseed the risk of outcrossing is by far less pronounced, nevertheless careful monitoring of related *Brassica* species, namely *Brassica campestris*, is advisable. *Brassica napus* is a anphidiploid hybrid of *B. campestris* and *B. oleracea*. The crosses of *B. napus* with *B. campestris* are partially fertile and the herbicide resistance gene could therefore introgress in the *B. campestris* gene pool (Scheffer and Dale, 1994). If a transformant is chosen, in which the inserted DNA is not located in the *campestris* genome this probability can be excluded.

Besides the efficacy of weed control of the complementary herbicide the degree of selectivity in the modified crop will determine the acceptance of the technology by the farmer. Only if the modified crop does not show even subtle signs of herbicide phytotoxicity leading to delayed flowering, reduced yield or vulnerability to diseases, herbicide tolerant crops are acceptable for them. If transgenic crops are considered it is necessary that such crops express the transgene even under adverse environmental conditions and careful testing of the varieties must exclude even a remote chance of gene silencing. Silencing of the transgene could lead to crop varieties with a partial or complete loss of the transgenic phenotype like herbicide tolerance. A careful selection of the transformants chosen for breeding programmes and strict quality control of the produced seeds are prerequisites to avoid surprises.

The potential risk of spreading the herbicide resistance into related wild species depends mainly on the crop species. For a crop like maize this risk is zero. The by far bigger probability is the spontaneous occurrence and spread of herbicide resistant weed mutants for herbicides for which it is known that spontaneous mutants can be selected by the repeated use of active ingredients with the same mode of action and mode of resistance. If such herbicide tolerant crops regardless if they carry a natural mutation or a foreign resistance gene in their genomes are grown in large areas especially in rotation with another crop in which the same class of herbicides is already routinely used, special care is highly advisable. Herbicide rotation or herbicide mixtures together with the combined efforts of integrated weed management will help to reduce the probability of a rapid adaptation of the weed flora. Transgenic crops conferring resistance to a herbicide with a unique mode of action and which express a unique mode of resistance which prevents any cross resistance to other herbicides will be one additional building block in the future for the integrated weed management approach (Duke, this volume). Substances like glyphosate and glufosinate will help to reduce the risk of an undesirable spread of resistant weeds.

Conclusions

Herbicide resistant crops are not miraculous new solutions for old problems nor the evil of modern biotechnology (Duke *et al* 1991; Dyer *et al.*, 1993). It depends on the careful consideration of the herbicides their properties, the mode of resistance, the crops and their genetic relations to wild species as well as their role in crop rotation, if herbicide resistant crops will help to stabilize integrated approaches of weed management or if they will make themself obsolete after several years of unreflected and unbalanced overuse.

References

Anderson, A.D. and Georgeson, M. (1989) Herbicide tolerant mutants of corn. *Genome* **31**: 994-999.

Barry, G., Kishore, G., Padgette, S., Taylor, M., Kolacz, K., Weldon, M., Re, D., Eichholtz, D.A., Fincher, K. and Hallas, L. (1992). *In: Biosynthesis and Molecular Recognition of Amino Acids in Plants* (B.K. Singh, H.E. Flores and J.C. Shannon, eds.), 139-145. American Society of Plant Physiology.

Bayley, C., Trolinder, N., Ray, C., Morgan, M., Quisenberry, L.E. and Ow, D.W. (1992) Engineering 2,4-D resistance into cotton. *Theoretical Applied Genetics* **83**: 645-649.

Buchanan-Wollaston, V., Snape, A. and Cannon, F. (1992) A plant selectable marker gene based on the detoxification of the herbicide dalapon. *Plant Cell Reports* **11**: 627-631.

Chaleff, R.S. and Parsons, M.F. (1978) Direct selection *in vitro* for herbicide resistant mutants of *Nicotiana tabacum. Proceedings National Academy Sciences USA* **75**: 5104-5107.

Chaleff, R.S. (1980) Further characterization of picloram tolerant mutants of *Nicotiana tabacum*. *Theoretical Applied Genetics* **58**: 71 -95.

Creason, G.L. and Chaleff. R.S. (1988) A second mutation enhances resistance of a tobacco mutant to sulfonylurea herbicides. *Theoretical Applied Genetics* **76**: 177-182.

Deak, M., Donn, G., Feher, A. and Dudits, D. (1988) Dominant expression of a gene amplification related herbicide resistance in *Medicago* cell hybrids. *Plant Cell Reports* **7**: 158-161.

De Block, M., Botterman, J., Vandewiele, M., Dockx, J., Thoen, C., Gossele, V., Rao Movva, N., Thompson, C., Van Montagu, M. and Leemans, J. (1987) Engineering herbicide resistance in plants by expression of a detoxifying enzyme. *EMBO Journal* **6**: 2513-2518.

D'Halluin, K., Bossut, M., Bonne, E., Mazur. B., Leemans, J. and Botterman, J. (1992) Transformation of sugar beet (*Beta vulgaris* L.) and evaluation of herbicide resistance in transgenic plants. *Biotechnology* **10**: 309-314.

Donn, G. and Eckes, P. (1992) Basta-vertragliche Kulturpflanzen durch übertragung eines synthetischen phosphinothricin-acetyl-transferase Gens. *Z. PflKrankh. PflSchutz. Sonderheft* **XIII**: 499-504.

Donn, G., Nilges, M. and Morocz, S. (1990) Stable transformation of maize with a chimeric, modified phosphinothricin-acetyl-transferase gene from *Streptomyces viridochromogenes*. *Abstracts VIIth International Congress on Plant Tissue and Cell Culture*, 53.

Donn, G., Tischer, E., Smith, J.A. and Goodman, H.M. (1984) Herbicide resistant alfalfa cells: an example of gene amplification in plants. *Journal Molecular Applied Genetics* **2**: 621-635.

Duke, S.O., Holt. J.S., Hess, F.D. and Christy, A.L. (1991) Herbicide resistant crops. *Comments from Council for Agriculture Science and Technology (CAST)*, 1991-1, 24 pp.

Dyer, W.E., Hess, F.D., Holt, J.S. and Duke, S.O. (1993) Potential benefits and risks of herbicide resistant crops produced by biotechnology. *Horticultural Review* **15**: 367-408.

Eckes, P., Schmitt, P., Daub, W. and Wengenmayer, F. (1989) Overproduction of alfalfa glutamine synthetase in transgenic tobacco plants. *Molecular General Genetics* **217**: 263-268.

Gasser, C.S. and Fraley, R.T. (1989) Genetic engineering plants for crop improvement. *Science* **244**: 1293-1299.

Gressel. J. (1992) Indiscriminate use of selectable markers-sowing wild oats. *Trends Biotechnology* **10**: 382.

Hughes, K.W., Negrotto, D., Daub, M.E. and Meeusen, R.L. (1984) Free radical stress response in paraquat sensitive and resistant tobacco plants. *Environmental Experimental Botany* **24**: 151-157.

Kishore, G.M., Padgette, S.R. and Fraley, R.T. (1992) History of herbicide tolerant crops, method of development and current state of the art. Emphasis on glyphosate tolerance. *Weed Technology* **6**: 626-634.

Klee, H.J., Muskopf, Y.M. and Gasser, C.S. (1987) Cloning of an *Arabidopsis thaliana* gene encoding 5-enolpyruvylshikimate-3-phosphate synthase sequence analysis and manipulation to obtain glyphosate tolerant plants. *Molecular General Genetics*. **210**: 437-442.

Li, Z., Hayashimoto, A. and Murai, U. (1992) A sulphonylurea herbicide resistance gene from *Arabidopsis thaliana* as a new selectable marker for production of fertile transgenic rice plants. *Plant Physiology* **100**: 662-668.

Marshall, L.C., Somers, D.A, Dotray, P.D., Gengenbach, B.G., Wyse, D.L. and Gronwald, J.W. (1992) Allelic mutation in acetyl-coenzyme A carboxylase confer herbicide tolerance in maize. *Theoretical Applied Genetics* **83**: 435-452.

Miki, B.L., Labbe, H., Hattori I., Ouellet, T., Gabard, J. Sunohara, G.. Charest, P.J. and Iyer, V.N. (1991) Transformation of *Brassica napus* canola cultivars with *Arabidopsis thaliana* acetohydroxyacid synthase genes and analysis of herbicide resistance. *Theoretical Applied Genetics* **80**: 449-459.

Misawa, N., Yamano, S., Linden, H., de Felipe, M.R., Lucas, M., Ikenga, H. and Sandmann, G. (1992) Functional expression of the *Erwinia uredovora* carotenoid biosynthesis gene crt_I in transgenic plants showing an increase of β-carotene biosynthesis activity and resistance to the bleaching herbicide norflurazon. *Plant Journal* **4**: 833-840.

Nehra, N., Chibbar, R.N., Leung, N., Caswell, K., Mallard, C., Steinhaver, L., Baga, M. and Kartha, K. (1994) Self-fertile transgenic wheat plants regenerated from isolated scutellar tissues following microprojectile bombardment with two distinct gene constructs. *Plant Journal* **5**: 285-297.

Newhouse, K.M., Singh, B., Shaner, D. and Stidham, M. (1991) Mutations in corn (*Zea mays* L.) conferring resistance to imidazolinone herbicide. *Theoretical Applied Genetics* **83**: 65-70.

Padgette, S.R., Re, D.B., Gasser, C.S., Eichholtz, D.A, Frazier, R.B., Hironaka, C.M., Levine, E.B., Shah, D.M., Fraley, R.T. and Kishore, G.M. (1991) Site directed mutagenesis of a conserved region of the 5-enol pyruvyl shikimate-3-phosphate synthase active site. *Journal Biological Chemistry* **266**: 22364-22369.

Parker, W.B., Marshall, L.C., Burton, J.D., Somers, D.A, Wyse, D.L., Gronwald, J.W. and Gengenbach, B.G. (1990) Dominant mutations causing alterations in acetyl-coenzyme-A-carboxylase confer tolerance to cyclohexanedione and aryloxyphennoxypropionate herbicides in maize. *Proceedings National Academy Sciences USA* **87**: 7175-7179.

Radin, D.N. and Carlson, P.S. (1978) Herbicide tolerant tobacco mutants selected *in situ* and recovered via regeneration from cell culture. *Genetical Research* **32**: 85-89.

Rudelsheim, P. (1993) Engineering crops for tolerance to specific herbicides: A valid alternative. *Brighton Crop Protection Conference Weeds*, 265-272.

Scheffer, J.A. and Dale, P.J. (1994) Opportunities for gene transfer from transgenic oilseed rape (*Brassica napus*) to related species. *Transgenic Research* **2**: 263-278.

Singer, S.S. and McDaniel, C.N. (1984) Selection of amitrole tolerant tobacco calli and the expression of this tolerance in regenerated plants and progeny. *Theoretical Applied Genetics* **67**: 427-432.

Spencer, T.M., Gordon-Kamm, W.J., Daines, R.J., Start, W.G. and Lemaux, P.G. (1990) Bialaphos selection of stable transformants from maize cell culture. *Theoretical Applied Genetics* **79**: 625-631.

Stalker, D.M., Hiatt, W.R. and Comai, L. (1985) A single amino acid substitution in the enzyme 5-enolpyruvylshikimate-3-phosphate synthase confers resistance to the herbicide glyphosate. *Journal Biological Chemistry* **260**: 4724-4728.

Stalker, D.M., McBride, K.E. and Malyi, L.D. (1988) Herbicide resistance in transgenic plants expressing a bacterial detoxification gene. *Science* **242**: 419-423.

Steinrücken, M.C., Schulz, A., Amrhein, N., Porter, C.A and Fraley, R.T. (1986) Overproduction of 5-enol-pyruvylshikimate-3-phosphate-synthase in a glyphosate tolerant *Petunia hybrida* cell line. *Archives Biochemistry Biophysics* **244**: 169-174.

Streber, W.R. and Willmitzer, L. (1989) Transgenic tobacco plants expressing a bacterial detoxifying enzyme are resistant to 2,4-D. *Biotechnology* **7**: 811-815.

Streber, W.R., Kutschka, U., Thomas, F. and Pohlenz, H.D. (1994) Expression of a bacterial gene in transgenic plants confers resistance to the herbicide phenmedipham. *Plant Molecular Biology* **25**: 977-987.

Swanson, E.B., Caumans, M.P., Brown, G.L., Patel, J.D. and Beversdorf, W.D. (1988) The characterization of herbicide tolerant plants in *Brassica napus* L. after *in vitro* selection of microspores and protoplasts. *Plant Cell Reports* **7**: 83-87.

Terakawa, T. and Wakasa, K. (1992) Rice mutant to the herbicide bensulfuronmethyl (BSA) by *in vitro* selection. *Japanese Journal Breeding* **42**: 267-275.

Thomas, B.R. and Pratt, D. (1982) Isolation of paraquat tolerant mutants from tomato cell cultures. *Theoretical Applied Genetics* **63**:169-176.

Vasil, I.K. (1994) Molecular improvement of cereals. *Plant Molecular Biology* **25**: 925-937.

Vasil, V., Srivastava, V., Castillo, A.M., Fromm, M.E. and Vasil, I.K. (1993) Rapid production of transgenic wheat plants by direct bombardment of cultured immature embryos. *Biotechnology* **11**: 1553-1558.

Ward, R., Baldwin, G., Stanton, J., Panter, D. and Kiser, J. (1993) Yield potential of Bromotol™ cotton strains containing Buctril® resistance. *Proceedings Beltville Cotton Production Research Conference* **617**.

White, R., Padgette, S., Delannay, X., Bradshaw, L., Wells, B. and Kishore, G. (1995) Development of glyphosate tolerant crops and perspectives on the potential for weed resistance to Herbicides.

Wohlleben, W., Arnold, W., Broer, I., Hillemann, D., Strauch, E. and Pühler. A. (1988) Nucleotide sequence of the phosphinothricin N-acetyl transferase gene from *Streptomyces viridochromogenes* and its expression in *Nicotiana tabacum*. *Gene* **70**: 27-37.

Wrather, J.A. and Freytag, A.H. (1991). Selection of atrazine tolerant soybean calli and expression of that tolerance in regenerated plants. *Plant Cell Reports* **10**: 44-47.

CHAPTER 25
MANIPULATION OF CROP TOLERANCE TO HERBICIDES WITH SAFENERS

K.K. HATZIOS
Department of Plant Pathology, Physiology and Weed Science, Virginia Polytechnic Institute and State University, Blacksburg, Virginia 24061-0330, USA

Introduction

Born out of keen observations of antagonistic herbicide interactions, the concept of improving crop tolerance to herbicides with safeners has evolved into a useful agricultural practice (Hatzios and Hoagland, 1989). Safeners (formerly known as antidotes) are chemical compounds that protect crop plants from herbicide injury without reducing weed control (Hatzios and Hoagland, 1989; Kreuz, 1993). Successful safeners are desirable because they allow us to: a) selectively control weeds in botanically related crops; b) improve the margin of crop selectivity of newly developed herbicides; and c) expand the uses of older herbicides (Hatzios and Hoagland, 1989).

Commercialized safeners are synthetic organic chemicals that resemble structurally the herbicides that they antagonize on crop plants. It should be noted, however, that rhizosphere-competent bacteria (e.g. *Pseudomonas* spp.) capable of degrading soil-applied herbicides are currently exploited as potential microbial safeners (Krueger *et al.*, 1991).

The possible biochemical mechanisms of action of herbicide safeners have been investigated extensively in the last two decades. Most of this research has been reviewed previously (Farago *et al.*, 1994; Hatzios and Hoagland, 1989; Hatzios, 1991; Kreuz 1993). The purpose of the present review is to summarize briefly current uses and recent developments in the biochemical aspects of the protective action of herbicides safeners.

Uses of Herbicide Safeners

Chemical safeners exhibit a high degree of botanical and chemical specificity protecting mainly large-seeded grass crops against injury caused by specific classes of herbicides. In the past, most of the developed safeners were used for the protection of maize (*Zea mays* L.), grain sorghum [*Sorghum bicolor* (L.) Moench] and wet-sown rice (*Oryza sativa* L.) against pre-plant incorporated or pre-emergence applications of carbamothioate and chloroacetanilide herbicides (Hatzios and Hoagland, 1989). More recently, safeners have been developed to protect winter cereal crops against post-emergence applications of the aryloxyphenoxypropionate herbicides (Kreuz, 1993). The safening of maize and rice against sulfonylurea, imidazolinone, cyclohexanedione, and isoxazolidinone herbicides is well documented (Hatzios and Hoagland, 1989; Kreuz, 1993).

Marketed safeners are members of diverse chemical groups including naphthopyranones (e.g., naphthalic anhybride or NA); dichloroacetamides (e.g., dichlormid, benoxacor, BAS-145138, R-29148, furilazole); dichloromethyldioxolans (e.g., MG-191); oxime ether derivatives (e.g., oxabetrinil, fluxofenim); substituted thiazoles (e.g., flurazole); phenylpyrimidines (e.g., fenclorim); triazole carboxylates (e.g., fenchlorazole-ethyl); and quinolinoxyacetates (e.g., chloquintocet-mexyl).

In most of the successful crop-herbicide-safener combinations, the selective improvement of crop tolerance to a given herbicide is achieved easily with safeners applied as prepackaged mixtures with the herbicide (Hatzios and Hoagland, 1989). A selective placement of the safener, however, is sometimes desirable for achieving the best results under field conditions. Thus, all safeners used with grain sorghum and the safener NA in maize or other crops are applied mainly as seed treatments (seed safeners) to avoid the safening of weeds that are botanically related to the crop (Hatzios and Hoagland, 1989; Kreuz, 1993).

Mechanisms of Action of Herbicide Safeners

An analysis of the currently available literature supports the view that herbicide safeners may act either as "antagonists" of herbicidal effects at a common target site or as "bioregulators" influencing the amount of a herbicide that reaches its target site in an active form (Hatzios, 1991).

Interactions of Herbicides and Safeners at Target Sites

The chemically diverse groups of herbicides that are antidoted by safeners on grass crops exert their action by a variety of biochemical mechanisms, which are well-defined for sulfonylureas, imidazolinones, isoxazolidinones, aryloxyphenoxypropionates and cyclohexanediones, but still speculative for carbamothioates and chloroacetanilides (Devine *et al.*, 1993).

Safeners could conceivably compete with respective herbicides at a common target site. However, the contribution of such a mechanism in the protective action of the currently available safeners is very unlikely. Safeners that protect grass crops from sulfonylureas and imidazolinones or from aryloxyphenoxypropionates and cyclohexanediones do not counteract the inhibitory effects of these herbicides on target enzymes such as acetolactate synthase (ALS, EC 4.1.3.8) and acetyl-CoA carboxylase (ACCase, EC 6.4.1.2), respectively (Hatzios and Hoagland, 1989; Kreuz, 1993). Recently, Boger and Miller (1994) showed that while safeners counteracted the accumulation of protoporphyrin IX induced by peroxidizing herbicides in maize plants, they did not alleviate the inhibition of protoporphyrinogen oxidase by these herbicides.

The antagonism of carbamothioate and chloroacetanilide effects on selected physiological processes by specific safeners has been reported (Hatzios and Hoagland, 1989; Hatzios, 1991). The biological basis for these interactions has been attributed partially to the existing structural similarity between some safeners and their respective herbicides. Studies, employing the use of computer assisted molecular modelling (CAMM) have shown that the most successful chemical safeners are structurally similar to their respective herbicides (Yenne and Hatzios, 1990). The safener dichlormid, for example, is structurally very similar to EPTC and other carbamothioate herbicides (Yenne and Hatzios, 1990). Allidochlor or CDAA, a chloroacetamide derivative whose chemistry is almost identical to that of the safener dichlormid, acts both as a safener and as a herbicide, depending on the concentration used (Hatzios and Hoagland, 1989; Yenne and Hatzios, 1990). However, chemical safeners do not protect grass crops against the herbicide dalapon, whose injury symptoms on grasses resemble those caused by the carbamothioate and chloroacetanilide herbicides (Hatzios and Hoagland, 1989). In addition, the safener-to-herbicide doses in prepackaged formulated mixtures of these herbicides and safeners ranges from 1:6 to 1:30. Such ratios do not favor the "antagonist" theory of safener action since very little safener will be available at the target site to compete effectively with its respective herbicide, which is applied at considerably higher concentrations.

Regulation of Herbicide Effects by Safeners

Safeners may reduce the amount of a herbicide reaching its site of action by either reducing the rate of its uptake and/or translocation or by enhancing the rate of its metabolic detoxification.

I) Safener Effects on Herbicide Uptake and Translocation

A search of current literature suggests that safener-induced effects on herbicide uptake/translocation are variable and cannot explain fully the protective action of herbicide safeners (Hatzios, 1991). A recent report by Omokawa *et al.* (1995) showed that the protective action of the rice safeners daimuron and its optically active S-enantiomer against injury from bensulfuron-methyl results from a significant reduction in the uptake of this herbicide by safened rice seedlings.

II) Enhancement of Herbicide Metabolism by Safeners

The inherent ability of plants to detoxify herbicides by certain metabolic reactions is not evenly distributed among various plant species and it is responsible for the observed selectivity of herbicides (Devine *et al.*, 1993; Sandermann, 1992). In general, the grass crops protected by safeners are moderately tolerant to the antagonized herbicides and safeners are most effective when applied prior to or simultaneously with their respective herbicides.

A safener-induced enhancement of herbicide detoxification in protected plants is currently accepted as the most likely mechanism explaining the protective action of all commercialized safeners. Safeners have been shown to enhance the conjugation of acetanilide and sulfoxidized carbamothioate herbicides with glutathione either by elevating the levels of reduced glutathione (GSH) or by inducing the activity of glutathione transferase (GST) enzymes. A safener-induced enhancement of the activity of other detoxifying enzymes such as cytochrome P450-dependent monooxygenases (Cyt P450s), UDP-glucosyl-transferases (GTs) and carboxylesterases appears to be involved in the protection of grass crops against sulfonylureas, imidazolinones, and aryloxyphenoxypropionates.

i) Effects of Safeners on Glutathione Levels and Biosynthesis

Glutathione, found primarily in its reduced form (GSH), is a very important non-protein thiol participating in key metabolic processes of plants such as protein synthesis, adaptation to environmental stresses (e.g., peroxidative damage, drought, etc.) and detoxification of selected herbicides (Farago *et al.*, 1994). Safeners may elevate GSH levels in protected plants either directly or indirectly.

The biosynthesis of free cysteine and GSH from inorganic sulfate in roots of maize seedlings was enhanced by dichlormid and benoxacor, which caused an increase in the extractable activity of ATP-sulfotransferase (EC 2.7.7.4) or of adenosine-5'-phosphosulfate sulfotransferase (APSSTase), two key enzymes in the assimilatory sulfate reduction in plants (Farago *et al.*, 1994).

A direct activation of maize GSH synthetase II (EC 6.3.2.3) by dichlormid and of glutamyl cysteine synthetase or GSH synthetase I (EC 6.3.2.2) by the safener flurazole in maize and grain sorghum has been reported (Breaux *et al.*, 1989; Farago *et al.*, 1994). Flurazole itself conjugates with GSH in maize and grain sorghum (Breaux *et al.*, 1989) and it is likely that the GS-conjugate of this safener may bind to GSH synthetase I and override the feedback inhibition of GSH synthesis by the end product of the pathway, GSH. Such a mechanism has been well documented in regulatory studies of GSH synthesis in mammalian systems (Pickett and Lu, 1989). Several other safeners such as dichlormid, benoxacor and fenclorim are also known to conjugate with GSH and they may increase GSH levels in protected plants by a similar mechanism (Hatzios, 1991).

Safeners may also elevate GSH levels in protected plants indirectly by inducing the activity of glutathione reductase (GR, EC 1.6.4.2), which mediates the reduction of oxidized glutathione (GSSG) to GSH. An induction of GR by safeners favors a high GSH/GSSG ratio in plant cells compensating for GSH used in the formation of the GS-conjugates of chloroacetanilide and sulfoxidized carbamothioates. The safeners dichlormid, MG-191, fenclorim and fluxofenim have been shown to stimulate GR activity in maize, rice and grain sorghum tissues (Hatzios and Hoagland, 1989; Hatzios, 1991).

ii) Induction of Glutathione Transferases by Safeners

The conjugation of the thiol group of GSH to a variety of electrophilic substrates is catalyzed by a superfamily of ubiquitous enzymes known as glutathione S-transferases (GSTS, EC 2.5.1.18) (Pickett and Lu, 1989; Timmerman, 1989).

Plant GSTs have been characterized mainly in maize, where they account for as much as 1-2% of soluble proteins in etiolated tissue (Jepson *et al.*, 1994; Timmerman, 1989). As in mammalian species, the maize GSTs are homo- or hetero-dimers of subunits ranging from 26 to 29 kDa. Four isoforms of maize GST known as GSTI, GSTII, GSTIII and GSTIV have been characterized based on their molecular weight, substrate specificity and responsiveness to safeners (Jepson *et al.*, 1994; Timmerman, 1989). GSTI and GSTIII are constitutively expressed, whereas GSTII and GSTIV are induced by herbicide safeners of the dichloroacetamide type (e.g., R-29148, dichlormid and benoxacor). The safener inducible 27 kDa subunit present in GSTIV is identical to that previously described in GSTII (Jepson *et al.*, 1994).

While little is known about their biological function, maize GSTs have been shown to conjugate GSH to a number of herbicides leading to their detoxification (Farago *et al* 1994; Jepson *et al.*, 1994; Timmerman, 1989). Apart from their involvement in herbicide detoxification, maize GSTs can also mediate the bioactivation of certain pro-herbicide molecules. Nicolaus *et al.* (1995) showed recently that a maize GST converts irreversibly specific thiadiazolidinine analogues to their respective triazolidines, which act as peroxidizing herbicides.

Treatment with selected herbicide safeners enhances the conjugation of chloroacetanilide and sulfoxidized carbamothioate herbicides to GSH in protected grass crops (Farago *et al.*, 1994). The efficacy of chemical safeners as protectants of sorghum against acetanilide herbicides was dependent on their ability to increase GST activity (Gronwald *et al.*, 1987). The exact mechanism of the safener-induced enhancement of GST activity is not known. It has been suggested that safeners act by an enzyme induction rather than activation process since dichlormid and oxabetrinil did not alter in vitro GST activity (Jepson *et al.*, 1994; Wiegand *et al.*, 1986).

iii) Effects of Safeners on Oxidative Metabolism of Herbicides.

Safeners may act also by inducing the activity or the *de novo* synthesis of cytochrome P450-dependent monooxygenases (EC 1.14.14.1), which are involved in the metabolic detoxification of carbamothioate, aryloxyphenoxypropionate, sulfonylurea, and imidazolinone herbicides in protected crops (Hatzios, 1991; Kreuz, 1993).

Preliminary evidence for such safener effects was provided by many indirect studies conducted with the use of selected antioxidants or insecticide synergists, which act as inhibitors of Cyt P450s in plants (Hatzios and Hoagland, 1989). More recently, several direct studies of safener effects on the activity of specific Cyt P450 enzymes have been reported. Thus, the Cyt P450-mediated aryl hydroxylation of the herbicides bentazon, diclofop, chlorsulfuron, triasulfuron, imazaquin, and flumetsulam was shown to be enhanced by NA and several dichloroacetamide safeners (Hatzios, 1991; Kreuz 1993).

iv) Effects of Safeners on Glucose Conjugation of Herbicides

Hydroxylated metabolites of selected herbicides, produced as a result of oxidative metabolism, are known to conjugate rapidly with glucose or other plant sugars (Devine *et al.*, 1993; Sandermann, 1992). The glycosylation of hydroxylated herbicides is mediated by UDP-glucosyltransferases (EC 2.4.1.71).

Lamoureux and Rusness (1991) showed that the safener BAS-145138 partially protects maize from chlorimuron-ethyl injury by increasing the rate of herbicide metabolism by hydroxylation, glucosylation and glutathione conjugation. Kreuz *et al.* (1991) have also reported that the safener cloquintocet-mexyl enhanced the rate of hydroxyldation and O-glucosylation reactions involved in the metabolism of the aryloxyphenoxypropionate herbicide clodinafop-propargyl in wheat.

v) Effects of Safeners on Hydrolytic Metabolism of Herbicides

A recent report (Hatzios, 1993) has suggested that the safener NA may protect maize against injury from the sulfonylurea herbicide thifensulfuron-methyl by inducing the activity of hydrolytic enzymes such as carboxylesterases. NA enhanced the deesterification of thifensulfuron-methyl causing a 30-50% increase in the formation of the parent acid , thifensulfuron, in maize coleoptiles (Hatzios, 1993).

The safener fenclorazole-ethyl has also been reported to enhance the deesterification of fenoxaprop-ethyl to the herbicidally active free acid in both wheat and crabgrass seedlings (Yaacobi *et al.,* 1991). Further metabolism and detoxification of the herbicide, however, was only enhanced in wheat but not in the weed. Thus, fenoxaprop-ethyl behaved as a wheat safener and as synergist of fenoxaprop-ethyl in crabgrass (Yaacobi *et al.*, 1991).

III) Effects of Safeners on Secondary Metabolism and Compartmentation of Herbicides

Glutathione conjugates of herbicides in plants are catabolized to their cysteine and thiolactic acids derivatives, which are further acylated with malonic acid (Sandermann, 1992). Similarly, the glucose conjugates of several herbicides are known to form secondary metabolites with malonic acid (Devine *et al.*, 1993; Sandermann, 1992). These terminal metabolites may be stored as soluble metabolites in the vacuoles of plant cells or deposited as "bound residues" into cell wall components.

Studies on the potential effects of safeners on the formation of secondary conjugates of herbicide metabolites as well as on the compartmentation of such metabolites have been limited. The safener BAS 145138 did not alter significantly the formation of soluble secondary metabolites and bound residues from the initial glutathione conjugates of propachlor and metolachlor in maize (Kreuz, 1993). Recently, Gaillard *et al.*, (1994) showed that apart from inducing the activity of herbicide detoxifying enzymes, the safener cloquintocet-mexyl stimulated the vacuolar transport of the GS-metolachlor and primisulfuron glucoside conjugates.

Molecular Aspects of Safener Action

Alteration of structural genes, gene regulation and gene amplification are currently recognized as significant genetic factors conferring herbicide tolerance or resistance on agronomic crops (Boger and Miller, 1994; Devine *et al.*, 1993). Our knowledge about the structure and function of specific plant enzymes that mediate the metabolic detoxification of herbicides in higher plants has advanced considerably in recent years.

As mentioned earlier, four isoforms of GST have been characterized in maize (Jepson *et al.*, 1994). GSTI and GSTII are constitutively expressed, however GSTI mRNA levels can be elevated three to four fold in etiolated maize seedlings by herbicide safener treatment (Jepson *et al.*, 1994; Wiegand *et al.*, 1986). Nucleotide sequences encoding the 29kDa and 26 kDa subunits have been reported (Jepson *et al.*, 1994). The isoforms GSTII and GSTIV are induced by dichloroacetamide safeners such as dichlormid, benoxacor and R-29148 (Jepson *et al.*, 1994). A cDNA GST sequence form rice has been reported recently by Wu *et al.* (1995).

Wiegand *et al.* (1986) showed that the safener flurazole may act at the transcriptional level inducing a 3- to 4-fold increase in the steady state level of mRNA encoding for the GST I gene in maize tissues grown from flurazole-treated seeds. The expression of a rice GST gene was induced by the safener fenclorim (Wu *et al.*, 1995). Several maize cDNA clones that are induced by treatment with substituted benzensulfonamide herbicide safeners have been also identified and studied by Hershey and Stoner (1991). One of these safener-induced clones in maize, In2-1, contains five of the six absolutely conserved residues seen in plant and animal GSTs. Furthermore, In2-1 shares 25% identity with the known tobacco GST NT103 gene, but only 0.4 identity with the maize clone GST-27 (Jepson *et al.*, 1994).

Because of the low levels of Cyt P450s in plants and the inherent difficulties for their purification, our knowledge of the biochemistry and molecular biology of plant Cyt P450s is rather poor (Donaldson and Luster, 1991; Durst and Benveniste, 1993). cDNA sequences encoding some plant Cyt P450s have just been published (Durst and Benveniste, 1993), but nothing is currently known about genes encoding plant P450s involved in herbicide detoxification as well as their regulation by herbicide safeners. There is some indirect evidence that induction of cytochrome P450 isoforms by safeners, like the safener induction of GSTs, might be exerted through enhanced gene expression. The induction of sulfonylurea oxidation in maize leaves required *de novo* protein synthesis. Published information on the enzymology and molecular biology of plant glucosyl transferases and carboxylesterases as well as on the proposed induction of these enzymes by safeners is also very limited.

Therefore, it appears that the "gene activation" theory which has been implicated in the action of plant hormones is also promising for explaining the protective action of herbicide safeners. Undoubtedly, further research is needed to elucidate the molecular mechanisms of safener action.

Concluding Remarks

A safener-induced enhancement of herbicide detoxification in safened plants seems to be the major mechanism involved in the protective action of the currently used safeners. Most safeners resemble structurally their respective herbicides and they induce the synthesis of cofactors or the expression of enzyme activity needed for their own metabolism as well as that of the antidoted herbicides in safened plants.

Treatment of plants with herbicides and their safeners appears to resemble the situation of other biotic or abiotic stresses that are known to induce, either directly or indirectly, a broad array of defense mechanisms generating plant resistance to such stresses. A better understanding of the mechanisms involved in stress recognition, signal transduction, and defense gene activation will facilitate our views on how safeners protect plants against herbicide injury. Continuing advancements in molecular biology techniques will undoubtedly contribute towards the achievement of these goals.

Nevertheless, metabolic enzymes mediating the detoxification of herbicides in plants may be regulated by many factors. For example, known modulators of GST activity in certain plants include safeners, fungal pathogens, ethylene, heat shock, heavy metals, and auxin (Farago *et al.*, 1994; Hatzios, 1993). Based on the broad nature of GST inducers in plants, one can speculate that the safener-induced expression of GST genes may be a part of a general stress response.

References

Boger, P. and Miller, R. (1994) Protoporphyrin accumulation by peroxidizing herbicides is counteracted by safeners. *Zeischrift Naturforschung* **49c**: 775-780.

Breaux, E.J., Hoobler, M.A., Patanella, J.E. and Leyes, G.A. (1989) Mechanisms of action of thiazole safeners. *In: Crop Safeners for Herbicides: Development, Uses, and Mechanisms of Action* (K.K. Hatzios and R.E. Hoagland, eds.), 163-175. Academic Press, San Diego.

Devine, M., Duke, S.O. and Fedtke, C. (1993) *Physiology of Herbicide Action*. Prentice Hall, Englewood Cliffs, NJ.

Donaldson, R.P. and Luster, D.G. (1991) Multiple forms of plant cytochromes P-450. *Plant Physiology* **96**: 669-674.

Durst, F. and Benveniste, I. (1993) Cytochrome P450 in plants. *In: Cytochrome P450* (J.B. Schenkman and H. Greim, eds.), 293-310. Springer Verlag, New York.

Farago, S., Brunold, C. and Kreuz, K. (1994) Herbicide safeners and glutathione metabolism. *Physiologia Plantarum* **91**: 537-542.

Gaillard, C., Dufaud, A., Tommasini, R., Kreuz, K., Amrhein, N. and Martinoia, E. (1994) A herbicide antidote (safener) induces the activity of both the herbicide detoxifying enzyme and of a vacuolar transporter for the detoxified herbicide. *FEBS Letters* **352**: 219-221.

Gronwald, J.A., Fuerst, E.P, Eberlein, C.V. and Egli, M.A. (1987) Effect of herbicide antidotes on glutathione content and glutathione transferase activity of sorghum shoots. *Pesticide Biochemistry Physiology* **29**: 66-76.

Hatzios, K.K. and Hoagland, R.E. (Eds.) (1989). *Crop Safeners for Herbicides: Development, Uses, and Mechanisms of Action.* Academic Press, San Diego.

Hatzios, K.K. (1991) An overview of the mechanisms of action of safeners. *Zeischrift Naturforschung* **46c**: 819-827.

Hatzios, K.K. (1993) Mode of action of naphthalic anhydride as a maize herbicide safener for thifensulfuron-methyl. *Proceedings Brighton Crop Protection Conference Weeds* **3**: 1259-1266.

Hershey, H.P. and Stoner, T.D. (1991) Isolation and characterization of cDNA clones for RNA species induced by substituted benzenesulfonamides in corn. *Plant Molecular Biology* **17**: 679-690.

Jepson, I., Lay, V.J., Holt, D.C., Bright, W.J. and Greenland, A.J. (1994) Cloning and characterization of maize herbicide safener-induced cDNAs encoding subunits of glutathione S-transferase isoforms I, II, and IV. *Plant Molecular Biology* **26**: 1855-1866.

Kreuz, K. (1993) Herbicide safeners: Recent advances and biochemical aspects of their mode of action. *Proceedings Brighton Crop Protection Conference Weeds* **3**: 1249-1258.

Kreuz, K., Gaudin, J., Stingelin, J. and Ebert, E. (1991) Metabolism of the aryloxyphenoxypropanoate herbicide CGA 184927 in wheat, barley and maize: Differential effects of the safener, CGA 185072. *Zeischrift Naturforschung* **46c**: 901-905.

Krueger, J.P., Butz, R.G. and Cork, D.C. (1991) The use of dicamba degrading microorganisms to protect dicamba susceptible plant species. *Journal Agricultural Food Chemistry* **39**: 1000-1003.

Lamoureux, G.L. and Rusness, D.G. (1991) The effect of BAS-145138 safener on chlorimuron ethyl metabolism and toxicity in corn. *Zeischrift Naturforschung* **46c**: 882-886.

Nicolaus, B., Sato, Y., Iida, T., Senoo, S., Wakabayashi, K. and Boger, P. (1995) Enzymatic conversion of isoimide-type peroxidizing herbicides into active imides. *Abstracts Weed Science Society America* **35**: 71.

Omokawa, H., Wu, J. and Hatzios, K.K. (1995) Mechanism of action of (S)-l-(a-methylbenzyl)-3-p-tolyurea, a rice safener against bensulfuron methyl. *Abstracts Weed Science Society America* **35**: 71.

Pickett, C.B. and Lu, A.Y.H. (1989) Glutathione S-transferases: Gene structure, regulation and biological function. *Annual Review Biochemistry* **58**: 743-764.

Sandermann, H. (1992). Plant metabolism of xenobiotics. *Trends Biochemical Sciences* **17**: 82-84.

Timmerman, K.P. (1989) Molecular characterization of corn glutathione S-transferase isozymes involved in herbicide detoxification. *Physiologia Plantarum* **77**: 465-471.

Wiegand, R.C., Shah, D.M., Mozer, T.J., Harding, E.I., Diaz-Collier, J., Sounders, C., Jaworski, E.G. and Tiemeier, D.C. (1986) Messenger RNA encoding a glutathione S-transferase responsible for herbicide tolerance in maize is induced in response to safener treatment. *Plant Molecular Biology* **7**: 235-243.

Wu, J., Meredith, S.A. and Hatzios, K.K. (1995) Investigations on the action of the rice safener fenclorim against pretilachlor injury. *Abstracts Weed Science Society America* **35**: 72.

Yaacoby, T., Hall, J.C. and Stephenson, G.R. (1991) Influence of fenclorazole-ethyl on the metabolism of fenoxaprop-ethyl in wheat, barley and crabgrass. *Pesticide Biochemistry Physiology* **41**: 296-304.

Yenne, S.P. and Hatzios, K.K. (1990) Molecular comparisons of selected herbicides and their safeners by computer-aided molecular modeling. *Journal Agricltural Food Chemistry* **38**: 1950-1956.

CHAPTER 26
GENE FLOW BETWEEN CROPS AND WEEDS: RISK FOR NEW HERBICIDE RESISTANT WEEDS ?

H. DARMENCY
Laboratoire de Malherbologie, INRA, BV1540, 21034 Dijon, France

Introduction

Literature reports illustrate the diversity of genetic relationships between crops and their wild relatives. Interspecific hybridization and introgression are described in allogamous species such as sunflower, maize, goosefoots, radish, sugar beet, sorghum, pearl millet and oilseed rape, and in autogamous species such as foxtail millet, rice, wheat and potatoes. These cases, although not widespread, provide a basis for discussion of the impact of releasing herbicide resistant crops and the potentiality to create new herbicide resistant weeds.

Genetic engineering in plants is currently used to develop new herbicide resistant crops (HRCs). The reasons and fundamentals of this new strategy have been recently discussed (Burnside, 1992; Gressel, 1995). The release of such HRCs is the genetic corollary of the use of selective herbicides for which genes of resistance are already present in the crop. Crop germplasms showing resistance to glyphosate, glufosinate, bromoxynil, sulfonylureas and 2,4-D are now available and some already got commercial agreement. However, there is some public and scientific concern about long term agronomic benefits and environmental hazards of the commercial release of HCRs (Rissler and Mellon, 1993). One frequent discussion is that we may lose control over engineered genes if wild plants acquire the resistance through pollen exchange with the crop (Ellstrand, 1988). Thereafter, weeds and wild plants expressing resistance might infest fields and be so troublesome that they might cancel the use of the herbicide.

Interspecific hybridization and introgression among wild species as well as between native and invading plant species have been described repeatedly (Heiser, 1973; Abbott, 1992). Such hybridization and introgression has led to new ecotypes, subspecies and allopolyploid species. Some of them survive in restricted habitats, but others invaded large areas (Thompson, 1991). The question I address here is to determine to what extent gene flow is possible between crops and wild relatives. I review reports of the tools used to show evidence of introgression, of those allogamous and autogamous crops and wild species already involved in introgression, and implications for HRCs.

Introgression versus Gene Pool Variability

It is tempting to attribute to introgressive hybridization all intermediate forms betweeen wild and cultivated forms within a gene pool. However, examples for which interspecific gene exchanges can be clearly identified are scarce because morphology varies as a continuum from wild forms to weeds and crops. For instance, introgressive hybridization may be suspected within and between the diploid and tetraploid levels of alfalfa, *Medicago sativa* L. (Small and Bauchan, 1984). Adaptive differences between wild and domesticated forms are quite small and some cultivated alfalfas survive very well outside fields. One may also assume that genes are exchanged between cultivated tomatoes, *Lycopersicum esculentum* Mill, and the wild species *L. pinpinellifolum* (Jusl.) Mill in their South American ranges (Rick, 1958), as well as within the bean gene pool, *Phaseolus vulgaris* L. (Debouck and Tohme, 1989; Gepts, 1993), but present data cannot allow to identify clear-cut phylogenetic origins.

This confusing situation should be an incentive for cytogenetic and molecular studies involving several types of markers. Gene flow can be traced and introgression quantified by karyotype analysis, C-banding and nuclear markers (isozymes, DNA). Cytoplasmic markers can help determine the maternal ancestors of introgressants. For instance, the origin of *Helianthus holanderi* Gray by introgression of *H. exilis* Gray and *H. annuus L.* has long been given as an example, although one could not assert which of either introgression or variability within a gene pool was involved (Heiser, 1973). Molecular investigations provided no evidence for introgression because *H. bolanderi* differed from both other taxa (Rieseberg *et al.*, 1988). On the other hand, molecular markers did suggest that introgression has occurred in *H. annuus* ssp. *texanus* Heiser (Rieseberg *et al.*, 1990).

Similarly, it was initially believed that mimicry between the teosinte weed form, *Zea mays* (L.) ssp *mexicana* Iltis and Doebley, and cultivated maize was caused by introgression due to close proximity in fields (Wilkes, 1977). Ssp *mexicana* grows only in arable fields and hybridizes easily with maize. Hybrids do occur, but they flower at the same time as maize and are thus partly isolated from the teosinte. Moreover, the genome structure of these two taxa differ considerably from each other (Smith *et al.*, 1982), which therefore reduces the likelihood of recombination and creates sterility thereby protecting ssp *mexicana* from gene pollution by maize. Indeed, leaf isozymes analysis showed that *mexicana* populations differ greatly from sympatric maize populations.

Paradoxically, populations of the wilder teosinte type, *Z. mays* (L.) ssp *parviglumis* Iltis and Doebley, look more like maize than the *mexicana* populations do (Doebley *et al.*, 1987). The ssp *parviglumis* taxon is present on stony slopes, near or in fields. Its morphology makes it easy to differentiate from maize. In fields containing some ssp *parviglumis* individuals, forms may grow which are luxuriant, tall, weakly tillering, broad-leafed and which bear a large infructescence all making them morphologically similar to plants descending from artificial crosses (Wilkes, 1977). This argues in favour of introgression and explains why ssp *parviglumis* has some similarity with maize at the molecular level. Nevertheless, straightforward direct origin of local maize from these teosintes cannot be excluded.

Experimental evidence of the possibility of intercrossing between two taxons may also provide light on complex questions. In South America and Mexico, the cultivated goosefoots, *Chenopodium quinoa* Willd and *C. nuttalliae* Aellen, form gene pools with the wild taxa *C. hircinum* Schrad. and *C. berlandieri* Moq., respectively. Hybrids within these groups are usually fertile, unlike those across the groups (Wilson, 1980). Intermediate populations exist in nature which botanists regard as varieties, such as *C. berliandieri* var *sinuatum* which has a more compact inflorescence, less dehiscent fruit and no dormancy, as in the cultivated species (Wilson and Heiser, 1979). There is no firm evidence yet that this variety originated from introgression. However, it has been shown that under experimental sympatry conditions, the spontaneous hybridization rate of *C. berliandieri* by *C. quinoa* was 33% (Wilson and Manhart, 1993). The hybrids seemed more robust than the parents and were fertile enough to give rise to introgressants which were able to survive in nature. This high potential for introgression would explain the lack of molecular differentiation between sympatric Andean domesticated and freeliving populations.

Introgression in Allogamous Species

Mating system is the most obvious important factor in introgression. It determines the possibility of exchange of genes at distant or close locations. *Raphanus* species display self-incompatibility systems that promote hybridization, and there is no isolation barriers between species. Population analysis showed that the pollen from a cultivated variety, bearing an isozyme marker absent in the area, may fertilize wild populations up to 1 km away (Klinger *et al.*, 1991). Hybrids obtained through artificial pollination showed 57% pollen fertility and 81% seed fertility. Cytological examination at meiosis showed seven bivalents and a quadrivalent. The analysis of populations of wild radish established along the Californian coast showed plants with swollen root, reduced fertility, and a quadrivalent formation, that indicates recent introgression between cultivated and wild radish (Panestos and Baker, 1968). This situation provides a good example of hybrid derivatives that have colonized habitats outside the range of the parental species.

Sugar beet provide another case because the weed forms which have developed and caused weed difficulties in sugar beet crops since the 1970's are probably not competitive enough to establish themselves in other crops or elsewhere (Fredshavn and Poulsen, 1996). Cultivated varieties, *Beta vulgaris* ssp *vulgaris* L., are usually biennial, as are wild forms along the shores of the eastern North Atlantic and of the North Sea, *B. vulgaris* ssp *maritima* (L.) Arcangeli. Certificated seed is produced in southern France, where the seed plants may sometimes spontaneously hybridize with annual inland wild plants. *B. vulgaris* is highly allogamous and the two forms are interfertile. Hybrids with a cultivated form cytoplasm are unwittingly exported, within sugar beet seed lots, to the production areas in Northern France, where they soon bolt and disperse their seeds from which new adventitious populations develop (Boudry *et al.*, 1993). Indeed, these new weed populations are annual and have a cultivated type cytoplasm, in contrast to local northern wild beet. One may fear that gene exchange between cultivated and adventitious sugar beet will carry on even if breeders take severe action now.

There is reason to believe that part of diversity created by man in cultivated varieties of sorghum, *Sorghum bicolor* (L.) Moench, has been transmitted back to wild plants belonging to group *S. arundinaceum* Roem. and Schult. Allogamy varies between 5 and 10% in crop and 18 to 30% in feral populations. Spontaneous hybrids between the various species of *Sorghum* are quite common. In a one case study, up to 18% of the descendants of one wild plant were identified as hybrids on the basis of their morphology. Unlike weed sugar beet, intermediate types *Sorghum* survive neither in fields where farmers remove them, although it does survive in waste land (Dogget and Majisu, 1968). The use of male sterile varieties for commercial production of hybrid varieties leads to another situation which is not strictly introgressive. Hybrids between *S. bicolor* and the tetraploid *S. halepense* (L.) Pers. do occur, giving rise to very vigorous triploid plants which are a serious nuisance in crops, but are luckily sterile (Baker, 1972).

Although highly allogamous, some species may develop several partial barriers which may provide isolation enough to protect them from extensive introgression. The various subspecies of pearl millet, *Pennisetum americanum* L. (Leeke), which are all allogamous annual plants with the same number of chromosomes, all interbreed readily. African farmers regularly find in their fields hybrid forms which are very hard to distinguish from cultivated pearl millet until flowering or even ripening. It seems that the wild pearl millet growing in and around fields *(P. violaceum* (Lam.) L. Rich.) is better protected from cultivated pearl millet than wild pearl millet growing away from cultivated areas (*P. mollissimum* Hochst) is. Cultivated characteristics are less susceptible to appear among the descendants of hybrids between the crop and the former taxon than among those of hybrids of crop and the latter taxon (Pernes, 1985). Seed malformation may affect the F_2 generation (Amoukou and Marchais, 1993). One may wonder whether these related varieties managed to maintain each their own characteristics sympatrically just because they had the appropriate genetic systems. The differences between wild and cultivated pearl millets are indeed arranged into linkage groups of coherent sets of characters (Pernes, 1985).

With millions of flowers each year and an outcrossing rate around 20%, oilseed rape, *Brassica napus* L., is also a good candidate for introgression. However, there has been only one report of naturally occurring interspecific hybridization. In the British Isles, *B. harmsiana* is given as hybrid between *B. napus* and *B. rapa* L., and shows intermediate chromosome number and sterility (Stace, 1975). A recent study shows that the hybridization rate may be as high as 93% and that introgression of *B. napus* does occur in wild *B. rapa* populations (Jorgensen and Andersen, 1994). Various studies using interspecific or intergeneric hybridizations (Mizushima, 1950) have shown that rape is partly homologous to various *Brassiceae*. Some of these crosses can only be obtained by culturing plants *in vitro*. These methods probably indicate the maximum potential for crossing with other taxa. For instance, up to at most one hybrid to every 100 flowers can be obtained by crossing rape with the wild species *Sinapis arvensis* L., *Raphanus raphanistrum* L., *Hirschfeldia incana* L. (Lagreze-Fossat), *B. oleracea* L. or *Diplotaxis erucoides* (L.) D.C. (Kerlan *et al.*, 1992; Delourme *et al.*, 1989). However, one cannot work out how frequent exchanges are spontaneously. By planting sterile male rape along alternating rows or in cages containing hives, spontaneous hybrids were obtained with *R. raphanistrum* and *H. incana* at a rate up to 2 per 100 flowers, but very few with *S. arvensis* (Darmency, 1994). The ploidy of the hybrids was usually intermediate between those of the parents. They developed well but had very low fertility. Artificial hybrids grown amongst the wild parent for backcrossing purposes produced fewer than two seeds per hybrid, which indicates a high sterility barrier to introgression.

Introgression in Autogamous Species

Introgression also occurs in partially to highly autogamous species. A well documented case is foxtail millet, *Setaria italica* (L.) Beauv., a highly autogamous cereal most probably the result of the domestication of the green foxtail, *S. viridis* (L.) Beauv. Hybrids between the wild and the domesticated species are fertile enough to produce abundant offspring which often resemble taxon *S. viridis* var. *major* (Gaud) Posp., the giant green foxtail known to botanists. These plants retain both wild characteristics (flowering, seed shedding) and domesticated characteristics after several generations of selfing, which make them even more noxious weeds of agriculture (Darmency *et al.*, 1987). In our experiments simulating millet cultivating conditions, 0.2% of the seeds produced by wild plants were hybrids. There were 100 times fewer hybrids among the descendants of the crop. Appearance of hybrids in these species which produce thousands of seeds per plant is unavoidable, however autogamous they may be. The very existence of the *major* taxon confirms this. Hybrid descendants in millet fields in France provide another indication that introgression is a occurrent process. These plants had a *major*-like morphology and segregated for several simple traits differing between the cultivated and the wild plants. Analysis of chloroplast DNA indicated that the female parent in spontaneous crosses was either wild or cultivated. Therefore, one should consider pollen exchanges in both directions, not just pollen flow from the cultivated plant (Till-Bottraud *et al.*, 1992).

Oryza sativa L., a highly autogamous species, is isolated from other cultivated and wild species either by sterility of the hybrids, as with *O. glaberrima* Steud. and *O. brevigulata* Chev. and Roehr, or by a lethal gene system, as with *O. perennis* Moench. *O. sativa* can hybridize with its own spontaneous forms, genes flowing more readily from the crop to the wild plant than the other way round (Oka and Chang, 1959, 1961). Crossing success varies widely from one variety to another. Late flowering enables the flowering of several species to be synchronized, and may lead to gene pollution rates as high as 52% in a feral population (Langevin *et al.,* 1990). Offtype plants have also been identified among African populations of *O. perennis*, a wild allogamous species that often develops rhizomes. Their morphology, lethal system, pollen fertility and isozymes all clearly show that they descend from crosses between taxa growing sympatrically (Chu and Oka, 1970). Controlled crosses show that the hybrids are highly but not totally sterile and without rhizomes. Fertility is restored in the next generation by backcrossing, with a drift towards the parent types.

A well known example of gene flow in rice was provided by rice growers in an Indian province that started growing in the 1940's a variety of rice with a red pigmentation. This was performed in order to identify easily wild rice (unpigmented) for uprooting (Dave, 1943). This weeding strategy was thwarted by introgression: wild rice picked up the pigmentation gene which had an extremely high selective value since it was used as the weeding criterion. The morphology of the *O. perennis* plants was intermediate and so was their autogamy rate and seed dormancy (Oka and Chang, 1959).

Attempts have been made to transfer genes from the cultivated tetraploid wheat (*Triticum turgidum* var *durum*) to the wild diploid level (*Aegilops speltoides* and *A. longissima* Scheinf and Muschli). Triploid hybrids are almost totally sterile, but backcrossing with pollen from the diploid wild parent does occasionally lead to the development of some of the seeds. This second generation is partly fertile. The third generation stabilizes at the diploid level and is fertile. Wheat genomes are soon eliminated and only very few traits actually get transmitted. Nevertheless, triploid hybrids can bridge across taxa and lead to the formation of individuals quite close to the wild taxon within few generations (Vardi, 1974). On the other hand, fertility and even vigour of hybrids obtained by crossing with the diploid *T. monococcum* could not be restored, indicating an evolutionary dead end (Sharma and Waines, 1981).

Similarly, recent work shows that it is virtually impossible to cross potatoes (*Solanum tuberosum* L.) with black nightshade (*S. nigrum* L.) or bittersweet (*S. dulcamara* L.). Only five hybrid embryos out of the over 20,000 cultured could be rescued. Three of these were not viable, the other two developed and produced tubers but were completely sterile (Eijlander and Stiekema, 1994). These data, together with limited pollen migration, i.e. a few meters (Tynan *et al.*, 1990; Dale *et al.*, 1992), suggest that gene escape from transgenic potatoes is unlikely in Europe. Spontaneous crosses seem to be possible only between wild and cultivated potatoes in their original range in South America, where cases of introgression have been described (Hawkes, 1962).

Introgression and HRCs

These examples are far from being exhaustive. Many other species that man has domesticated for the food value of their grain or roots, for their fibres or for their sheer beauty may be affected, but detailed studies of them are not available. The cases reported here show that intergeneric and interspecific hybridizations have occurred in the past, both with allogamous and autogamous species, and have produced descendants which may become established in new habitats or as weeds. This event does not necessarily mean that introgression occurs frequently. The chance of introgression depends on four major parameters: sympatry, hybridization rate, fitness of hybrids, and advantage provided by herbicide resistance genes.

The likelihood of hybridization depends on the coexistence of crops and wild relatives within a distance pollen may travel (Raybould and Gray, 1993). Indeed, the spread of genes from maize and sunflower is unlikely in most countries as there is no closely related wild counterpart growing sympatrically. In other cases, as with oilseed rape, there might be a higher chance of crosses with wild relatives. Such crosses must be kept rare events under agricultural conditions since weeds within the crop are being controlled by herbicides. In addition, the various species may have different time of flowering (Lefol *et al.*, 1996). However, wild relatives growing outside cultivated fields and volunteer crop escaped into uncultivated habitats are potential risks.

Study of pollen flow and potentiality of intercrossing allow to predict high or low hybridization rate in fields. For instance, potato hardly spreads pollen and intercross with other *Solanum* species in Europe. In contrast, sugar beet widely spreads pollen by wind and easily crosses with wild forms of the same species. The special case of areas producing hybrid seed varieties must be carefully examined. Male sterile varieties, which are necessary to produce these high yield hybrid varieties, can catch every kind of pollen from the neighbourhood, even from several kilometers away via the wind or insects. In the case of oilseed rape, this provides a new risky opportunity to cross.

In cases where hybridization is not so difficult, hybrids may survive in fields as well as, and even better than, their wild parent. However, the reproduction potential of hybrids may be very low compared to that of the wild parent, as for oilseed rape x wild *Brassiceae* hybrids. That means that hybrids have a very low fitness. The odds that the offspring will establish themselves amongst a normal population are therefore quite low. This suggests that even if hybrids appear in the wild, then introgression would seldom occur. Things are different in the case of sugar beet and foxtail millet as descendants proved to be fertile enough to establish weed populations and to be recognised as botanical variants.

The type of genes that may possibly circulate by introgression is a matter of great importance. Genetic engineering can produce genes that have no counterpart in nature (Gliddon, 1994). Resistances to bromoxynil, glyphosate, glufosinate and 2,4-D were engineered from microorganisms, which means that a higher plant would never get them through mutation. With HRCs, the fitness of hybrids and wild plants which have captured a resistance gene will be much higher than previous spontaneous hybrids observed to date, and it is likely that new hybrids will be quickly selected for in the field. Their fitness in herbicide treated fields should be higher than that of their wild parent, so that they could occur at high frequency within populations and be able to transfer the resistance gene to the wild species. The behaviour of such plants is predicted to be similar to that of herbicide resistant weeds (Darmency, 1994), with known troublesome problems of weed control and even less possibility of finding an alternative chemical strategy.

Acknowledgements

We thank Ph. Butler for translating the paper into English.

References

Abbott, R.J. (1992) Plant invasions, interspecific hybridization and the evolution of new plant taxa. *TREE* **7**: 401-405.

Amoukou, A.I. and Marchais, L. (1993) Evidence of a partial reproductive barrier between wild and cultivated pearl millets (*Pennisetum glaucum). Euphytica* **76**: 19-26.

Baker, H.G. (1972) Human influences on plant evolution. *Economic Botany* **32**: 26-43.

Boudry, P., Morchen, M., Saumitou-Laprade, P., Vernet, Ph. and Van Dijk, H. (1993) The origin and evolution of weed beets: consequences for the breeding and release of herbicide resistant transgenic sugar beets. *Theoretical Applied Genetics* **87**: 471-478.

Burnside, O.C. (1992) The impetus for a symposium on herbicide-resistant crops. *Weed Technology* **6**: 611-612.

Chu, Y.E. and Oka, H.I. (1970) Introgression across isolating barriers in wild and cultivated *Oryza* species. *Evolution* **24**: 344-355.

Dale, P.J., Mc Partlan, H.C., Parkinson, R., Mackay, G.R. and Scheffler, J.A. (1992) Gene dispersal from transgenic crops by pollen. *In: Proceedings 2nd International Symposium Biosafety Results of Field Tests of Genetically Modified Plants and Microorganisms* (R. Casper and J. Landsmann, eds.), 54-60. Goslar.

Darmency, H. (1994). The impact of hybrids between genetically modified crop plants and their related species: introgression and weediness. *Molecular Ecology* **3**: 37-40.

Darmency, H., Zangre, G.R. and Pernes, J. (1987). The wild weed-crop complex in *Setaria* a hybridization study. *Genetica* **75**: 103-107.

Dave, B.B. (1943) The wild rice problem in the central provinces and its solution. *Indian Journal Agricultural Science* **13**: 46-53.

Debouck, D.G. and Tohme, J. (1989) Implications for bean breeders of studies on the origins of common beans, *Phaseolus vulgaris* L. In: *Current Topics in Breeding Common Bean* (S. Beebe, ed.), 3-42. CIAT, Cali.

Delourme, R., Eber, F. and Chevre, A.M. (1989) Intergeneric hybridization of *Diplotaxis erucoides* with *Brassica napus.* I: cytogenetic analysis of Fl and BCI progeny. *Euphytica* **41**: 123-128.

Doebley J., Goodman, M.M. and Stuber C.W. (1987) Patterns of isozyme variation between maize and mexican annual teosinte. *Economic Botany* **41**: 234-246.

Dogget, H. and Majisu, B.N. (1968) Disruptive selection in crop development. *Heredity* **23**: 1-23.

Eijlander, R. and Stiekema, W.J. (1991) Biological containment of potato (*Solanum tuberosum*) outcrossing to the related wild species black nightshade (*Solanum nigrum*) and bittersweet (*Solanum dulcamara*). *Sexual Plant Reproduction* **7**: 2940.

Ellstrand, N.C. (1988) Pollen as a vehicle for the escape of engineered genes? *TREE* **3**: S30-S32.

Fredshavn, J.R. and Poulsen, G.S. (1996) Growth behaviour and competitive ability of transgenic crops. *Field Crops Research* **45**: 11-18.

Gepts, P. (1993) The use of molecular and biochemical markers in crop evolution studies. *Evolutionary Biology* **27**: 51-94.

Gliddon, C. (1994) The impact of hybrids between genetically modified crop plants and their related species: biological models and theoretical perspectives. *Molecular Ecology* **3**: 41-44.

Gressel, J. (1996) The potential role for herbicide resistant crop in world agriculture. *In: Herbicide Resistant Crops: Agricultural, Economic Environmental, Regulatory and Technological Aspects* (S.O. Duke, ed.), 231-250. Lewis Publishers, Boca Raton.

Hawkes, J.G. (1962) Introgression in certain wild potato species. *Euphytica* **11**: 26-35.

Heiser, C.B. (1973) Introgression re-examined. *Botanical Review* **39**: 347-366.

Jorgensen, R.B. and Andersen, B. (1994) Spontaneous hybridization between oilseed rape (*Brassica napus*) and weedy *B. campestris* (*Brassicaceae*): a risk of growing genetically modified oilseed rape. *American Journal Botany* **81**: 1620-1626.

Kerlan, M.C., Chevre, A.M., Eber, F., Baranger, A. and Renard, M. (1992) Risk assessment of outcrossing of transgenic rapeseed to related species: I; interspecific hybrid production under optimal conditions with emphasis on pollination and fertilization. *Euphytica* **62**: 145-153.

Klinger, T., Elam, D.R. and Ellstrand, N.C. (1991) Radish as a model system for the study of engineered gene escape rates via crop-weed mating. *Conservation Biology* **5**: 531-535.

Langevin, S.A., Clay, K. and Grace J.B. (1990) The incidence and effects of hybridization between cultivated rice and its related weed red rice (*Oryza sativa* L.). *Evolution* **44**: 1000-1008.

Lefol, E., Danielou, V. and Darmency, H. (1996) Predicting hybridization between transgenic oilseed rape and wild mustard. *Field Crops Research* **45**: 153-161.

Mizushima, U. (1950) On several artificial allopolyploids obtained in the *Brassiceae* of *Crucifera. Tohoku Journal Agricultural Research* **1**: 15-27.

Oka, H.I. and Chang, W.T. (1959) The impact of cultivation on populations of wild rice, *Oryza sativa* f. *Spontanea. Phyton* **13**: 105- 117.

Oka, H.I. and Chang, W.T. (1961) Hybrid swarms between wild and cultivated rice species, *Oriza perennis* and *O. sativa. Evolution* **15**: 418-430.

Panestos, C.P. and Baker, H.G. (1968) The origin of variation in "wild" *Raphanus Sativus (Cruciferae)* in California. *Genetica* **38**: 243-274.

Pernes, J. (1985) Evolution des plantes cultivees: l' exemple des cereales. *La Vie des Sciences* **2**: 429-447.

Raybould, A.F. and Gray, A.J. (1993) Genetically modified crops and hybridization with wild relatives: a UK perspective. *Journal Applied Ecology* **30**: 199-219.

Rick, C.R. (1958) The role of natural hybridization in the derivation of cultivated tomatoes of Western South America. *Economic Botany* **12**: 348-367.

Rieseberg, L.H., Soltis, D.E. and Palmer, J.D. (1988) A molecular re-examination of introgression between *Helianthus annuus* and *H. bolanderi. Evolution* **42**: 227-238.

Rieseberg, L.H., Beckstrom-Sternberg, S. and Doan, K. (1990). *Helianthus annuus* ssp *texanus* has chloroplast DNA and nuclear ribosomal RNA genes of *Helianthus debilis* ssp *Cucumerifolius. Proceedings National Academy Sciences USA* **87**: 593-597.

Rissler, J. and Mellon, M. (1993) *Perils amidst the promise: Ecological risks of transgenic crops in a global market.* Union of Concerned Scientists, Cambridge.

Sharma, H.C. and Waines, J.G. (1981) Attempted gene transfer from tetraploids to diploids in *Triticum. Canadian Journal Genetics Cytology* **23**: 639-645.

Small, E. and Bauchan, G.R. (1984) Chromosome numbers of the *Medicago sativa* complex in Turkey. *Canadian Journal Botany* **62**: 749-752.

Smith, J.S.C., Goodman, M.M. and Kato, T.A. (1982) Variation within teosinte. II. Numerical analysis of chromosome knob data. *Economic Botany* **36**: 100-112.

Stace, C.A., (1975) *Hybridization and the flora of the Britishes Isles.* Academic Press, London.

Thompson, J.D. (1991) The biology of an invasive plant: what makes *Spartina anglica* so successful? *BioScience* **41**: 393-401.

Till-Bottraud, I., Reboud, X., Brabant, P., Lefranc, M., Rherissi, B., Vedel, F. and Darmency, H. (1992) Outcrossing and hybridization in wild and cultivated foxtail millets: consequences for the release of transgenic crops. *Theoretical Applied Genetics* **83**: 940-946.

Tynan, J.L., Williams, M.K. and Conner, A.J. (1990) Low frequency of pollen dispersal from a field trial of transgenic potatoes. *Journal Genetics Breeding* **44**: 303-306.

Vardi, A. (1974) Introgression from tetraploid durum wheat to diploid *Aegilops Longissima* and *Aegilops speltoldes. Heredity* **32**: 171-181.

Wilkes, H.G. (1977) Hybrization of maize and teosinte, in Mexico and Guatemala and the improvement of maize. *Economic Botany* **31**: 254-293.

Wilson, H.D. (1980) Artificial hybridization among species of *Chenopodium* section *Chenopodium. Systematic Botany***5**: 253-263.

Wilson, H.D. and Heiser, C.B. (1979) The origin and evolutionary relationships of "Huauzontle" (*Chenopodium muttalliae* Safford) domesticated chenopod of Mexico. *American Journal Botany* **66**: 198-206.

Wilson, H.D. and Manhart, J. (1993) Crop/Weed gene flow: *Chenopodium quinoa* Willd. and *C. berlandieri* Moq. *Theoretical Applied Genetics* **86**: 642-648.

CHAPTER 27
TECHNOLOGICAL, ECOLOGICAL AND SOCIAL ASPECTS OF HERBICIDE RESISTANT CROPS

W.E. DYER
Department of Plant, Soil and Environmental Sciences, Montana State University, Bozeman, Montana

Introduction

Recent advances in the technologies of gene transfer and crop plant regeneration now permit the creation of transgenic herbicide-resistant cultivars for essentially all major crop species. Several cultivars have already been commercialized and more will be marketed within the next few years. These crops have the potential to provide a valuable strategy for weed management if they are used appropriately and in ongoing crop and herbicide rotations. However, several controversies surrounding their development and use may limit or slow their widespread adoption. One of the most widely publicized concerns, the likelihood of herbicide resistance gene escape from resistant cultivars into weedy relatives, is addressed in an accompanying paper in this section. In addition, other problems may arise in transgenic plants, including the loss of expression from the transgene over time and unpredictable pleiotropic effects of the transgene and the enzyme it encodes on physiological aspects of the resistant plant. The genetic mechanisms underlying these phenomena are not completely understood but are under active study in a number of laboratories. Other agronomically important concerns of HRCs include the possibility that genetic modifications for resistance will confer weediness to the cultivar (ie. volunteer plants in subsequent crops) and selecting additional herbicide resistant weeds in the new crop/herbicide regime. Of these concerns, the potential for selecting new resistant weeds may have the highest likelihood of affecting the long-term success of herbicide resistant crops.

Herbicide resistant crops (HRCs) represent one of the first and most highly publicized applications of plant biotechnology. Most HR cultivars have been created through the stable integration of a foreign gene (transgene) using recently developed techniques of molecular biology and plant transformation. This approach takes advantage of the potential ability of every plant cell to be regenerated into a whole plant after receiving a foreign gene. However, it is important to note that HRCs have also been developed using traditional plant breeding methods as well as hybridization with conspecific HR relatives. In this paper, the term HRC is reserved for cultivars created by hybridization, mutagenesis, somaclonal variation, protoplast fusion, gene transfer or other novel technologies to be resistant to a previously toxic herbicide. All crops are naturally resistant to certain herbicides because they possess insensitive target enzymes or metabolic capabilities, but are not included in this paper because there do not appear to be new technological or ecological questions about this kind of resistance. Most of the following discussion applies to those HRCs created through introduction of novel genes, either through biotechnology or conventional breeding.

As of 31 December 1994 the United States Department of Agriculture/Animal and Plant Health Inspection Service (USDA/APHIS) had issued 1,803 permits or notifications for field trials of transgenic organisms in 35 states (Anonymous 1994). Worldwide, at least 14 countries are involved in field testing of transgenic plants. These plants contain genes conferring a wide variety of agriculturally important traits, including modification of food quality, expression of valuable proteins, disease and insect resistance, and herbicide resistance. HRC permits represent about 57% of the total, demonstrating the strong emphasis being placed on this particular trait. The technology and development of HRCs have been reviewed previously (Botterman and Leemans, 1988; Dyer *et al.*, 1993b; Gressel, 1993; Mazur and Falco, 1989; Mullineaux, 1992; Oxtoby and Hughes, 1989; Quinn, 1990; Schulz *et al.*, 1990; Stalker, 1991).

Technological Aspects of HRC Development

Most HRCs created to date have been developed by introducing a foreign gene that confers resistance (Dyer *et al.*, 1993b). The gene may encode a herbicide-resistant form of an endogenous enzyme or a novel enzyme that alters and thereby inactivates the herbicide. In recent years, the major focus of transgenic crop research has been on optimizing transformation and regeneration methods for particular crop species. Crops like *Nicotiana* and *Petunia* spp. were the first to be transformed because these species are highly responsive to tissue culture and regenerate easily. Concurrent advances in plant transformation protocols, especially using *Agrobacterium tumefaciens*, allowed early and rapid success with many dicotyledonous species (Horsch *et al.*, 1985). However, transformation of agronomically important graminaceous monocots such as wheat (*Triticum aestivum* L.), rice (*Oryza sativa* L.) and maize (*Zea mays* L.) has been slower because they are not hosts for *A. tumefaciens*. Therefore, new techniques, most notably particle bombardment using a «biolistic» (biological ballistic) device, were developed to transform cereals.

One of the early successful applications of particle bombardment was the transformation and regeneration of fertile maize plants, a species notoriously recalcitrant to these manipulations. Cells from embryogenic suspension cultures were bombarded with the *bar* gene and transgenic plants with heritable glufosinate resistance were recovered (Gordon-Kamm *et al.*, 1990). Maize transformation has since been accomplished in many laboratories and is now routine. Similar methodology was used to obtain stable glufosinate-resistant sugarcane (*Saccharum* spp.) calli (Chowdhury and Vasil, 1992) and rice plants (Cao *et al.*, 1992) after bombarding embryogenic cell suspension cultures with the *bar* gene. Glufosinate-resistant oat plants were likewise obtained from bombarding embryogenic callus (Somers *et al.*, 1992). Regeneration of fertile glufosinate-resistant wheat plants was obtained from bombarded embryogenic calli, using a system that required the identification and maintenance of a specific regenerable callus type (Vasil *et al.*, 1992). A potentially more universal technique was demonstrated by bombarding calli from immature wheat embryos with the *bar* gene and regenerating glufosinate-resistant fertile wheat plants (Weeks *et al.*, 1993). Most recently, particle bombardment of several barley (*Hordeum vulgare* L.) tissues was successfully used to produce fertile herbicide-resistant transgenic plants (Wan and Lemaux, 1994). Since all important monocot crops have now been transformed, research priorities are shifting away from transformation/regeneration techniques and are moving towards other potential concerns about transgenic crops.

Unpredictable Changes in Gene Expression

Gene expression levels from transgenes can vary as much as 100-fold among independent plants or tissues resulting from the same transformation experiment (Flavell, 1994). This phenomenon is thought to result from the fact that, regardless of the transformation method used (*A. tumefaciens*, particle bombardment, microinjection, etc.), transgene insertion occurs in random locations within the host plant cell genome. Therefore, transgenes may be inserted into chromatin regions of widely differing transcriptional activity, resulting in variable expression levels. In addition, transgene sequences can be rearranged, altered or, most often, duplicated during the transformation process. The underlying mechanisms controlling these sequence alterations are not completely understood, although they are under active study. In practice, such phenomena are not regarded as particularly problematic, because it is often simply a matter of selecting among regenerated transformants to obtain individual plants with the desired levels of gene expression.

In contrast, several other fates of transgenes in host plant cells can cause unpredictable and undesirable side effects. Because the location of transgene insertion in the host plant genome is essentially random, the insertion of transgenic sequences may occur within the coding or regulatory regions of important genes, most often resulting in gene inactivation. Naturally, if such a target gene encodes an enzyme responsible for a critical function in the cell, gene inactivation will be lethal and that particular transformant will never be recovered. However, if the enzyme encoded by an inactivated gene is involved in a regulatory activity manifested only under certain conditions, then the resulting phenotype may go undetected until the transgenic plant experiences those particular conditions. For example, a minor gene controlling a subtle aspect of cold stress adaptation may not be essential under normal growing conditions, but its inactivation by transgene insertion would cause the plant to experience severe stress and loss of vigor in a cold environment. For HRCs, these effects may not be observed unless the cultivars are tested under a wide range of possible environments during field screening and before commercialization. The important point is that these undesirable side effects of transgene insertion are unpredictable, especially since relatively little is known about the interrelationships of genes controlling quantitative or other multigenic traits.

i) Trans-Inactivation in Transgenic Plants

Another possible effect occurring in transgenic plants is the loss of gene expression over time. This phenomena occurs most often if the introduced transgene is homologous to an endogenous gene. For HRCs, examples would include insertion of a gene encoding a herbicide-resistant acetolactate synthase (ALS), the target enzyme of sulfonylurea and imidazolinone inhibition, or a resistant 5-enolpyruvylshikimate 3-phosphate (EPSP) synthase, the target of glyphosate inhibition. Since the transgenes are essentially identical in DNA sequence to the native gene(s), unusual and incompletely understood hybridization events can occur that result in loss of expression from both the transgene and the endogenous genes. This phenomenon has been termed «cosuppression» (Napoli *et al.*, 1990) and more recently gene silencing or trans-inactivation. Loss of expression can occur rapidly after transformation or alternatively may not appear until regenerated transformants have gone through several generations. Although the phenomenon has not been reported specifically for HRCs, it is well documented for many plant species transformed with other genes, and thus represents a possible unpredictable and undesirable side effect in transgenic HRCs.

One of the first reports of cosuppression in plants involved the loss of gene expression in petunia plants after insertion of an additional copy of the chalcone synthase gene (Napoli *et al.*, 1990). Chalcone synthase is an important enzyme in the pathway responsible for anthocyanin pigment production, which is especially substantial in petunia corolla and anther tissues. Flower color phenotypes of regenerated plants ranged from partial white sectoring to completely white flowers. The lack of pigment was correlated with very low levels of mRNA from transgenic and endogenous copies of chalcone synthase genes. Interestingly, some branches on white-flowered plants reverted to purple or purple sectoring, indicating that the loss/reversion of expression is under developmental control and that different somatic states of transgene activity can occur. These phenotypes were heritable through sexual reproduction and seed development/germination. Subsequent reports provide evidence for the following commonalities: cosuppression and trans-inactivation of genes are dependent on sequence homology, can be epigenetically reversible and break down to various extents when the homologous loci are segregated away from one another. Other observations that do not apply in all situations include: trans-inactivation is sometimes dependent on transcription or a specific promoter, is sometimes under developmental control, does not occur with all copies of transgenes, is sometimes associated with changes in the cytosine methylation pattern of genes, all or only one copy can be trans-inactivated and loss of gene expression appears to be due to inhibition of transcription in some cases and degradation of mRNA in others. Although these phenomena have been described only recently in transgenic plants, many of the phenotypes are similar to a well-characterized example of allelic interaction in plants known as paramutation (Matzke and Matzke, 1993), and the mechanisms proposed below may apply to both instances.

Several hypotheses have been put forth to explain these trans-inactivation phenomena (reviewed in Flavell, 1994). Current theories of nuclear structure and transcription propose that transcription is initiated after DNA is decondensed from chromatin and becomes attached to the nuclear matrix. If, in the presence of a transgene, homologous DNA sequences interact to form novel hybrid DNA duplex or triplex structures, several unusual consequences could arise. The new hybrid would create a different DNA/chromatin structure, resulting in the possible exchange of chromatin proteins and novel cytosine methylation patterns in the participating DNA strands. These events could occur irregularly during cell division and differentiation, giving rise to unstable phenotypes and allelic variation. In contrast, a single changed pattern could be imprinted which would result in a stable allele chromatin structure, albeit different from the original transcriptional conformation of the endogenous gene(s). Any of these changes could result in aberrant (reduced) expression of some or all of the genes.

ii) Post-transcriptional Loss of Gene Expression

It is important to point out that changes in gene expression resulting from transgene insertion, such as the documented phenomena described above, are unpredictable in nature and may result in very subtle phenotypes. Alterations may not appear for several generations after transformed plants are regenerated. Such possibilities are included in this manuscript in order to point out that the creation of transgenic crops, including HRCs, may include some changes with unforseen consequences.

Pleiotropic Effects of Transgenes

In addition to the trans-inactivation phenomena described above, enzymes encoded by transgenes may exhibit altered and unusual characteristics. For example, an introduced ALS enzyme encoded by a mutant transgene may function apparently normally in its proper biosynthetic reaction. However, the mutation leading to herbicide resistance could also alter the enzyme's kinetic or regulatory properties in very subtle ways. This phenomenon is illustrated by field-selected chlorsulfuron-resistant *Kochia scoparia* L. Scop accessions. Field observations suggested that germination occurred earlier and/or faster than in susceptible types. Even thought greenhouse and field competition and fitness studies suggested that resistant and susceptible types were equally fit, germination tests at several temperatures showed that germination rates of resistant types were in fact faster than susceptibles, but only at low temperatures (Figure l; Dyer *et al.*, 1993a). Subsequent studies showed that free levels of the branched chain amino acids isoleucine, leucine and valine were about 2-fold higher in resistant accessions. The authors' tentative conclusion was that the mutation(s) conferring resistance had altered the feedback inhibition sensitivity of ALS such that the amino acids synthesized from this pathway were overproduced. Since isoleucine and possibly valine appear to be absolutely required for DNA synthesis and therefore cell division (Rost *et al.*, 1990), elevated levels of these amino acids in resistant types could lead to more rapid germination under stressful low temperatures (Dyer *et al.*, 1993a). An alternative explanation, and one that could apply more specifically to HRCs, is that the interaction of resistant and susceptible ALS subunits could create heteromultimers with altered kinetic and/or regulatory properties. The important point illustrated here is that

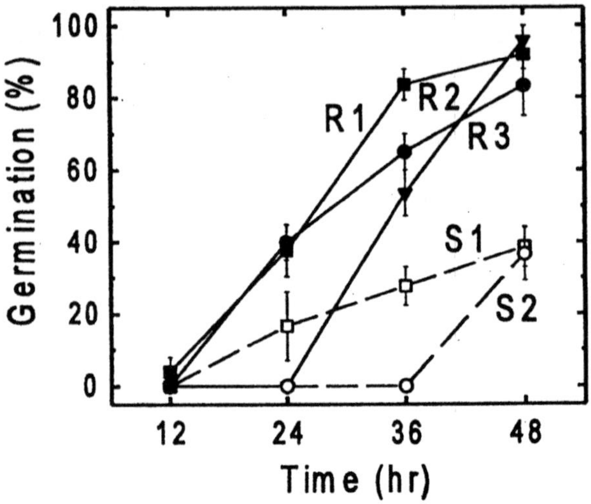

Figure 1. Germination (expressed as a percent of total germination achieved by each accession after 72 h) of seeds from sulfonylurea-resistant (Rl, R2, R3: solid lines) and susceptible (S1, S2; dashed lines) Kochia scoparia accessions at 4.6 °C. Vertical bars are standard errors of means.

such an alteration in germination physiology would not be predicted from a mutation conferring herbicide resistance. Similar subtle changes could occur in HRCs as a result of introducing a transgene for an altered enzyme.

For some herbicides that are not rapidly metabolized by plants, presence of an insensitive target enzyme in the HRC may lead to an accumulation of the parent herbicide in harvested portions of the crop, particularly after pre-harvest treatments. Other HRCs created by introducing microbial genes encoding herbicide metabolic enzymes may produce novel metabolites not encountered during the initial herbicide registration process. Transgenic carrot plants expressing a fungal gene for resistance to glufosinate (2-amino-4 [hydroxymethylphosphinyl] butanoic acid) contained different herbicide metabolites than those produced by nontransformed plants (Droge *et al.*, 1992). Such a possibility would be investigated during the HRC registration process, since all herbicide and metabolite residues are identified, quantified, and their toxicology determined before HRC release (Dyer *et al.*, 1993b).

Ecological Aspects of HRCs

Volunteer crop weeds (plants emerging from remnant seed in the subsequent season) are a common problem in most crop rotation systems. Choice of a weed management strategy must include measures that will control these plants. However, producers' choices may be limited if the volunteer plants are HRCs. For example, control of volunteer graminicide-resistant maize in a subsequent soybean (*Glycine max* L.) crop would require use of an alternative measure (another herbicide or a cultural practice) to control the HRC, possibly at an additional production cost. The possible development of HRCs with 'stacked genes' for multiple herbicide resistance would further complicate management of volunteers. Another consideration is the existence of a persistent HRC seed bank, particularly for crops with hard seed such as alfalfa (*Medicago sativa* L.) which could further limit options in succeeding years. Thus, incorporation of HRCs into a crop production system can offer valuable weed management options, but HRC persistence may conversely limit options in some crop rotations.

Can HRCs Become New Weeds?

The possibility exists that introduction of a novel gene for herbicide resistance could sufficiently change a plant's growth characteristics or fitness (survival and reproductive success) so that it becomes a weedy pest in noncultivated or feral habitats. The traits traditionally used to define weediness result from the action of many genes, so it is unlikely that introduction of one new gene could cause a crop to become a weed. Cultivated crops generally cannot survive and reproduce without human intervention so the potential for a HRC to invade feral environments seems extremely low. In experiments comparing the invasiveness and ecological performance of transgenic and conventional oilseed rape (*Brassica napus* subsp. *oleifera*), the herbicide-resistant rape displayed equal or less invasiveness than its non transformed counterparts in 12 different non-agricultural environments and in the presence or absence of various stresses (Crawley *et al.*, 1993).

Scenarios can be conceived in which introduced herbicide resistance could provide a plant with a competitive advantage under certain conditions. Novel enzymes for herbicide metabolism could be sufficiently promiscuous to recognize and inactivate an insect or pathogen toxin and thus render the HRC resistant to this stress. Although such added resistance could be considered a production advantage in the target HRC, its presence could also increase a persistent HRC's potential for weediness.

Selection for Resistant Weeds

Intensive use of herbicides over the last 40 years has led to the selection of resistant weedy plants and subsequent populations within formerly susceptible species. This phenomenon is rapidly becoming an important agricultural problem in many countries and the subject has been recently reviewed (Caseley *et al.*, 1991; Holt *et al.*, 1993; Moss and Rubin, 1993; Powles and Holtum, 1994). Primary factors controlling the incidence and evolution of resistance include the initial frequency of the resistance trait in a population, extent of selection pressure, and the comparative fitness of resistant and susceptible types. Of these factors, the only one that can be controlled by agricultural producers is the extent of selection pressure, by manipulating the type and efficacy of weed management methods they employ. Current recommendations for resistance prevention and management focus on crop and herbicide rotations and herbicide mixtures to interrupt selection pressures imposed by continuous use of chemicals having the same mechanism of action (Gressel and Segel, 1990; LeBaron and McFarland, 1990).

HRCs can fulfill a valuable role in such a resistance prevention strategy by allowing producers the flexibility to use alternative herbicides not traditionally used in the crop. The commercial development of triazine resistant canola (Beversdorf and Kott, 1987) has provided a useful means of controlling certain troublesome weeds. Ideally, the novel herbicide used on a HRC would have a mechanism of action substantially different from herbicides previously used on that species. HRCs thus incorporated into ongoing crop and herbicide rotations will undoubtedly provide a significant weed management advantage to producers. However, the likelihood of this scenario becoming a reality depends entirely on how HRCs are perceived and marketed. If HRCs prove to be very successful in some situations, particularly in continuous cropping areas with few available rotations, producers' natural tendency will be to continue using the same HRC cultivar for several consecutive years. Producers growing HRCs may be tempted to make multiple herbicide applications during one season to control successive weed flushes. Both of these situations will surely help set the stage for selecting resistant weeds by increasing the effective selection pressure. This possibility is particularly disturbing for HRCs resistant to environmentally benign herbicides such as glyphosate and glufosinate, for which the evolutlon of weed resistance has not been reported (Dyer, 1994). Multiple applications over several consecutive years would compensate for these herbicides' nonresidual characteristics and undoubtedly increase the chances of selecting resistant weeds, as has been shown for other nonresidual herbicides like paraquat (1,l' -dimethyl-4,4'-bipyridinium ion) (Holt *et al.*, 1993).

Social and Economic Aspects of HRCs

Only one study has been published examining the possible economic effects of HRC use (Tauer and Love, 1989). In particular, the study focused on the adoption of transgenic herbicide-resistant corn in the midwestern U.S. Producers first adopting the new cultivars would gain from higher yields and higher profits, but as the technology became more widely diffused, increased production would reduce market prices and consumers would eventually capture most of the benefits. The authors estimated that total benefits to producers and consumers from HRC use would be $1.9 to $3.8 billion annually after the technology was fully adopted. In addition, regions that previously suffered the most from uncontrolled weed problems would benefit the most from the improved weed control technology. If HRC technology was priced at $ 13 per acre, aggregate corn production was projected to increase 2% to 4% and corn prices would decrease about $0.30 per bushel. Falling corn prices would subsequently result in production shifts to more soybean acreage. Overall, the authors predicted that HRC effects on corn acreage and total changes in net income would be relatively small.

Conclusions

The long-term goal of agriculturalists should be to work towards developing environmentally compatible systems for crop production. In the interim, we must continue to optimize and attempt to minimize agrichemical use for weed control HRCs represent a potentially valuable tool for weed management in our current agricultural systems. However, there are enough concerns about the technology, ecology and social aspects of this new technology that due caution should be exercised before they are widely distributed across the landscape. In particular, the potential for unpredictable transgene side effects to alter HRC behavior suggests that these phenomena should be carefully studied under a variety of environments. It appears that commercialization of this technology may be proceeding at such a rapid pace that questions like these will be overlooked. Likewise, the likelihood of selecting for additional resistant weeds with widespread HRC use seems high enough that some restraint should be exercised in their market penetration. Perhaps a long-term perspective on careful HRC development and product stewardship will prevail in this situation so that HRCs can be successfully used for successful weed management now and in the future.

Acknowledgements

Research in the author's laboratory is supported by the National Science Foundation (OSR-9350546), USDA/NRICGP (92-37303-7614), and the Montana Agricultural Experiment Station.

References

Anonymous (1994) USDA, APHIS Biotechnology Permits Unit, database release 31 December 1994.

Beversforf, W.D. and Kott, L.S. (1987) Development of triazine resistance in crops by classical plant breeding. *Weed Science* **35** (Suppl. 1), 9-11.

Botterman, J. and Leemans, J. (1988) Engineering herbicide resistance into plants. *Trends Genetics* **4**: 219-232.

Cao, J., Duan, X., McElroy, D. and Wu, R. (1992) Regeneration of herbicide resistant transgenic rice plants following microprojectile-mediated transformation of suspension culture cells. *Plant Cell Reports* **11**: 586-589.

Caseley, J.C., Cussans, G.W. and Atkin, R.K. (eds.) (1991) *Herbicide Resistance in Weeds and Crops*, 513 pp. Butterworth-Heinemann, Oxford.

Chowdhury, M.K.U. and Vasil, I.K. (1992) Stably transformed herbicide resistant callus of sugarcane via microprojectile bombardment of cell suspension cultures and electroporation of protoplasts. *Plant Cell Reports* **11**: 494-498.

Crawley, M.J., Hails, R.S., Rees, M., Kohn, D. and Buxton, J. (1993) Ecology of transgenic oilseed rape in natural habitats. *Nature* (London) **363**: 620-623.

Dröge, W., Broer, I. and Pühler, A. (1992) Transgenic plants containing the phosphinothricin-N-acetyltransferase gene metabolize the herbicide L-phosphinothricin (glufosinate) differently from untransformed plants. *Planta* **187**: 142-151.

Dyer, W.E. (1994) Resistance to glyphosate. *In: Herbicide Resistance in Plants: Biology and Biochemistry* (S.B. Powles and J.A.M. Holtum, eds.), 229-241. CRC Press, Oxford.

Dyer, W.E., Chee, P.W. and Fay, P.K. (1993a) Rapid germination of sulfonylurea-resistant *Kochia scoparia* accessions is associated with elevated seed levels of branched chain amino acids. *Weed Science* **41**: 18-22.

Dyer, W.E., Hess, F.D., Holt, J.S. and Duke, S.O. (1993b) Potential benefits and risks of herbicide-resistant crops produced by biotechnology. *Horticultural Review* **15**: 367-408.

Flavell, R.B. (1994) Inactivation of gene expression in plants as a consequence of specific sequence duplication. *Proceedings National Academy Sciences USA* **91**: 3490-3496.

Gordon-Kamm, W.J., Spencer, T.M., Mangano, M.L., Adams, T.R., Daines, R.J., Start, W.G., O'Brien, J.V., Chambers, S.A., Adams, W.R., Willetts, N.G., Rice, T.B., Mackey, C.T., Krueger, R.W.,

Kausch, A.P. and Lemaux, P.G. (1990) Transformation of maize cells and regeneration of fertile transgenic plants. *Plant Cell* **2**: 603-609.

Gressel, J. (1993) Advances in achieving the needs for biotechnologically-derived (B-D) herbicide-resistant crops. *Plant Breeding Review* **11**: 155-183.

Gressel, J. and Segel, L.A. (1990) Herbicide rotations and mixtures. Effective strategies to delay resistance. *American Chemical Society Symposium Series* **421**: 430-458.

Holt, J.S., Powles, S.B. and Holtum, J.A.M. (1993) Mechanisms and agronomic aspects of herbicide resistance. *Annual Review Plant Physiology Plant Molecular Biology* **44**: 203-229.

Horsch, R.B., Fry, J.E., Hoffmann, N.L., Eichholtz, D., Rogers, S.G. and Fraley, R.T. (1985) A simple and general method for transferring genes into plants. *Science* **227**: 1229- 1232.

LeBaron, H.M. and McFarland, J. (1990) Herbicide resistance in weeds and crops. An overview and prognosis. *American Chemical Society Symposium Series* **421**: 336-352.

Matzke, M. and Matzke, A.J.M. (1993) Genomic imprinting in plants: parental effects and *trans*-inactivation phenomena. *Annual Review Plant Physiology Plant Molecular Biology* **44**: 53-76.

Mazur, B.J. and Falco, S.C. (1989) The development of herbicide resistant crops. *Annual Review Plant Physiology Plant Molecular Biology* **40**: 441-470.

Moss, S.R. and Rubin, B. (1993). Herbicide-resistant weeds: a worldwide perspective. *Journal Agricultural Science* **120**: 141-148.

Mullineaux, P.M. (1992) Genetically engineered plants for herbicide resistance. *In: Plant Genetic Manipulation for Crop Protection* (A.M.R. Gatehouse, V.A. Hilder and D. Boulder, eds.). *Biotechnology in Agriculture Series* **7** : 75-107.

Napoli, C., Lemieux, C. and Jorgensen, R. (1990) Introduction of a chimeric chalcone synthase gene into petunia results in reversible co-suppression of homologous genes *in trans. Plant Cell* **2**: 279-289.

Oxtoby, E. and Hughes, M.A. (1989) Breeding for herbicide resistance using molecular and cell techniques. *Euphytica* **40**: 173-180.

Powles, S.B. and Holtum, J.A.M. (eds.) (1994) *Herbicide Resistance in Plants: Biology and Biochemistry.* CRC Press, Oxford, 353 pp.

Quinn, J.P. (1990) Evolving strategies for the genetic engineering of herbicide resistance in plants. *Biotechnology Advances* **8**: 321-333.

Rost, T.L., Gladish, D., Steffen, J. and Robbins, J. (1990) Is there a relationship between branched amino acid pool size and cell cycle inhibition in roots treated with imidazolinone herbicides? *Journal Plant Growth Regulation* **9**: 227-232.

Schulz, A., Wengenmayer, F. and Goodman, H.M. (1990) Genetic engineering of herbicide resistance in higher plants. *CRC Critical Review Plant Sciences* **9**: 1 -34.

Somers, D.A., Rines, H.W., Gu, W., Kaeppler, H.F. and Bushnell, W.R. (1992) Fertile, transgenic oat plants. *Bio/Technology* **10**: 1589-1593.

Stalker, D.M. (1991) Developing herbicide resistance in crops by gene transfer technology. *Plant Biotechnol*ogy **1**: 82-104.

Tauer, L.W. and Love, J.M. (1989) The potential economic impact of herbicide-resistant corn in the USA. *Journal Production Agriculture* **2**: 202-207.

Vasil, V., Castillo, A., Fromm, M. and Vasil, I. (1992) Herbicide resistant fertile transgenic wheat plants obtained by micro-projectile bombardment of regenerable embryogenic callus. *Bio/Technology* **10**: 667-673.

Wan, Y.C. and Lemaux, P.G. (1994) Generation of large numbers of independently transformed fertile barley plants. *Plant Physiol*ogy **104**: 37-42.

Weeks, J.T., Anderson, O.D. and Blechl, A.E. (1993) Rapid production of multiple independent lines of fertile transgenic wheat (*Triticum aestivum*). *Plant Physiol*ogy **102**: 1077-1084.

CHAPTER 28
UNEXPECTED FALLOUT FROM HERBICIDE RESISTANCE; RESISTANCE TO DROUGHT, OXIDANTS AND SOME FUNGAL STRESSES

G. M. PASTORI
Laboratorio de Fisiología Vegetal, Facultad de Ciencias Exactas, Físicas y Naturales, Universidad de Córdoba, 5000 Córdoba, Argentina.

Introduction

Higher plants are sometimes subjected to environmental stress situations which involve the generation of active O_2 species. This "oxygen" stress may be caused by many factors, such as ionising and UV-radiation, flooding, drought, high salt concentrations, dramatic temperature changes (heating, chilling or freezing), air pollutants (SO_2, O_3), herbicides (paraquat, diquat, acifluorfen), heavy metals and pathogens. In this work, the mechanisms of plant resistance to herbicides, drought, oxidants and fungi will be discussed from the point of view of the active oxygen species.

Active O_2 species are formed in normal cell metabolism and their production and destruction is a regulated cellular phenomenon. Oxygen as an unreactive molecule in the triplet ground state ($^3O_2$), must be activated in order to react with atoms or molecules in the singlet ground state (for review, see Elstner and Osswald, 1994). Oxygen activation may occur by two different mechanisms:

1. Oxygen can be activated by photodynamic reactions through pigments, leading the formation of singlet oxygen ($^1O_2$). $^1O_2$ reacts rapidly with most organic molecules, especially at double bonds, producing hydroperoxides.

2. Oxygen can be reduced by one, two, three or four electrons, yielding superoxide anion radical ($O_2^{\cdot-}$), hydrogen peroxide (H_2O_2), hydroxyl radical ($^{\cdot}OH$) and water, respectively. $O_2^{\cdot-}$ and H_2O_2 are produced enzymatically in the normal cell metabolism; no enzyme is yet known capable of yielding $^{\cdot}OH$. Active O_2 species are also generated by non-enzymatic reactions even in normal conditions. In the presence of iron complexes, $O_2^{\cdot-}$ and H_2O_2 can be generated, react with each other and give rise to the highly toxic $^{\cdot}OH$. This is the main source of $^{\cdot}OH$ and, since no enzyme is capable of removing it, its formation is dangerous for the cell.

Active O_2 species react with all the cell components, causing breakdown of nucleic acids, enzyme inactivation, lipid peroxidation, photosynthesis inhibition, proteolysis, and chlorophyll degradation (for review, see Halliwell an Gutteridge, 1989).

Mechanisms of Active O_2 Detoxification

Plant survival against active O_2 depends on the presence of highly efficient antioxidant system, continued by enzymatic and non-enzymatic components. Non-enzymatic components which are small molecules like glutathione, ascorbate, α-tocopherol and carotenoid pigments are very well-distributed in the cell compartments and play an essential role against toxic O_2 species. Enzymatic components are superoxide dismutase, catalase, ascorbate peroxidase and other peroxidases, and enzymes that synthesize and regenerate antioxidants, such as glutathione reductase (for review, see Foyer *et al.*, 1994). These antioxidant enzymes play a significant role in many physiological and stress situations. Ascorbate is present in chloroplasts, cytosol, vacuole and apoplastic space of leaf cells at high concentrations and, closely linked to glutathione (GSH) plays a very important role, as a H_2O_2-detoxifying system (Figure 1). In this detoxification mechanism, H_2O_2 is reduced to water by the enzyme ascorbate peroxidase (AP), which is present in the chloroplast and in the cytosol and exhibits a high degree of specificity for ascorbate as the electron donor. Ascorbate is oxidized in two steps, producing monodehydroascorbate (MDA) and dehydroascorbate (DHA). Ascorbate is regenerated by two chloroplastic enzymes, monodehydroascorbate reductase (MDAR) and dehydroascorbate reductase (DHAR), the first one depending on the presence of NADPH. Glutathione is the major low molecular weight thiol compound in most plants. It has an important role in sulphur transport, it can react with singlet O_2 and $^{\cdot}OH$ and protects the thiol groups of enzymes. It participates in the regeneration of ascorbate from dehydroascorbate, via dehydroascorbate reductase. Glutathione is synthesized in both the chloroplast and in the cytosol, where it is maintained in its reduced form (GSH) by the enzyme glutathione reductase (GR) in a NADPH-dependent reaction. GR is located mainly in the chloroplast stroma, but is also found in the mitochondria and cytosol.

Superoxide dismutases (SODs) are a family of metalloenzymes that catalyze the disproportionation of superoxide anion ($O_2^{\cdot-}$) to molecular oxygen and H_2O_2, decreasing the risk of $^{\cdot}OH$ formation by the Haber-Weiss reaction. There are three forms of SOD, depending on the metal cofactor: copper-zinc, manganese and iron. SOD is an ubiquitous enzyme located in all cellular compartments, with a very important function in the $O_2^{\cdot-}$ detoxification process (for a review, see Bowler *et al.*, 1992).

Catalase is a tetrameric heme-containing enzyme which protects from the toxic effects of H_2O_2, by converting H_2O_2 to water and molecular oxygen, although it has a poor affinity for its substrate and is a light-sensitive enzyme exclusively located in peroxisomes (Feierabend *et al.*, 1992). All these mechanisms of detoxification must ensure that $O_2^{\cdot-}$ and

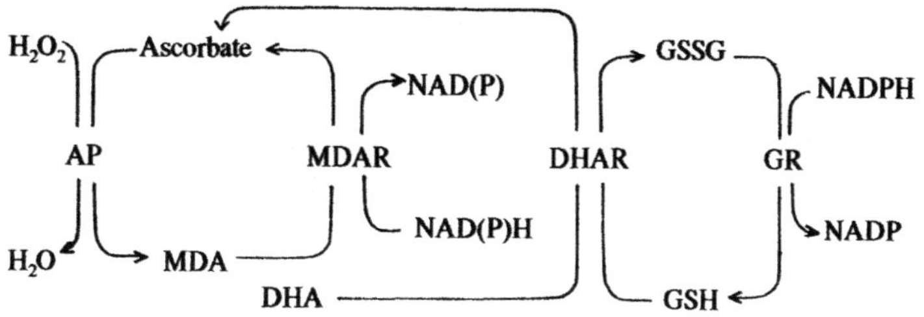

Figure 1. H_2O_2 detoxifying system (taken from Foyer *et al.*, 1994).

H_2O_2 do not come into contact with each other, in order to avoid the risk of formation of the highly toxic $^{\cdot}OH$ by the Haber-Weiss reaction.

Response of Antioxidant Enzymes to Stress

Herbicide Stress

Bipyridyl herbicides such as paraquat or diquat provoke oxidative stress directly by generating active O_2 species. Paraquat is a non-selective, contact and photosynthesis-inhibiting herbicide extensively used for total weed control. In light-exposed plant it exerts its phytotoxic effect by accepting electrons from PSI and transferring them to molecular oxygen producing $O_2^{\cdot-}$ and other toxic O_2 species such as H_2O_2 and $^{\cdot}OH$ (Dodge, 1971). With the constant regeneration of oxidized paraquat by O_2, the herbicide is always avaible to shunt electrons to O_2 and, therefore, is effective in catalytic amounts inside the chloroplasts. Paraquat can also be toxic in non-photosynthetic tissues following its reduction in the mitochondrial electron transport chain (Harvey and Harper, 1982). Paraquat-generated $O_2^{\cdot-}$ or its decomposition products, $O_2^{\cdot-}$ and H_2O_2, produce increases in lipid peroxidation, membrane permeability and proteolysis, as well as chlorophyll degradation (Pastori and Trippi, 1992, 1993a).

There are many examples showing the response of antioxidant enzymes to the effect of paraquat. SOD and catalase activities increase in maize by the effect of paraquat (Matters and Scandalios, 1986). Increases in SOD and GR activities were also found in paraquat-stressed wheat leaves (Baisak *et al.*, 1994).

Correlations between antioxidant enzymes and paraquat resistance have been demonstrated. Harper and Harvey (1978) analyzed antioxidant activities in paraquat-tolerant and paraquat-susceptible cultivars of *Lolium perenne* and found that constitutive activities of SOD and CAT were higher in the paraquat-tolerant lines than in the susceptible ones. SOD, GR and AP were also induced in a paraquat-resistant biotype of *Conyza bonariensis* when subjected to atrazine or acifluorfen; in this biotype, there were elevated constitutive levels of these enzymes (Shaaltiel and Gressel, 1986). A *Conyza canadiensis* biotype with resistance to paraquat also showed high SOD, GR and AP activities (Matsunaka and Ito,

1991). In an ozone-tolerant cultivar of tobacco which was also resistant to paraquat, increases in GR and SOD activities were found (Shaaltiel *et al.*, 1988). The effect of this herbicide was studied in drought-resistant and drought-sensitive maize strains (Table 1) (Pastori and Trippi, 1992, 1993a). Results showed that SOD, GR and AP activities were increased in the drought-resistant strain, even in senescent tissues. In the drought-sensitive strain there were stronger symptoms of oxidative damage by decreasing of chlorophyll and protein content and increasing lipid peroxidation and membrane permeability.

Table 1. Antioxidant enzyme activities in leaves of drought-resistant (LIZA) and drought-sensitive (LG11) maize strains, exposed to 1 mM paraquat and 1 mM H_2O_2 for 12 h (redrawn from Pastori and Trippi, 1992). Enzyme activities are expressed in U/cm^2 leaf area (for definition of U and statistics, see Pastori and Trippi, 1992).

	Glutathione reductase		Ascorbate peroxidase		Superoxide dismutase	
Assay	LIZA	LG11	LIZA	LG11	LIZA	LG11
Control	5.1	2.1	20.2	19.3	20.4	19.5
Paraquat	21.3	7.2	50.3	22.4	37.8	22.0
H_2O	20.5	6.3	43.7	20.3	35.3	20.6

Drought Stress

Stomatal closure takes place during drought stress, as a way of optimizing the water use efficiency of the plant. This results in a reduction of CO_2 assimilation and therefore, in CO_2 fixation while photosynthetic electron transport is operating at normal rates. Under these conditions, limited quantities of NADP are available to accept electrons, therefore O_2 can functions as an alternative electron acceptor (Gamble and Burke, 1984). Reduction of O_2 and formation of $O_2^{\cdot -}$ and together with high concentrations of iron give rise to the highly reactive hydroxyl radical $^{\cdot}OH$ during drought (Price *et al.*, 1989), which can cause severe damage to membranes, lipids and proteins (Pastori and Trippi, 1993b).

In wheat leaves, high GR and catalase activities were found by the effect of water stress (Gamble and Burke, 1984). Burke *et al.* (1985) showed increases in GR activity in cotton leaves exposed to drought. During water stress, a drought-tolerant moss *Tortula ruralis* showed increased levels of SOD and catalase (Dhindsa and Matowe, 1981). Drought-tolerant and-intolerant maize inbreds were studied by Malan *et al.* (1990); resistance was correlated with CuZn-SOD and GR activities. The drought-resistant variety was also tolerant to paraquat, acifluorfen and SO_2. In drought-resistant wheat leaves, SOD, GR and AP activities increased when subjected to water stress and showed more capacity to overlead oxidative damage to chlorophylls, proteins and membranes (Pastori and Trippi, 1993b). The drought-resistant strain also exhibited resistance to high O_2 stress.

Oxidant Stress

Many types of oxidants have been used to provoke stress or are produced by environmental contaminations. Several works have been carried out analysing the effect of high O_2 pressures and H_2O_2. H_2O_2, as a reduced form of O_2, causes oxidative damage. Due to its diffusible properties, it reaches a wide range of cellular components provoking lipid peroxidation, chlorophyll degradation and proteolysis (Pastori and Trippi, 1992, 1993a). The effect of H_2O_2 was analyzed in *Ricinus communis* endosperm in which catalase activity was inhibited by aminotriazole, yielding H_2O_2 accumulation. AP activity was increased by this treatment (Klapheck *et al.*, 1990). The response to H_2O_2 was studied in drought-resistant and drought-sensitive maize strains. Significant increases in GR, AP and SOD activities were found in leaves of the drought-resistant strain (Table 1; Pastori and Trippi, 1992), and also in senescent tissues (Pastori and Trippi, 1993a). Decreases in chlorophyll and protein content and increases in lipid peroxidation and membrane permeability were stronger in the drought-sensitive strain.

The effects of high O_2 pressures could be similar to those provoked by O_3. Ozone enters plant leaves through open stomata. Under high humidity, stomatal opening increases and plants are more sensitive to air pollution. Once O_3 has entered the leaf, reduced forms of O_2, like $O_2^{\cdot-}$, H_2O_2 and $^{\cdot}OH$, alter cell integrity by reacting with components of cell walls and plasma membranes (Melhorn *et al.*, 1990). It has been shown that atmospheres of 100% O_2 cause oxidative damage to membranes either by increasing lipid peroxidation and permeability or provoking important changes in lipid composition, as well as decreasing chlorophyll contents (Pastori and Trippi, 1993b, 1995). Responses of SOD and GR were analyzed in cotton and maize leaves exposed to 75% O_2 (Foster and Hess, 1980, 1982). SOD and GR activity were increased by the treatment in cotton whereas only GR activity was induced in maize. An atmosphere containing 90% O_2 provoked increases in the cytosolic SOD activity (Matters and Scandalios, 1986). In leaves of drought-resistant and drought-sensitive wheat strains subjected to 100% O_2, high levels of GR, AP and SOD activities were found but only in the drought-resistant strain (Pastori and Trippi, 1993b).

Fungal Stress

Many plant pathogens produce toxins which cause a rapid tissue necrosis. Fungi of the genus *Cercospora* are important pathogens of numerous crops, such as corn, rice, soybeans, coffee and tobacco. *Cercospora* species infect aerial portions of the plant by producing a light-induced, photoactivated polyketide toxin, cercosporin. By itself, cercosporin does no damage to the plant. However, in the presence of light the toxin molecule absorbs energy from the blue portion of the spectrum and is concerted to an excited triplet state, which reacts with molecular oxygen yielding both singlet O_2 and $O_2^{\cdot-}$ (Daub and Hangarter, 1993). These activated O_2 species are directly toxic to living tissues, causing oxidation of cellular components, including lipids, proteins carbohydrates and nucleic acids.

In maize leaves, high total catalase activity as well as changes in the steady-state RNA leavels were found as a consequence of exposure to cercosporin (Williamson and Scandalios, 1992). Transcript accumulation of maize SODs increased strongly whereas total SOD activity and isozyme protein patterns remained constant after cercosporin treatment. The authors suggest that protein turnover might play a role in the response of SODs to stress. A paraquat-resistant tobacco variety, also resistant to cercosporin, showed high levels of SOD activity (Furusawa *et al.*, 1984). Paraquat-tolerant tobacco plants were selected and regenerated *in vitro*, exhibiting cross tolerances to acifluorfen, freezing and other oxidative stresses. Paraquat-tolerant plants did not show necrotic symptoms after fungal stress (Barna *et al.*, 1993).

Cross Resistance and Antioxidant Enzymes

Active O_2 species are highly dangerous; therefore a regulated balance between oxygen radical production and destruction is required for metabolic efficiency and function either in normal or in stress conditions. The key role of antioxidant enzymes in the mechanism of defense against oxidative stress generated by paraquat, drought, oxidants and fungi has been shown, and points out that the efficiency of the antioxidant capacity could be at least one of the mechanisms of stress resistance. The phenomenon of cross resistances described previously between paraquat, drought, oxidants and fungal stresses could be the result of a common feature, which is the generation of active O_2 species. But concluding that a plant which shows resistance to any type of stress could be resistant to any other, would be a simplified and incomplete approach in order to define the mechanism of stress resistance. Each type of stress has its own mode of action and generates oxidative damage in specific sites and compartments of the cell. However, a plant with an efficient antioxidant capacity capable of reducing oxidative damage generated by any kind of stress may have an important advantage over a plant which has not got it. A good tool for evaluating this phenomenon in the overexpression of antioxidant enzymes in plants with agronomic relevance. There are many groups in the world transforming plants which overexpress any of the antioxidant enzymes with the aim to improve plant stress resistance. Bowler *et al.* (1991) found that a high level of overproduction of Mn-SOD in tobacco plants could give protection against the oxidative effect of paraquat. Potato plants transformed with tomato cDNAs from chloroplastic and cytosolic CuZn-SODs also showed increased tolerance to paraquat (Perl-Treves *et al.*, 1993). Tobacco plants overexpressing chloroplastic CuZn-SOD showed decreased levels of oxidative damage in the presence of paraquat and were less sensitive to photoinhibition (Gupta *et al.*, 1993). Foyer *et al.* (1991) demonstrated that tobacco plants transformed using the bacterial *gor* gene, which codifies for GR, showed increased GR activity under light and paraquat stresses. In the same way, Aono *et al.* (1991, 1993) observed a decrease in leaf oxidative damage in transformed tobacco plants with the *gor* gene when exposed to paraquat and photooxidative stress. All these evidences lead to suggest that the simultaneous overexpression of more than one component of the antioxidant system could be a way to improve substantially stress resistance. The selection of which antioxidant enzymes are going to be used to transform the plant, as well as the compartments in which these enzymes are going to be overexpressed are important things to take into account in order to reinforce the antioxidant capacity in the specific sites in which active O_2 species are being produced. This would be a good approach to the comprehension of the mechanism of plant stress resistance. The release of transgenic plants overexpressing antioxidant enzymes to the field in which environmental stress situations take place, will let us evaluate the role of antioxidant enzymes in plant stress resistance and its real agronomical value.

References

Aono, M., Kubo, A., Saji, H., Natori, T., Tanaka, H. and Kondo, N. (1991) Resistance to active oxygen toxicity of transgenic *Nicotiana tabacum* that expresses the gene for glutathione reductase from *Escherichia coli. Plant Cell Physiology* **32**: 691-697.

Aono, M., Kubo, A., Saji, H., Tanaka, H. and Kondo, N. (1993) Enhanced tolerance to photo-oxidative stress of transgenic *Nicotiana tabacum* with high glutathione reductase activity. *Plant Cell Physiology* **34**: 129-135.

Baisak, R., Rana, D., Acharya, P. and Kar, M. (1994) Alterations in the activities of active oxygen scavenging enzymes of wheat leaves subjected to water stress. *Plant Cell Physiology* **35**: 489-495.

Barna, B., Ádám, A. and Király, Z. (1993) Juvenility and resistance of a superoxide-tolerant plant to diseases and other stresses. *Naturwissenschaften* **80**: 420-422.

Bowler, C., Slooten, L., Vandenbraden, S., De Rycke, R., Botterman, J., Sybesma, C., Van Montagu, M. and Inzé, D. (1991) Manganese superoxide dismutase can reduce cellular damage mediated by oxygen free radicals in transgenic plants. *EMBO Journal* **10**: 1723-1732.

Bowler, C., Van Montagu, M. and Inzé, D. (1992) Superoxide dismutase and stress tolerance. *Annual Review Plant Physiology Plant Molecular Biology* **43**: 83-115.

Burke, J., Gamble, P., Hatfield, J. and Quisenberry, J. (1985) Plant morphological and biochemical responses to field water dificits. I. Responses of glutathione reductase activity and paraquat sensitivity. *Plant Physiology* **79**: 415-419.

Daub, M. and Hangarter, R. (1983) Production of singlet oxygen and superoxide by the fungal toxin, cercosporin. *Plant Physiology* **73**: 855-857.

Dhindsa, R. and Matowe, W. (1981) Drought tolerance in two mosses correlated with enzymatic defense against lipid peroxidation. *Journal Experimental Botany* **32**: 79-91.

Dodge, A. (1971) The mode of action of bipyridylium herbicides, paraquat and diquat. *Endeavour* **30**: 130-135.

Elstner, E. and Osswald, W. (1994) Mechanisms of oxygen activation during plant stress. *Proceedings Royal Society of Edinburgh* **102B**: 131-154.

Feierabend, J., Schaan, C. and Hertwing, B. (1992) Photoinactivation of catalase occurs under both high and low-temperature stress conditions and accompanies photoinhibition of photosystem II. *Plant Physiology* **100**: 1554-1561.

Foster, J. and Hess, J. (1980) Responses of superoxide dismutase and glutathione reductase activities in cotton leaf tissue exposed to an atmosphere enriched in oxygen. *Plant Physiology* **66**: 482-487.

Foster, J. and Hess, J. (1982) Oxygen effects on maize leaf superoxide dismutase and glutathione reductase. *Phytochemistry* **21**: 1527-1532.

Foyer, C., Lelandais, M., Galap, C. and Kunert, K. (1991) Effects of elevated cytosolic glutathione reductase activity on the cellular glutathione pool and photosynthesis in leaves under normal and stress coditions. *Plant Physiology* **97**: 863-872.

Foyer, C., Descourvières, P. and Kunert, K. (1994) Protection against oxygen radicals: an important defence mechanism studied in transgenic plants. *Plant Cell Environment* **17**: 507-523.

Furusawa, I., Tanaka, K., Thanutong, P., Mizuguchi, A., Yazaki, M. and Asada, K. (1984) Paraquat resistant tobacco calluses with enhanced superoxide dismutase activity. *Plant Cell Physiology* **25**: 1247-1254.

Gamble, P. and Burke, J. (1984) Effect of water stress on the chloroplast antioxidant system. I. Alterations in glutathione reductase activity. *Plant Physiology* **76**: 615-621.

Gupta, A., Heinen, J., Holady, A., Burke, J. and Allen, R. (1993) Increased resistance to oxidative stress in transgenic plants that over-express chloroplastic Cu/Zn superoxide dismutase. *Procedings National Academy Sciences USA* **90**: 1629-1633.

Halliwell, B. and Gutteridge, J., eds. (1989) *Free Radicals in Biology and Medicine*. Clarendon Press, Oxford.

Harper, D. and Harvey, P. (1978) Mechanism of paraquat tolerance in perennial ryegrass. II. Role of superoxide dismutase, catalase and peroxidase. *Plant Cell Environment* **1**: 211-215.

Harvey, B. and Harper, D. (1982) Tolerance to bipyridylium herbicides. *In: Herbicide Resistance in Plants* (H. LeBaron and J. Gressel, eds.), 215-223. John Wiley & Sons, Chichester.

Klapheck, S., Zimmer, I. and Cosse, H. (1990) Scavenging of hydrogen peroxide in the endosperm of Ricinus communis by ascorbate peroxidase. *Plant Cell Physiology* **31**: 1005-1013.

Malan, C., Greyling, M. and Gressel, J. (1990) Correlation between Cu-Zn superoxide dismutase and glutathione reductase, and environmental and xenobiotic stress tolerance in maize inbreds. *Plant Science* **69**: 157-166.

Matters, G. and Scandalios, J. (1986) Effect of free radical-generating herbicide paraquat on the expression of the superoxide dismutase (SOD) genes in maize. *Biochimica Biophysica Acta* **882**: 29-39.

Matsunaka, S. and Ito, K. (1991) Paraquat resistance in Japan. *In: Herbicide Resistance in Weeds and Crops* (J. Caseley, G. Cussans, and R. Atkin, eds.). Butterworths, Oxford.

Melhorn, H., Tabner, B. and Wlburn, A. (1990) Electron spin resonance evidence for the formation of free radicals in plants exposed to ozone. *Physiologia Plantarum* **79**: 377-383.

Pastori, G. and Trippi, V. (1992) Oxidative stress induces high rate of glutathione reductase synthesis in a drought-resistance maize strain. *Plant Cell Physiol* **33**: 377-383.

Pastori, G. and Trippi, V. (1993a) Antioxidative protection in a drought-resistant maize strain during leaf senescence. *Physiologia Plantarum* **87**: 227-231.

Pastori, G. and Trippi, V. (1993b) Cross resistance between water and oxidative stresses in wheat leaves. *Journal Agricultural Science* **120**: 289-294.

Pastori, G. and Trippi, V. (1995) Fatty acid composition in water-and oxygen-stressed leaves of maize and wheat strains. *Phytochemistry* (in press).

Perl, A., Perl-Treves, R., Galili, S., Shalgi, E., Malkin, S. and Galun, E. (1993) Enhanced oxidative-stress defense in transgenic potato expressing tomato Cu, Zn superoxide dismutases. *Theoretical Applied Genetics* **85**: 568-576.

Price, A., Atherton, N. and Hendry, G. (1989) Plants under drought-stress generate activated oxygen. *Free Radical Researchs Communications* **8**: 61-66.

Shaaltiel, Y. and Gressel, J. (1986) Mutienzyme oxygen radical detoxifying system correlated with paraquat resistance in *Conyza bonariensis. Pesticide Biochemistry Physiology* **26**: 22-28.

Shaaltiel, Y., Glazer, A., Bocion, P. and Gressel, J. (1988) Cross tolerance to herbicidal and environmental oxidants of plant biotypes tolerant to paraquat, sulfur dioxide and ozone. *Pesticide Biochemistry Physiology* **31**: 13-23.

Williamson, J. and Scandalios, J. (1992) Differential response of maize catalases and superoxide dismutases to the photoactivated fungal toxin cercosporin. *Plant Journal* **2**: 351-358.

CHAPTER 29
SYNERGIZED MYCOHERBICIDES FOR RESISTANCE MANAGEMENT

J. GRESSEL[1], Z. AMSELLEM[1], D. MICHAELI[1] and A. SHARON[2]
1. Plant Genetics, Weizmann Institute of Science, Rehovot, Israel. 2. Botany Deptartment, Tel-Aviv University, Tel-Aviv, Israel.

Why Mycoherbicides?

Despite the large number of excellent herbicides available, there are many cases where there are needs to control single weed species. These needs include:

a) Weeds that have evolved resistance to a still favored herbicide. Typically, only a single weed (initially) evolves resistance under monoherbicide monoculture, and the herbicide still adequately controls all other weeds.

b) There are weeds that are naturally resistant to the herbicide, and no selectivity is available (except by genetically engineering new resistance to the crop). This is typical when the weeds are within the same botanical family as the crops: e.g. *Cassia* in soybeans, *Abutilon* in cotton, barleys and bromes in wheat, etc. Parasitic weeds are often in this class; few selective herbicides control them. Recently, acetolactate synthase (ALS) inhibitors were shown to control parasites in legumes where there is a modicum of effective natural herbicide resistance (Gacia-Torres and López-Granados, 1991), as well as where there is genetically-engineered herbicide resistance in crops (Joel *et al.*, 1995). In such instances the parasitic weeds might join group (a), as weeds, including parasites rapidly evolve resistance to this herbicide group (Sibony *et al.*, 1995).

c) There are many cases where an especially high rate of a herbicide is needed to control one weed species and the rest of the weed spectrum can be controlled by lower rates. If this single weed spp. could be controlled by non-chemical means, then lower, more economical and less potentially polluting rates of the herbicide could be used.

d) There are often problems with single weed spp., where herbicides cannot economically or easily be used. These include: (i) pasture weeds such as spurge and bracken, where herbicides can be uneconomical; (ii) urban situations with allergenic weeds, but spraying witht chemicals is unacceptable for any of a variety of reasons; (iii) narcotic weeds (that fit the definition of being deleterious to humanity), which need aerial treatment without damage to nearby crops.

In all these cases biological control is appropriate because it can potentially afford selectivities unavailable cost-effectively with presently available chemicals. This review will not cover the following biological control agents: (a) manual labores; (b) insects; (c) "classical" biological control agents; i.e. agents that adequately keep a wild plant species from becoming a weed. Where such species are indigenous they are naturally suppressed, but where they have been introduced they became weeds. Only cases are discussed where a pathogen was initially found in biological balance with a weed. Pathogens are often specific to single host species, and it is such pathogens that can allow the necessary selectivity between crop and weed host. There are many reports about finding myco-herbicidal agents, as well as excellent books on the biocontrol of weeds (cf. TeBeest, 1991), but commercial successes are limited. The typical cases use indigenous agents often just make plants sick, but do not kill them. A truly "inundative" approach is initiated; only with heavy inoculation are weeds sometimes controlled (Table 1).

Reasons for High Inoculation Levels

From an evolutionary standpoint the inundation (Table 1) is understandable; if an organism was hyperpathogenic, both it and the weed host might have become extinct or at least the weed would be diseased enough to render it to be just another wild but not weedy species. Thus, ways must be found to circumvent the need for heavy inoculation, unless one accepts it to be as a "given", as many do. Most industrial efforts have been to produce and formulate the inocula less expensively, instead of finding ways to use less.

There are two groups of reasons for the need of such gargantuan inundations of fungal material: (a) There can be inherent problems in getting the organism to the right place and keeping it alive until an adequate infection is established. There are problems in getting soil and stem pathogens to their inaccessable targets. Leaf and stem pathogens often need long dew periods (6-18h) to germinate and become established. Such optimal conditions do not often exist in nature, which is why many crop diseases are sporadic and weather related; (b) Despite a pathogen being "compatible" with the weed, it does have to overcome the weed's inherent defense mechanisms. This is a "quantitative" battle; i.e .a function of the amount of soldiers (inoculum) used. The attacked weeds initiate various defenses to suppress spread of invading pathogens including: i) PR (pathogenicity related) proteins (chitinases and glucanases as well as other stress-related proteins). The chitinases and glucanases degrade fungal cell walls, preventing penetration and growth in the tissue; (ii) phytoalexins, non-protein, secondary metabolites of various chemistries that poison the fungus; (iii) physical barriers composed of polymers such as suberin, lignin, callose, or mixtures of some or all of these. Often papillae of callose rapidly form, blocking the advancing fungal hyphae. Such papillae can contain small amounts of lignin. Traces of lignification can severely suppress degradation of carbohydrate polymers such as callose and cellulose by fungal enzymes (Gressel *et al.*, 1983). Different species initiate different defenses, and the levels of induced and pre-existing defenses vary during weed growth and development. The outcome of this is that different quantities of inocula are needed at different stages of growth, a major complication to using biological control agents. Similar problems often exist with herbicides.

Table 1. Examples of inoculum levels of mycoherbicides.

Pathogen	Major target	Level used (spores/cm^2)	Refer.
Fusarium lateritium	*Abutilon* sp. & *Sida* sp.	7,500	e
Colletotrichum coccodes	*Solanum ptycanthum*	3,000	f
	Abutilon theophrasti	20,000	g
C. gloeosporiodes	*Aeschynomene virginica*	600	h
C. orbiculare	*Xanthium orbiculare*	2,000	i
C. dematium	*Crotolaria spectabilis*	10,000	j
Alternaria macrospora	*Anoda cristata*	200	k
		1155 kg/ha[b]	l
Alternaria cassiae	*Cassia obtusifolia*	500[a]	m
Phoma aquilina	*Pteridium aquilinum*	40,000	n
Cercospora rodmanii	*Eichhornia crassipes*	11kg/ha[c]	o
Puccinia expansa	*Senecio* spp.	500,000[d]	p
Fusarium udum	*Crotolaria spectabilis*	10,000	j

[a]calculated from stated spore concetration and typical spray volume of 115 l/ha, unless stated as "to run off" and then 1000 l/ha.; [b]Granular formulation containing 10^5 spores/g.; [c]blended mycelium; [d]per seedling at 3 leaf stage; [e]Boyette and Walker 1985; [f]Anderson and Walker 1985; [g]Wymore *et al.*, 1988; [h]Daniel *et al.*, 1973; [i]Klerk *et al.*, 1985; [j]Charudattan, 1986; [k]Walker 1981; [l]Walker and Sciumbato 1973; [m]Walker and Riley, 1982; [n]Burge and Irvine, 1985; [o]Charudattan *et al.*, 1985; [p]Alber *et al.*, 1986.

Synergizing Mycoherbicides by Formulation

Getting a biocontrol agent to the right place and keeping it there is the task of formulants. Initially, mycoherbicides that attack roots were at rates of hundreds of kg per hectare together with the solid substrate in which they were cultured. Organisms for infecting weed leaves and stems were formulated in water, sometimes with some thickening agents to keep spores from settling out, some detergent to prevent clumping, and sticking agents to hold the inoculum on the leaves and stems.

The need for extended dew periods was addressed by developing formulants having an oil based component and an emulsifier, such that water and medium would surround the propagule (spore or mycelium) with an oil wrapper to prevent loss of water (cf. Boyette *et al.*, 1991). One such invert emulsion very efficient in lowering the inoculum level required (Amsellem *et al.*, 1990), but also abolished selectivity (Amsellem *et al.*, 1991). Organosilicone surfactants also show promise (Zidack *et al.*, 1992).

Several novel solid substrates, and pelleted or beaded formulations are discussed in Boyette *et al.* (1991) for preplant or pre-emergence soil applications. They somewhat decrease inoculum requirements from gargantuan to very large. We are testing a special case; the biocontrol of root-attacking parasitic weeds. We have some success at the greenhouse level in treating transplant plugs of tomatoes with a pathogen of broomrape prior to planting out. Presumably the pathogen continues growing along the rhizoplane, attacking the broomrape as or after it attaches to roots. The ideal general situation would be to coat crop seeds with a weed-specific pathogen that would spread out in the soil near the developing crop, and kill nearby weeds.

Synergizing Mycoherbicides by Suppressing Weed Defenses

Chemicals, especially herbicides, have been used, some seemingly "off the shelf", to ascertain if they will synergize mycoherbicides and bacterial agents used for biocontrol of weeds (Caulder and Stowell, 1988; Gressel *et al.*, 1993; Christy *et al.*, 1992). Random testing in this manner with the few hundred registered herbicides seems like a daunting task. There are also many non-herbicidal chemicals that also might prevent the inception of induced defenses in the weeds. A little forethought should clearly allow testing only more relevant compounds as synergists. This requires an understanding of the defenses being used by the pathogens. This approach can work successfully, at least at the laboratory level. We chose a model legume weed *Cassia obtusifolia*, a great problem in soybeans. Attack by *Alternaria cassiae*, a rather specific pathogen induced the synthesis of a large number of phenolic compounds. Most did not inhibit pathogen growth but one was clearly a phytoalexin. This compound was isolated (Sharon *et al.*, 1992a) and its structure determined by NMR and MS to be 2-(p-hydroxyphenoxy)-5,7 dihydrochromone (Figure 1, insert). The B ring of this compound came from the shikimate pathway, as ascertained by feeding leaves with ^{14}C-labeled phenylalanine. Leaves were treated with a spore suspension containing various concentrations of glyphosate, an inhibitor of the shikimate pathway. As the level of glyphosate was increased towards 50 μM (20 fold less than is phytotoxic) the level of phytoalexin decreased and the intensity of infection increased (Figure 1). The number of conidia required for equal infection also dropped >5 fold when glyphosate used (Sharon *et al.*, 1992b).

Abutilon theophrasti produces callose when attacked by specific pathogens. Various chemicals inhibit callose synthase *in vitro* (Bayles and Aist, 1987), but do not penetrate plant cuticles. Our colleagues A. Warshawsky and V. Kampel have synthesized analogs that penetrate cuticles, and one such compound increased infectivity of a strain of *C. coccodes* that is specific for *Abutilon*, but does not affect cotton. Such synergists could even be envisaged to be used to increase the host range of non-specific pathogens, when the crop does not use callose synthesis as part of its defense mechanisms. We hope this line of

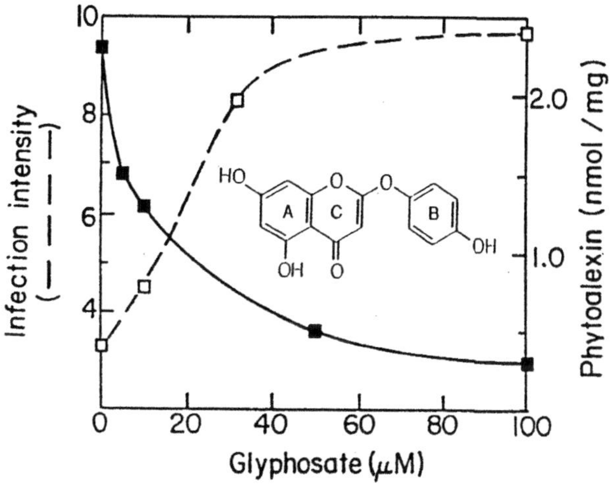

Figure 1. Suppression of the production of the phytoalexin [2-(p-) (hydroxyphenoxy)-5,7dihydroxychromone (insert)], and enhancement of virulence by glyphosate. Virulence (according to Sharon, 1991) and phytoalexin levels (Sharon and Gressel, 1991) were measured a week after infection. Source: previously unpublished.

research continues to show promise, and they are synthesizing more of these compounds to optimize structure.

Biologically Synergizing Biocontrol Agents

Genetic Manipulations

Organisms can be potentially modified to increase pathogenicity by genetic transformation with genes for virulence from other species, by increasing the endogenous expression of genes, or by transfer from other organisms by protoplast fusion (cf. Kistler, 1991; Harman and Stasz, 1991). Increasing virulence, especially by gene transfer, requires extreme care due to environmental impact, i.e. the possibility of increasing the host range to include other crops. Groups are considering engineering "fail safe devices" into their pathogens. Conversely, highly virulent, not too specific organisms can be "disarmed" to allow them to be used where there is crop selectivity. For example a widespread broad-spectrum pathogen *Sclerotinia* was mutated to auxotrophy and to sclerotia minus, i.e. it could only reproduce vegetatively in culture, with the missing nutrients added. It could still infect weeds having the missing nutrients, but "commits suicide" when the weed dies (Sands and Miller, 1993).

One genetic manipulation of mycoherbicides will often be called for; to fungicide resistance. This would allow farmers to treat crops with a fungicide of the types, and at times that would otherwise have suppressed the biocontrol agent. There are alleles for resistance to many fungicides in most fungal populations, and with a little selection pressure in the laboratory, resistant strains can be made available.

Interorganism Synergies

Just as herbicide mixtures often provide synergistically cost-effective weed control, the use of more than one organism can effect better biocontrol. How some of the cases work is unclear. Some show ingenuity and mimic nature, e.g. the use of arthropods to spread microorganisms (cf. Wapshere *et al.*, 1989).

There is a possibility that scientists have overextended Koch's postulates (that to be considered a pathogen, an organism must always be present, isolatable in pure culture, causes disease when reinoculated into a healthy plant, reisolatable) into Koch's "axiom", despite contradictions. The pathogenic *Agrobacterium* disappears from crown galls. Some organisms are only pathogenic when a subsidiary, non-pathogen is present. Many microorganisms are typically found in diseased lesions. Most are there for a free meal, but some may facilitate the primary pathogen. This was clearly shown with a rust that infected but did not control a burr. A *Colletortrichum* that also by itself was ineffective, enters the rust lesions, killing the weed (Morin *et al.*, 1993). Far more research is needed to ascertain synergisms by such facilitory organisms.

Concluding Remarks

We have discussed how it should be possible to synergize biocontrol agents to lower inoculum levels. There are theoretically 2 to 5 orders of magnitude for improvement (Table 1). One order of magnitude renders most biocontrol agents cost-effective.

Impermanent synergies (the use of chemicals and adjuvants) may have advantages over many of the permanent synergies (genetic manipulations without suicide, or organism mixtures that may reproduce in concert). One may have valid fears of both changing host-specificity as well as "persistent residues" of biocontrol agents. Impermanent synergies provide weed control when applied, and the level of the organism should dissipate, just like an environmentally sound chemical herbicides.

It has been said "the human mind is like a parachute; both work best when open". It is hoped that those used to chemical weed control will have an open mind about using synergized biocontrol agents to augment chemical control of weeds, where chemicals fail.

Acknowledgements.

The work of Z. A. and J. G. on the biocontrol of parasitic weeds is supported by grant number HNE-0158-G-00-3060-00 of the trilateral USA-Egypt-Israel program of the U. S. Agency for international Development (A.I.D.). J. G. has the Gilbert de Botton Chair of Plant Sciences.

References

Alber, G., Defago, G., Kern, H. and Sedlar, L. (1986) Host range of *Puccinia expansa*, a possible fungal biocontrol agent against *Senecio* weeds. *Weed Research* **26**: 69-74.

Anderson, R.N. and Walker, H.L. (1985) *Colletotrichum coccodes*: A pathogen of eastern black nightshade *(Solanum ptycanthum)*. *Weed Research* **26**: 69-74.

Amsellem, Z., Sharon, A., Gressel, J. and Quimby, P.C. (1990) Complete abolition of high inoculum threshold of two mycoherbicides (*Alternaria cassiae* and *A. crassa*) when applied in invert emulsion. *Phytopathology* **80**: 925-929.

Amsellem, Z., Sharon, A. and Gressel, J. (1991) Abolition of selectivity of two mycoherbicidal organisms and enhanced virulence of avirulent fungi by an invert emulsion. *Phytopathology* **81**: 985-988.

Bayles, C.J. and Aist, J.R. (1987) Apparent calcium mediation of resistance of an *ml-o* barley mutant to powdery mildew. *Physiological Molecular Plant Pathology* **30**: 337-345.

Boyette, D.C. and Walker, H.L. (1985) Factors influencing biocontrol of velvetleaf (*Abutilon theophrasti*) and prickly sida *(Sida spinosa)* with *Fusarium lateritium*. *Weed Science* **33**: 209-211 .

Boyette, C.D., Quimby, P.C., Connick, W.J., Daigle, D.J. and Fulgham, F.E. (1991) Progress in the production formulation, and application of mycoherbicides. *In: Microbial Control of Weeds* (D.O. TeBeest, ed.), 209-222. Chapman and Hall, New York.

Burge, M.N. and Irvine, J. A. (1985) Recent studies on the potential for biological control of bracken using fungi. *Proceedings Royal Society Edinburg* **86B**: 187-194.

Caulder, J. D. and Stowell, L. (1988) *U.S. Patent* **4**: 775-405.

Charudattan, R., Linda, S.B., Kluefel, M. and Osman, Y.A. (1985) Biocontrol efficacy of *Cercospora radmanii* on water hyacinth. *Phytopathology* **75**: 1263-1269.

Charaduttan, R. (1986) Biological control of showy crotolaria. *Weed Science Society America Abstracts* **26**: 137.

Christy, A.L., Herbst, K., Kostka, S.J., Mullen, J.P. and Carlson, P.S. (1992) Synergizing weed biocontrol agents whith chemical herbicides. *In: Pest Control with Enhanced Environmental Safety* (S.O. Duke, J. J. Menn and J.R. Plimmer, eds.), 87-100. American Chemical Society, Washington, DC.

Daniel, J.T., Templeton, G.E., Smith, R.J. Jr. and Fox, W.T. (1973) Biological control of northern jointvetch in rice with an endemic fungal disease. *Weed Science* **21**: 303-307.

Garcia-Torres, L. and López-Granados, F. (1991) Control of broomrape (*Orobanche crenata* Forsk.) in broad bean (*Vicia faba* L.) with imidazolinones and other herbicides. *Weed Research* **31**: 227-235.

Gressel, J., Vered, Y., Bar-Lev, S., Milstein, O. and Flowers, H.M. (1983) Partial suppression of cellulase action by artificial lignification of cellulose. *Plant Science Letters* **32**: 349-353.

Gressel, J., Shaaltiel, Y., Sharon, A. and Amsellem, Z. (1993) Biorational *in vitro* screening for herbicide sygnergists. *In: Herbicide Bioassay* (J.C. Streibig and P. Kudsk, eds.), 217-252. CRC Press, Boca Raton.

Harman, G.E. and Stasz, T.E. (1991) Protoplast fusion for the production of superior biocontrol fungi. *In: Microbial Control of Weeds*, (D.O. TeBeest, ed.), 171-188. Chapman and Hall, New York.

Joel, D.M., Kleifeld, Y., Losner-Goshen, D., Herzlinger, G. and Greseel, J. (1995) Transgenic crops against parasites. *Nature* **374**: 220-221.

Kistler, H.C. (1991) Genetic manipulation of plant pathogenic fungi. *In: Microbial Control of Weeds*, (D.O. TeBeest, ed.), 152-170. Chapman and Hall, New York.

Klerk, R.A., Smith, R.J. Jr. and TeBeest, D.O. (1985) Integration of a microbial herbicide into weed and pest control programs in rice *(Oryza sativa)*. *Weed Science* **33**: 95-99.

Morin, L., Auld, B.A. and Brown, J. F. (1993) Synergy between *Puccinia xanthi* and *Colletotrichum orbiculare* on *Xanthium occidentale*. *Biological Control* **3**: 293-310.

Sands, D.C. and Miller, R. V. (1993) Altering the host range of mycoherbicides by genetic manipulation. *In: Pest Control with Enhanced Environmental Safety* (S.O. Duke, J.J Menn and J.R. Plimmer, eds.), **99:** 101-109. American Chemical Society, Washington, DC.

Sharon, A. (1991) Suppression of elicited defense responses in Cassia obtusifolia: increased susceptibility to pathogens. *Ph.D. Dissertation.* Weizmann Institute of Science, Rehovot.

Sharon, A. and Gressel, J. (1991) Elicitation of a flavonoid phytoalexin accumulation in *Cassia obfusilifia* by a mycoherbicide: determination by $AICI_3$-spectrofluorimetry. *Pesticide Biochemistry Physiology* **41**: 142-149.

Sharon, A., Ghirlando, R. and Gressel, J. (1992a) Isolation, purification and identification of 2-(p-hydroxyphenoxy)-5,7-dihydroxychromone: a fungal induced phytoalexin from *Cassia obtusifolia. Plant Physiology* **98**: 303-308.

Sharon, A., Amsellem, Z. and Gressel, J. (1992b) Glyphosate suppression of induced defense responses: increased susceptibility of *Cassia obtusifolia* to a mycoherbicide. *Plant Physiology* **98**: 654-659.

Sibony, M., Lior, E., Fishler, A. and Rubin, B. (1995) Resistance of *Amaranthus* spp., horseweed (*Conyza canadensis*) and field dodder (*Cuscuta campestris*) to acetolactate synthase (ALS) inhibitors. *Weed Science Society America Abstracts* **35**: 261.

TeBeest, D.O., Ed. (1991) *Microbial Control of Weeds.* Chapman and Hall, New York.

Walker, H.L. (1981) Granular formulation of *Alternaria macrospora* for control of spurred anoda (*Anoda cristata*). *Weed Science* **29**: 342-344.

Walker, H.L. and Riley, J.A. (1982) Evaluation of *Alternaria cassiae* for the biocontrol of sicklepod (*Cassia obtusifolia*). *Weed Science* **30**: 651-654.

Walker, H.L. and Sciumbato, G.L. (1979) Evaluation of *Alternaria macrospora* as a potential biocontrol agent for spurred anoda (*Anoda cristata*). Host range studies. *Weed Science* **27**: 612-614.

Wapshire, A.J., Delfosse, E.S. and Cullen, J.M. (1989) Recent developments in the biocontrol of weeds. *Crop Protection* **89**: 227-250.

Watson, A.K., Gotlieb, A.R. and Wymore, L.A. (1986) Interaction between a mycoherbicide *Colletotrichum coccodes* and herbicides for the control of velvetleaf. *Weed Science Society America Abstracts,* **26**: 143.

Wymore, L.A. and Poirier, C. (1988) *Colletotrichum coccodes*, a potential bioherbicide for control of velvetleaf (*Abutilon theophrasti*). *Plant Disease* **72**: 534-538.

Zidack, N.H., Backman, P.A. and Shaw, J.J. (1992) Promotion of bacterial infection of leaves by an organosilicone surfactant: implications for biological weed control. *Biological Control* **2**: 111-117.

CHAPTER 30
PARASITIC WEED CONTROL USING TRANSGENIC HERBICIDE-RESISTANT CROPS

D.M. JOEL[1], Y. KLEIFELD[1] and J. GRESSEL[2]
1. Department of Weed Research, ARO, Newe-Ya'ar Research Center, Haifa 31900, Israel. 2. Department of Plant Genetics, Weizmann Institute of Science, Rehovot 76100, Israel

Abstract

The parasitic flowering plants *Orobanche* spp. (broomrapes) and *Striga* spp. (witchweeds) parasitize the roots of important crops, heavily reducing yields. The development of genetically-engineered herbicide-resistant crops was tested as solution to this intractable problem. Here we show that the use of three target-site resistances decimated *Orobanche*, demonstrating the potential of transgenic herbicide-resistant crops in the control of parasitic weeds.

The Problem

Two of the most devastating pests in agricultural fields, mainly in Africa, the middle east and southern Europe, are the broomrapes (*Orobanche* spp.) and the witchweeds (*Striga* spp.). These flowering plants parasitize roots of many crop plants and exploit photosynthate and water, significantly reducing and damaging the yields (Musselman, 1980).

The broomrapes devastate the legumes that serve as an important source of protein in middle-eastern societies, and cause heavy damage to vegetables, sunflower and fodder crops in Mediterranean areas in the Middle-East, southern Europe and parts of Asia, gradually spreading to new areas also in other continents including America. The witchweeds affect the grains of millions of people in sub-Saharan Africa (Musselman, 1986). Similarly, other parasitic plants like the dodders (*Cuscuta* spp.) and mistletoes cause considerable damage to forests and crops.

While effective control methods are avaible for most weeds, there are so far only limited ways to control the parasitic weeds (Parker and Riches, 1993). The impact of parasitic weeds on the world economy is therefore tremendous, particularly in developing countries where farmers have no alternatives to the few crops they grow for their living.

There are few herbicides that are capable of killing parasitic weeds, but none is sufficiently selective to leave the crops unaffected (Foy *et al.*, 1989).

The Hypothesis

The development of genetically engineered herbicide-resistant crops in recent years (De Block *et al.*, 1987; Della-Cioppa *et al.*, 1987; Guerineau *et al.*, 1990; Lee *et al.*, 1988; Mazur and Falco, 1989) might signify a milestone in our ability to control parasitic weeds. It was assumed that once we introduce herbicide-resistances to host plants, exclusion of this kind of parasitism is possible by treating infested fields with the respective herbicides (Foy *et al.*, 1989; Gressel, 1992).

Two main kinds of resistance to herbicides can be conferred on crop species: metabolic resistance, where the resistant plant degrades the herbicide to non-toxic products, and target-site resistance, where the target enzyme affected by the herbicide is modified to preclude herbicide binding without changing the normal function of the enzyme. We assumed that when root parasites such as *Orobanche* are to be controlled, target-site resistances are more relevant because only they permit movement of the intact herbicide from the foliage of the host crop, to the parasite that is connected to its roots. Crops with metabolic resistance would be useful only if the herbicide is directly applied to the parasite underground, because the crop degrades the foliar-applied herbicide before it gets to the parasite through host tissues (Gressel *et al.*, 1994).

Materials and Methods

We have examined the idea of applying herbicides to genetically-engineered herbicide-resistant crops, grown in soil infested with *Orobanche* seeds. Four transgenic resistances were tested as models to examine the hypothesis:

a) Transgenic rapeseed plants, engineered with the *aro* A gene that encodes a modified enolphosphate-shikimate phosphate (EPSP) synthase gene (Della-Cioppa *et al.*, 1987) conferring target-site resistance to glyphosate.
b) Transgenic tobacco plants containing the SURB-Hra gene, a modified acetolactate synthase (ALS) gene conferring target-site resistance to chlorsulfuron (Lee *et al.*, 1988).
c) Tobacco plants engineered with the *sul* I gene that encodes a modified dihydropteroate synthase (DHPS) conferring target-site resistance to asulam (Guerineau *et al.*, 1990).
d) Transgenic tomato plants with the *bar* gene conferring metabolic resistance to glufosinate-ammonium (De Block *et al.*, 1987).

Seedlings of these transgenic plants were selected with the respective herbicides. The resulting young plants were then transplanted into pots filled with soil infested with seeds of *Orobanche aegyptiaca*. During the development of the plants they were treated with various doses of the respective herbicides. Their development was followed and broomrape infection was documented. As controls we used treated non-transgenic plants as well as untreated transgenic plants, with and without broomrape.

Results

As expected, the application of all three target site resistances proved effective in controlling the parasite (Table 1).

Table 1. Broomrape control on different herbicide-resistant crops.

Crop	Resistant to	Foliar treatment (g ha^{-1})	Yield (% of control)	% reduction of broomrape spikes
Tobacco	Chlorsulfuron	15	100	100
Rape	Glyphosate	720	100	100
Tobacco	Asulam	8000	80	70
Tomato	Glufosinate	1200	0	0

A single application of 720g ha^{-1} glyphosate to glyphosate-resistant rapeseed infected with *Orobanche*, resulted in a complete prevention of broomrape emergence, with concomitant normal growth and flowering of the transgenic host. Non-treated control plants were heavily infected with the parasite, and did not devlop any flowers (Joel *et al.*, 1995). The infected plants died soon after the emergence of the parasite.

Broomrape was similarly controlled on transgenic tobacco plants resistant to chlorsulfuron, by foliar application of 15g ha^{-1} chlorsulfuron 10, 20, or 30 days after planting in broomrape-infested soil (Joel, 1992). A single application of the herbicide resulted in normal growth and flowering of the transgenic host. All controls were heavily infected with the parasite, did not flower and reached not more than 50% of the normal height and shoot dry weight.

With asulam resistance there was a considerable reduction in broomrape infection. Foliar application of 8 kg ha^{-1} asulam resulted in a 70% reduction of the number of broomrape plants parasitizing a single host plant (Joel *et al.*, 1995).

Conversely, the application of glufosinate ammonium to transgenic tomato plants with the metabolic resistance to this herbicide failed to control *O. aegyptiaca* that parasitized its roots (Joel *et al.*, 1992). The treated plants, like the controls, were heavily infected with the parasite and the yield was drastically reduced.

Conclusions

It was possible to achieve nearly complete control of broomrape and to prevent its damage using transgenic herbicide-resistant crops. As presumed, applying the respective herbicides to target-site herbicide-resistant host plants alleviated parasitism. World agriculture could gain control over parasitic weeds if effort is put in the transfer of the available target-site resistance into local varieties of the many crops that are susceptible to these parasites. Because the parasites might evolve resistances to the herbicides, this approach should serve as a temporary, though powerful, solution in the control of parasitic weeds, until other means are found.

Acknowledgements

We would like to thank J. Leemans and P. Rüdelsheim (PGS) for the glufosinate-resistant tomatoes, B. Mazur and F. Lichtner (Dupont) for the chlorsulfuron-resistant tobacco, G. Kishore and S. Padgette (Monsanto) for the glyphosate-resistant oilseed rape, and G. Freyssinett and B. Pelissier (Rhone Poulenc) for the asulam-resistant tobacco. J.G. has the Gilbert de Botton Chair of Plant Sciences. This work was supported by the United States Agency for International Development (Trilateral Egypt-USA-Israel Program grant number HNE-0158-G-00-3060-00), by PGS, and by the Israel Ministry of Agriculture.

References

De Block, M., Botterman, J., Vandewiele, M., Dockx, J., Thoen, C., Gossele, V., Movva, N.R., Thompson, C., Van Montagu, M. and Leemans, J. (1987) Engineering herbicide-resistance in plants by expression of a detoxifying enzyme. *EMBO Journal* **6**: 2513-2518.

Della-Cioppa, G., Bauer, S.C., Taylor, M.L., Rochester, D.E., Klein, B.K., Shah, D.M., Fraley, R.T. and Kishore, G. M. (1987) Targeting a herbicide resistant enzyme from *Esherichia coli* to chloroplasts of higher plants. *Bio/Technology* **5**: 579-584.

Foy, C.L., Jain, R. and Jacobsohn, R. (1989) Recent approaches for chemical control of boomrape (*Orobanche* spp.). *Reviews Weed Science* **4**: 123-152.

Gressel, J. (1992) The need for new herbicide-resistant crops. *In: Resistance '91: Achievements and Developments in Combatting Pesticide Resistance* (I. Denholm, A.L. Devonshire and D.W. Holloman, eds.), 283-294. Elsevier, London.

Gressel, J., Kleifeld, Y. and Joel, D. M. (1994) Genetic engineering can help control parasitic weeds. *In: Biology and Management of Orobanche* (A.H. Pieterse, J.A.C. Verkleij and S.J. ter Borg, eds), 406-418. Royal Tropical Institute, Amsterdam.

Guerineau, F., Brooks, L., Meadows, J., Lucy, A., Robinson, C. and Mullineaux, P. (1990) Sulfonamide resistance gene for plant transformation. *Plant Molecular Biology* **15**: 127-136.

Joel, D.M. (1992) Control of broomrape (*Orobanche aegyptiaca* Pers.) whith chlorsulfuron on a transgenic sulfonylurea-resistant crop. *Weed Science Society America Abstracts* **32**: 65.

Joel. D.M., Losner-Goshen, D. and Herzlinger, G. (1992) Use of transgenic plants for control of *Orobanche. Phytoparasitica* **20**: 346.

Joel, D. M., Kleifeld, Y., Losner-Goshen, D., Herzlinger, G., and Gressel, J. (1995) Transgenic crops to fight parasitic weeds. *Nature* **374**: 220-221.

Lee, K.Y., Townsend, J., Tepperman, J., Black, M., Chui, C.F., Mazur, B., Dunsmuir, P. and Bedbrook, J. (1988) The molecular basis of sulfonylurea herbicide resistance in tobacco. *EMBO Journal* **7**: 1241-1248.

Mazur, B.J. and Falco, S.C. (1989) The development of herbicide resistant crops. *Annual Review Plant Physiology Plant Molecular Biology* **40**: 441-470.

Musselman, L.J. (1980) The biology of *Orobanche, Striga,* and other root parasitic weeds. *Annual Review Phytopathology* **18**: 463-489.

Musselman, L.J. (1986) Parasitic weeds and their impact in south-west Asia. *Proceedings Royal Society of Edinburgh* **89B**: 283-288.

Parker, C., and Riches, C. (1993) *Parasitic Weeds of the World: Biology and Control.* CAB International, London.

PART V

MANAGING OR AVOIDING HERBICIDE RESISTANCE: INTEGRATED MECHANICAL, CHEMICAL AND BIOLOGICAL METHODS FOR WEED CONTROL

CHAPTER 31
STRATEGIES FOR THE PREVENTION AND CONTROL OF HERBICIDE RESISTANCE IN ANNUAL GRASS WEEDS

S. MOSS
IACR-Rothamsted, Harpenden, Herts AL5 2JQ,, UK.

Introduction

Practical prevention and control strategies usually need to be implemented before the biochemical and genetic basis of resistance are totally understood. Consequently strategies must be robust, and capable of controlling populations in which several different resistance mechanisms may be present. An integrated weed management (IWM) approach is required, in which as many tactics as possible are used to combat weeds. This paper reviews the many cultural and herbicidal methods available, and stresses the importance of effectively conveying advice to farmers.

Prediction of Resistance

The underlying ecological and genetic factors determining selection for resistance are well documented (Gressel, 1991; Maxwell & Mortimer, 1994). To predict whether resistance will become a problem in any individual situation, some form of risk assessment is required. This needs to consider:

1 *The weed* - what characteristics of the biology, agro-ecology and genetics of the weed influence the evolution of resistance? Is this a major weed?

2 *The cropping system* - how serious would resistance be?

3 *The herbicide* - how vulnerable to resistance is this herbicide? What is the intensity of use? What alternative herbicides are available?

4 *Cultural (non-herbicidal) control* - what methods are available? How effective are they? How much do they cost?

In the Netherlands, guidelines on resistance risk evaluation are being devised for all pesticides (T. Rotteveel, personal communication). These include a much more detailed appraisal of all the factors listed above. The ability to make an accurate evaluation will be influenced greatly by the amount of information already available and experience elsewhere. It is difficult to predict the chance of resistance developing in any individual species, as there are no clear relationships between plant families or genera and their tendency to evolve resistance (Le Baron, 1991). Resistance has usually developed in one, or at most a few, species in a weed community despite all being exposed to the same intensity of herbicide use.

Monitoring for Resistance

It is vital that resistance is detected at an early stage. If resistance becomes an acute, whole farm problem, then options are more limited and greater expense is almost inevitable. Worldwide experience has been that farmers tend to do nothing about resistance management until it is a problem on their own farm - or their neighbours. Farmers may be more prepared to implement anti-resistance strategies if these are simple, have no cost penalties and have other benefits.

Accurate observation is important so that any reduction in herbicide efficacy can be detected. This may indicate developing resistance. An essential pre-requisite for confirmation of resistance is a good diagnostic test. Ideally this should be rapid, accurate, cheap and accessible.

Occurrence of Resistance in Annual Grass Weeds

At least 25 annual grass weed species have evolved resistance to one or more herbicides worldwide. These include *Alopecurus myosuroides* (UK, Spain, Germany, Netherlands), *Avena* spp. (North America, Australia, UK), *Echinochloa colona* (Columbia, Costa Rica), *Eleusine indica* (USA), *Lolium multiflorum* (USA, UK), *Lolium rigidum* (Australia, Spain), *Phalaris minor* (India) and *Setaria viridis* (Canada) (LeBaron, 1991; Powles & Holtum, 1994).

Important characteristics of many annual grass weeds are a high competitive ability, the capacity for rapid population increase and the production of seeds which have relatively short persistence in the soil - commonly a 70 - 90% decline per annum. Consequently, populations of annual grass weeds tend to be very dynamic, and populations can increase or decrease dramatically during a period of a few years depending on the degree of weed control achieved.

A wide range of herbicides is available for the selective control of many annual grass weeds in various crops. Many of these herbicides are highly effective and consequently there has been reliance on herbicides to achieve the high levels of control required. It is in such situations, where annual grasses are major weeds, that herbicide resistance has become a serious problem.

Management of Resistance

Herbicide resistance is a consequence of the repeated use of herbicides, often in association with crop monoculture and reduced cultivation practices. It is necessary, therefore, to modify these practices in order to prevent or delay the onset of resistance, or to control existing resistant populations. Practical prevention and control strategies usually need to be implemented before the biochemical and genetic basis of resistance has been fully determined. Consequently, such strategies must be robust, and capable of coping with populations in which several different resistance mechanisms may be present.

An integrated weed management (IWM) approach is essential in which as many tactics as possible are used to combat weeds. These should include cultural, as well as herbicidal methods of control. In this way less reliance is placed on herbicides and selection pressure should be reduced. Differences in `fitness' may affect the relative degree of control achieved of resistant and susceptible plants by both cultural and herbicidal methods. However, this will be dependent on the nature and the scale of any such `fitness' differences.

Cultural Control methods

A wide range of non-herbicidal methods is available for reducing weed populations. These methods will not be appropriate in all situations and individual circumstances will dictate which are most practicable. Provided susceptible and resistant plants respond similarly, there should be no selection in favour of resistant individuals.

a) Cultivations

Non-inversion tillage maximizes the proportion of the weed population derived from seeds shed in the previous crop, because seeds are retained close to the soil surface, and so minimizes the probability of back-crossing with earlier, less selected generations derived from older, buried seeds. Inversion tillage, such as mouldboard ploughing, can bury most freshly shed seeds to a depth from which seedlings are unlikely to emerge (>5 cm).

As most annual grass weeds have seeds which are relatively non-persistent in the seed bank, usually fewer old buried seeds are brought back up to the surface.

Inversion tillage has two distinct benefits. Firstly it can substantially reduce the weed population and secondly, it can increase the proportion of susceptible individuals by increasing the number of seedlings derived from older, less selected seeds. The greater the seed longevity and depth of emergence, the lesser the beneficial effect on population size. However, greater seed persistence may improve changes to the susceptible: resistance ratio, especially if inversion tillage follows a period of minimum cultivation.

In some situations annual ploughing is not feasible, but rotational ploughing, once every 4-5 years, may be an acceptable alternative. Such a strategy may be a method of returning the genetic structure of the population to that which existed several years previously.

It should be recognized that this strategy could be detrimental in situations where resistance has been managed successfully for several years prior to ploughing.

In crop cultivations such as inter-row hoeing or harrowing are established methods of weed control. The feasibility and effectiveness of these techniques is dependent on the crop being grown.

b) Crop Rotation

Many annual weeds are strongly associated with specific crops, so crop rotations can reduce the intrinsic success of such weeds and may permit the use of alternative herbicides. The inclusion of a grass ley in an arable rotation can also be an effective means of reducing populations of many arable grass weeds, providing seed return is prevented. Crop rotation may reduce the overall usage of herbicides and extend the range of active ingredients available.

c) Setaside/Fallowing

Fallowing is a traditional method of weed control, although land is taken out of production. 'Setaside', in which farmers are paid not to crop the land, also provides an opportunity to reduce weed populations. However, failure to prevent seed return in fallow or 'setaside' land may lead to increased weed problems in subsequent crops.

d) Burning Crop Residues

Burning can destroy 50-80% of freshly shed seeds lying on the soil surface. In addition, burning can break the dormancy of some surviving seeds and the emerging seedlings may then be destroyed prior to sowing the next crop.

e) Stubble Hygiene

It is important that all weed seedlings that have emerged before sowing are destroyed, otherwise some may re-establish in the crop. This may be achieved either by effective seedbed cultivations or by using non-selective herbicides such as glyphosate.

f) Delayed Drilling

Delayed drilling allows a higher proportion of seedling emergence to occur before sowing the crop. As many annual grass weeds have relatively short periods of innate dormancy, this can be an effective means of reducing the potential weed infestation, provided there is sufficient moisture to permit seed germination.

g) Crop Competition

Some agronomic practices can favour the development of competitive crops which are better able to suppress weeds, e.g. higher than normal seed rates, more competitive crops or varieties, narrower row spacings, improved drainage, good seedbeds.

h) Prevention of Seed Return

Grazing or cutting weeds may be possible in some situations. This is an important component in strategies for controlling *Lolium rigidum*, which is a valuable constituent of the pasture phase of the rotation (Matthews, 1994). In setaside land too, cutting can prevent seed return. Removal of plants from within crops ('roguing') is feasible for low populations, or patches of tall weeds such as *Avena spp.*

i) Seed Recovery at Harvest

This is another method of reducing seed return which is being used in Australia for control of *Lolium rigidum* (Matthews, 1994). It involves modification to the harvesting operation so that weed seeds are diverted into a container rather than returned to the ground. The seeds can therefore be removed from the field and destroyed. The efficacy of such a technique is very dependent on the time of shedding of weed seeds relative to crop harvest.

j) Avoidance of the Introduction and Spread of Resistant Seeds and Plants

Avoid sowing contaminated seed and minimize the dissemination of seeds and plants in combine harvesters, cultivation equipment, straw or manure.

k) Manipulation of Gene Flow

The management of populations with the specific aim of manipulating the relative frequency of susceptible and resistant genes has been postulated as a management concept (Radosevich *et al*, 1991). This philosophy can be viewed as encouraging susceptibility, in contrast to the more common objective of reducing resistance. This concept has been studied experimentally in *Lolium multiflorum* (Ghersa *et al.*, 1994).

Other techniques such as use of mulches, allelopathy, seed dormancy breaking chemicals and biological control may be feasible alternatives to herbicides in some situations. These, and other alternatives to herbicides have been reviewed by Morgan (1989).

Herbicidal control

a) Alternative Herbicides

When resistance is first confirmed, the efficacy of alternatives is likely to be the first consideration. The use of alternative herbicides, which remain effective on resistant populations, can be a successful strategy, at least in the short term. The value of alternative herbicides will be highly dependent on the extent of cross-resistance. If there is resistance to a single class of herbicides, then use of herbicides from other classes may provide a simple and effective solution, at least in the short term. If resistance extends to more than one group, then choices are more limited.

It should not be assumed that resistance will automatically extend to all herbicides with the same mode of action. For example, some populations of *Lolium multiflorum* with target site resistance (insensitive ACCase) show resistance to `fops' aryloxyphenoxypropionate (APP) but not to `dims' cyclohexanedione (CHD), despite both classes being ACCase inhibitors (Gronwald *et al.*, 1992). This may be due to slight differences in the precise binding site for different ACCase inhibiting herbicides. Consequently, selection for different mutations of the ACCase gene may result in different patterns of cross-resistance. Some populations of *Alopecurus myosuroides* with an enhanced metabolism mechanism show resistance to pendimethalin but not to trifluralin, another dinitroaniline. This is due to differences in the vulnerability of these two herbicides to oxidative metabolism (James *et al.*, 1995).

b) Mixtures and Sequences

The use of two or more herbicides which have differing modes of action can potentially reduce the selection for resistant genotypes. Ideally each component in a mixture should have similar levels of efficacy; have activity at different target sites; be detoxified by different biochemical pathways; have similar soil persistence (if a residual herbicide); exert negative cross-resistance; synergise the activity of the other component. It is extremely unlikely that any mixture will have all these attributes and therefore the benefits of mixtures may be substantially reduced (Wrubel and Gressel, 1994). There is also a risk that mixtures will select for resistance to all components. One practical advantage of sequences compared with mixtures is that a better appraisal of the efficacy of each herbicide component is possible, provided sufficient time elapses between each application.

c) Herbicide Rotations

Rotating herbicides from different classes in successive years should reduce selection for resistance. This is a key element in most resistance prevention programmes. The value of this approach depends on the extent of cross-resistance, and whether multiple resistance occurs. A practical problem is the lack of awareness by farmers of the different classes of herbicide that exist. In Australia a scheme has been introduced recently in which identifying letters are included on the product label as a means of enabling farmers to distinguish products with different modes of action (Anonimous, 1995).

d) Herbicide Management

Optimizing herbicide input to the economic threshold level should avoid the unnecessary use of herbicides and reduce selection pressure. Herbicides should be used to their greatest potential by ensuring that the timing, dose, application method, soil and climatic conditions are optimal for good activity. Spraying only the badly infested areas of a field is another means of reducing total herbicide use.

e) Synergists and Safeners

The potential use of synergistic mixtures as a tool for managing resistance has been reviewed (Kemp and Caseley, 1991). Although the practical implementation of synergists in weed management programmes is very limited at present, this approach may be particularly appropriate in cases where resistance is due to enhanced metabolism, although it is important that herbicide selectivity in the crop is not compromised.

f) Spray Topping

This involves the application of herbicides at a late stage of weed growth with the objective of reducing viable seed return. In Australia, spray topping of *Lolium rigidum* is practised as a means of reducing seed return in pastures (Matthews, 1994).

g) Herbicide Resistant Crops

These could allow the use of total herbicides, such as glyphosate, for control of resistant weeds. To date, no confirmed cases of weeds with evolved resistance to glyphosate have been reported. Herbicide resistant crops could encourage the repeated use of the same herbicide type, which may favour the development of resistant biotypes. Crops with resistance to herbicides to which resistance has already been demonstrated in weeds, (e.g. sulphonyl-ureas) would be a higher risk. Herbicide resistant crops will need to be used with care if they are to reduce, rather than increase, the risk of evolution of herbicide resistance in weeds.

The list above demonstrates that a wide range of weed control techniques is available, both cultural and herbicidal. The Australian `WeedSmart' programme for *L. rigidum* and UK `Guidelines for the prevention and control of herbicide resistant black-grass *(A. myosuroides)'* (Moss and Clarke, 1994) are examples of attempts to provide advice on integrated weed management. No single blueprint strategy will be appropriate for every situation and the cropping system and biology of the weed must be considered as well as the herbicides affected and possible alternatives.

Conclusion

Herbicide resistance has evolved because too much reliance has been placed on herbicides to control weeds. It is clear that a broader based approach to weed control is needed in which herbicide use is integrated with non-chemical methods of weed control.

Strategies for resistance management are of no use unless implemented. Farmers receive information from a wide variety of sources. It is necessary to consider a variety of channels of communication if advice is to be conveyed effectively.

At present many farmers consider herbicides as the main method of weed control. They may well practice non-herbicidal methods of control, perhaps without being aware of their impact. A measure of the effectiveness in communicating the concept of integrated weed management (IWM) as an anti-resistance strategy, will be when farmers consider cultural control as their primary, and not secondary, method of weed control.

References

Anonimous (1995) Australia pioneers herbicide resistance labelling. *Agrow* **224**: 12-13.

Ghersa, C.M., Martinez-Ghersa, M.A., Brewer, T.G. and Roush, M.L. (1994) Use of gene flow to control diclofop-methyl resistance in Italian rye-grass (*Lolium multiflorum*). *Weed Technology* **8**: 139-147.

Gressel, J. (1991) Why get resistance? It can be prevented or delayed. *In: Herbicide Resistance in Weeds and Crops* (J.C. Caseley, G.W. Cussans and R.A. Atkin, eds.), 1-25. Butterworth-Heinemann, Oxford.

Gronwald, J.W., Eberlein, C.V., Betts, K.J., Baerg, R.J., Ehlke, N.J. and Wyse, D.L. (1992) Mechanism of diclofop resistance in an Italian rye-grass *(Lolium multiflorum)* biotype. *Pesticide Biochemistry Physiology* **44**: 126-139.

James, E.H., Kemp, M.S. and Moss, S.R. (1995) Phytotoxicity of trifluoromethyl and methyl-substituted dinitroaniline herbicides on resistant and susceptible populations of black-grass *(Alopecurus myosuroides). Pesticide Science* (in press).

Kemp, M.S. and Caseley, J.C. (1991) Synergists to combat herbicide resistance. *In: Herbicide Resistance in Weeds and Crops* (J.C. Caseley, G.W. Cussans and R.K. Atkin, eds.), 279-292. Butterworth-Heinemann, Oxford.

LeBaron, H.M. (1991) Distribution and seriousness of herbicide-resistant weed infestations worldwide. *In: Herbicide Resistance in Weeds and Crops* (J.C. Caseley, G.W. Cussans and R.K. Atkin, eds.), 27-43. Butterworth-Heinemann, Oxford.

Matthews, J.M. (1994) Management of herbicide resistance weed populations. *In: Herbicide Resistance in Plants - Biology and Biochemistry* (S.B. Powles and J.A.M. Holtum, eds.), 317-335. CRC Press, Boca Raton, Florida.

Maxwell, B.D. and Mortimer, A.M. (1994) Selection for herbicide resistance. *In: Herbicide Resistance in Plants - Biology and Biochemistry* (S.B. Powles and J.A.M. Holtum, eds.), 1-25. CRC Press, Boca Raton, Florida.

Morgan, W.C. (1989) Alternatives to herbicides. *Plant Protection Quarterly* **4**: 33-37.

Moss, S.R. and Clarke, J.H. (1994) Guidelines for the prevention and control of herbicide-resistant black-grass *(Alopecurus myosuroides* Huds.). *Crop Protection* **13**: 230-234.

Powles, S.B. and Holtum, J.A.M. (1994) *Herbicide Resistance in Plants - Biology and Biochemistry,* 353 pp. CRC Press, Boca Raton, Florida.

Radosevich, S.R., Maxwell, B.D. and Roush, M.L. (1991) Managing herbicide resistance through fitness and gene flow. *In: Herbicide Resistance in Weeds and Crops* (J.C. Caseley, G.W. Cussans and R.K. Atkin, eds.), 129-143. Butterworth-Heinemann, Oxford.

Wrubel, R.P. and Gressel, J. (1994) Are herbicide mixtures useful for delaying the rapid evolution of resistance? A case study. *Weed Technology* **8**: 635-648.

CHAPTER 32
RISKS AND BENEFITS OF WEED MANAGEMENT TECHNOLOGIES

M.D.K. OWEN
Iowa State University, Ames, IA 50011 USA

Introduction

Weeds are the most important pest complex that threaten world food and fiber production while herbicides represent the most prevalent pesticide used (Bridges, 1994). Currently herbicides are used on a majority of the crop acres and generally provide economically acceptable control of weedy pests. However, due to the success of herbicide use over the last three decades, weed science and management has evolved into herbicide science. Research on weed biology, weed ecology and cropping systems has not been a high priority nor has funding been readily available to those researchers who chose to emphasize this area. The high emphasis on herbicides for management of weeds has resulted in the limited use of alternative strategies and the general lack of biological information. Problems associated with this system include herbicides being perceived as major environmental contaminants, weed populations developing resistance to herbicides and the sustainability of agriculture being questioned. Importantly, it is estimated that less than 2% of the population in the United States is directly involved with production agriculture and thus do not understand nor appreciate the impact that food and fiber production has on daily existence. This paper will address a number of issues that reflect these areas of concern in weed management technology using crop production in the midwestern United States as an example.

Weed Management

Herbicides are used on over 95% of the row crop acres in Iowa and in the midwestern United States Corn Belt (Hartzler and Wintersteen, 1991). The herbicide classes of importance include the benzoic acids, chloroacetamides, dinitroanilines, imidazolinones, phenoxys , sulfonylureas, thiocaramates, triazines and others. Herbicide applications have shifted during the 1990's from a high percentage of acres with preplant incorporated herbicides to the use of early preplant applications, postemergence treatments and soil-applied/ postemergence sequential combinations. A majority of acres receive more than one herbicide application per season and are treated with more than one active ingredient. Chemical weed management has an increasingly important role in crops produced without seed bed preparation. For example, the success of no tillage crop production relies on herbicidal weed control.

The use of triazine herbicides, primarily atrazine and cyanazine, has increased in the midwest despite restrictions developed to minimize potential movement of these products into ground and surface water. More than 81% of the corn acres in Iowa receive triazine herbicides and atrazine use has increased from 49% of the corn acres in 1985 to over 61% in 1990 (Hartzler and Wintersteen, 1991). The triazine herbicides provide consistently acceptable control of many weeds in corn and the loss of these products represents a significant increase in weed management costs.

The recent development of herbicide classes that inhibit the acetolactate synthase (ALS) enzyme in plants has been of major importance to agriculture. Imidazolinone and sulfonylurea herbicides have provided broad spectrum weed control in a number of crops and have been rapidly accepted in the marketplace. Currently, an estimated 85% of the soybean acres in Iowa and 64% of the acres for the United States receive applications of ALS herbicides (Wrubel and Gressel, 1994). The ALS inhibitor herbicides have been developed by most of the agricultural chemical companies and are represented by 15 different chemical classes (Powles and Holtum, 1994). Given the toxicological and environmental characteristics of these herbicides, ALS inhibitor herbicides represent the most important group of agricultural chemicals for the foreseeable future.

Cultural and mechanical practices continue to be important weed management strategies, however the use of mechanical techniques has declined (Hartzler *et al.*, 1993). The reasons for the decline of mechanical weed management is attributable to the increasing number of acres the average grower manages, the efficacy of herbicides, herbicide marketing strategies and the adoption of conservation tillage practices (Gunsolus, 1990). However, cultural weed management may actually be increasing (Jordan, 1993). The average row width in corn production is narrower and soybeans seeded with grain drills in row widths less than 25cm are rapidly increasing in popularity. However, growers are typically planting crops earlier in the growing season and thus limiting the utility of delayed planting as a weed management strategy. Crop rotation is used on a majority of the row crop acres in the Midwestern United States and is perceived to be an important weed management strategy (Hartzler and Wintersteen, 1991). However, rotations of perennial and annual crops are not widely used.

Herbicide Resistant Crops

The recent introduction of crops that have altered genetic characteristics that confer resistance to herbicides has not been widely accepted as a weed management strategy. These products of biotechnological research have been represented to be a major threat to the environment, cause significant increase in herbicide use and facilitate the flow of genetic resistance to weed populations (Goldburg *et al.*, 1990). However, herbicide resistant crops can provide a number of important agronomic opportunities such as minimizing potential crop phytotoxicity to herbicides used in rotational crops or allowing the use of alternative herbicide classes thus providing management of difficult weeds (Owen, 1994b). It is anticipated that the availability of glyphosate resistance in soybeans will dramatically increase the utilization of no tillage production systems by providing assurance of consistent weed management. This potentially represents a significant reduction in soil erosion and thus lessening environmental degradation. The development of imidazolinone resistant corn hybrids and sulfonylurea resistant soybean varieties do however, potentially reduce herbicide class rotation thus increasing selection for ALS inhibitor herbicide resistant weed populations. Whether this represents a significant economic concern can not be determined currently. However, it is documented that selection for ALS inhibitor herbicide resistance is relatively rapid and has been accomplished in numerous locations on many different weed species in several different cropping systems (Powles and Holtum, 1994). It is important to recognize that cropping systems that allow alternative weed management strategies are represented in these locations and the speed at which the resistant weed populations developed is alarming. Thus, it can be assumed that the utilization of ALS resistant crops, at least, does represent a real problem. Arguments have been presented to support the position that glyphosate resistant crops and the resultant use of glyphosate does not represent a significant problem with regard to the development of glyphosate resistant weed biotypes (Bradshaw *et al.*, 1995).

Herbicide Resistant Weeds

Herbicide resistant weed populations are developing at an alarming rate. Reports of new weed populations that are resistant to various herbicides increase almost daily and compilations of weed species and locations of resistant populations are outdated upon publication. Resistant populations have been reported for most of the major herbicide classes, cross resistance to specific herbicide classes is common and multiple resistance has been reported (Powles and Holtum, 1994). Weed populations that are resistant to ALS inhibitor herbicides have developed within 3 to 5 years after initial selection and there are reports of ALS resistant *Amaranthus* spp. that were economically significant the first year of selection (Horak, personal communication).

In Iowa, ALS inhibitor herbicides are used extensively and are extremely important to crop production. Iowa production practices usually include a number of weed management strategies that lessen selection for resistant weed populations. These strategies include the use of herbicide tank mixtures, mechanical practices, crop rotation and rotation of herbicide classes. However, there are reports of triazine resistant *Chenopodium album* and *Amaranthus* spp. (Owen, unpublished data) and ALS resistant *Amaranthus rudis* (Hartzler, personal communication) in Iowa. Thus, overall weed management strategies must be carefully considered and integrated with other crop management practices to minimize the selection for resistant weed biotypes (Swanton and Murphy, 1995).

Another consideration about the development of resistant weed populations is perception that the non-agrarian public will have on this problem. They will perceive, and possibly with great accuracy, that resistant weed populations have been caused by the misuse of herbicides. There are a number of widely-circulated publications that emphasize this point (Goldburg *et al.*, 1990; Wiles *et al.*, 1993). Weed scientists have reported that resistant weed populations are a likely consequence of improper weed management strategies that rely primarily on herbicides (Owen, 1994a). Many weed scientists have stressed the need for research on weed biology, weed ecology and alternative management systems in an attempt to resolve the development of these problems (Holt, 1994; Wyse, 1994). The agricultural chemical industry has created several herbicide resistance working groups such as the ALS/AHAS Inhibitor Resistance Working Group and the ACC-ase Inhibitor Resistant Working Group with the objective to develop management strategies that will slow or eliminate the development of herbicide resistant weed populations. Thus far, the efforts of private and public sector scientists to curtail the development of herbicide resistant weeds have not been successful, despite the overall agreement of these different factions on the likely solutions.

Economics and Weed Control

The average cost for herbicides in Iowa represents an estimated $500,000,000 annual investment by growers. Growers recognize the critical need for weed management and thus treat a majority of the acres whether a herbicide treatment is biologically warranted by the weed population or not (Hartzler *et al.,* 1993). Generally, herbicides consistently provide control of more weeds than would cause an economic reduction in yield. This represents an opportunity for better weed management strategies that would potentially improve the efficiency of herbicide use. These strategies include utilizing weed thresholds, modeling weed seed bank dynamics, reducing herbicide rates or replacing herbicide applications with mechanical practices (Hartzler, 1993; Swanton and Murphy, 1995; Liebman and Dyck, 1993; Hamill *et al.*, 1994; Dexter, 1994). These practices rely heavily on an understanding of weed biology, timely management and an integrated approach to weed control. Generally, the research on weed biology has not been sufficient to support the ability of growers to utilize these strategies. Of considerable note is the fact that when growers employ these management practices, they incur additional risk for successful management of weeds. It can be questioned whether or not the benefits of these strategies outweigh the current level of risks. Thus, growers traditionally rely on herbicidal weed management and a majority of the crop acres are treated annually.

Given the potential profitability of the herbicide marketplace and the lack of untreated acres, the herbicide industry is extremely competitive. Due to this competition, advertising and promotion of herbicides has increased dramatically in the last 2 decades. Television, radio and the popular press have frequently and consistently presented a biased perspective on herbicide use, weed management strategies and although unintended, weed biology. These perspectives have been extremely effective and have convinced growers on how weed control should look; herbicides are expected to consistently provide 100% weed control regardless of other management inputs or environmental conditions. This obviously is not a realistic picture, yet growers have been convinced about the accuracy of their expectations. The current perception of weed control in agriculture is not appropriate.

When a new herbicide is introduced into the marketplace, the company must convince the customer that the new product has advantages over older products. The reality is that most of these new products are biologically and economically similar to existing products and thus the benefits of the new herbicide to growers are marginal. Companies have adopted a strategy of guaranteeing performance of these herbicides thus eliminating a major risk for their customers. Given that herbicides are typically deemed commercial when they provide 85 to 90% control of weeds (based on visual estimates) and that the expectations of growers are 100% weed control, the industry has created a situation where they have provided supplemental applications of herbicides on a large percentage of acres in the midwestern United States. Importantly, growers are not likely to utilize alternative weed management strategies because of the performance guarantees and the implied 100% control.

The result of these marketing strategies has been significant acres resprayed with herbicides. In 1994, an estimated 14 to 20% of the Iowa corn acres were resprayed with herbicides as the direct result of performance guarantees. Estimates for soybeans are similar or slightly higher. Specific companies may have retreated up to 40% of the acres initially treated with they proprietary products. While there is no means of determining the actual economic need of these reapplications, based on past experience, it is assumed that supplemental mechanical strategies would have resolved most of the potential loss of yield and in many situations, no action would have not reduced yields (Hartzler *et al.*, 1993).

The consequences of these performance guarantees may be serious. While growers may perceive these programs as eliminating risk, increased risk may actually result. This increased risk is in the form of weed competition, poor performance of resprays, potential crop injury and selection for resistant weeds. Increased weed competition and poor performance result from the lack of timely action by the growers; when performance is guaranteed, industry representatives are likely to delay visits to complaints thus allowing weeds to compete and increase in size beyond that which can be effectively treated. Delayed applications also allow the crop to develop to a stage that may be more sensitive to the herbicide. Finally, many of the alternatives for respray are ALS inhibitor herbicides. There is an high probability that the initial application was an ALS herbicide thus resulting in two ALS inhibitor herbicides in one season and dramatically increasing selection for ALS resistance.

These practices are not in keeping with those recommended by public sector weed scientists nor do they follow the guidelines proposed by the various herbicide resistant working groups. However, performance guarantees and subsequent herbicide resprays continue to drive the marketplace. Marketing strategies and concern for market share has taken precedence over scientific and environmental concerns. These resprays can realistically be positioned as misuse of herbicides and do not represent a positive environmental position for the industry. It is likely that these programs may result in restrictions and regulations on herbicide use in the United States similar to those mandated by other countries (Bellinder *et al.*, 1994).

Conclusions

Weed management relies primarily on herbicides and alternative strategies are viewed as less important for most producers. However, significant opportunities exist for mechanical and cultural weed management strategies when the research base for weed biology and ecology improves. Conservation tillage practices are becoming increasing important economically and environmentally. The success of these practices reflect concerns for weed management and herbicides represent the best opportunity for consistent weed management in these systems. Herbicide resistant crops also represent potentially important, albeit controversial weed management opportunities. Herbicide resistant weed populations continue to increase and will likely become economically and politically problematic. The industry has positioned herbicides and weed control in a questionable manner. Performance guarantees and subsequent resprays will potentially increase the development of herbicide resistant weeds, result in more crop phytotoxicity and possibly cause increased government regulation of these important agricultural tools.

References

Bellinder, R.R., Gunnar, G. and Christer K. (1994) Percentage-driven government mandates for pesticide reduction: the Swedish model. *Weed Technology* **8**: 350-359.

Bradshaw, L.D., Padgette, S.R. and Wells, B.H. (1995) Perspectives on the potential of the development of glyphosate-resistant weeds. *Abstracts WSSA* **35**: 66.

Bridges, D.C. (1994) Impact of weeds on human endeavors. *Weed Technology* **8**: 392-395.

Dexter, A.G. (1994) History of sugarbeet (*Beta vulgaris*) herbicide rate reduction in North Dakota and Minnesota. *Weed Technology* **8**: 334-337.

Goldburg, R., Rissler, J., Shand, H. and Hassebrook, C. (1990) Biotechnology's bitter harvest: herbicide-tolerant crops and the threat to sustainable agriculture. *Biotechnology Working Group Report.*

Gunsolus, J.L. (1990) Mechanical and cultural weed control in corn and soybeans. *American Journal Alternative Agriculture* **5**: 114-119.

Hamill, A.S., Surgeoner, A.S. and Roberts, W.P. (1994) Herbicide reduction in North America: In Canada, an opportunity for motivation and growth in weed management. *Weed Technology* **8**: 366-371.

Hartzler, R.G. and Wintersteen, W. (1991) A survey of pesticides used in Iowa crop production in 1990. *Iowa State University Extension Report,* IPM-1441.

Hartzler, R.G. (1993) Reducing herbicide use in field crops. *Iowa State University Extension Report,* IPM-36.

Hartzler, R.G., Van Kooten, B.D., Stoltenberg, D.E., Hall, E.M. and Fawcett, R.S. (1993) On-farm evaluation of mechanical and chemical weed management practices in corn (*Zea mays*). *Weed Technology* **7**: 1001-1004.

Holt, J.S. (1994) Impact of weed control on weeds: new problems and research needs. *Weed Technology* **8**: 400-402.

Jordan, N. (1993) Prospects for weed control through crop interference. *Ecological Applications* **3**: 84-91.

Liebman, M. and Dyck, E. (1993) Weed management: a need to develop ecological approaches. *Ecological Applications* **3**: 39-41.

Owen, M.D.K. (1994a) Assessment of strategies to manage herbicide resistant weeds. *Proceedings North Central Weed Science Society* **49**: 164-165.

Owen, M.D.K. (1994b) Impact of crop tolerance ot specific herbicides on weed management systems: corn and soybeans. *Proceedings North Central Weed Science Society* **49**: 167-168.

Powles, S.B. and Holtum, J.A.M. (1994) *Herbicide Resistance in Plants: Biology and Biochemistry*, 353 pp. CRC Press, Inc. Boca Raton.

Swanton, C.J. and Murphy, S.D. (1995) Weed science beyond the weeds: the role of integrated weed management (IWM) in agroecosystem health. Submitted for publication.

Wiles, R., Cohen, B., Campbell, C. and Elderkin, S. (1994) Tap water blues: herbicides in drinking water. *Environmental Working Group Publication.*

Wrubel, R.P. and Gressel, J. (1994) Are herbicide mixtures useful for delaying the rapid evolution of resistance? A case study. *Weed Technology* **8**: 635-648.

Wyse, D.L. (1994) New technologies and approaches for weed management in sustainable agriculture systems. *Weed Technology* **8**: 403-407.

CHAPTER 33
ROLE OF THE HERBICIDE RESISTANCE ACTION COMMITTEE IN WEED RESISTANCE MANAGEMENT

J.C. GRAHAM
Herbicide Resistance Action Committee, Monsanto Europe S.A., Avenue de Tervuren 270-272, 1150 Brussels, Belgium

Introduction

The Herbicide Resistance Action Committee (HRAC) is an industry-led group that sponsors initiatives to understand and manage weed resistance to herbicides more effectively. The efforts of the Committee have been directed at funding research to understand mechanisms of weed resistance and its prevention, funding literature reviews covering current knowledge of weed resistance, supporting travel of scientists to address resistance problems or attend key scientific meeting, and encouraging exchange of weed resistance information among interested parties.

The paper will summarize the key accomplishments of the HRAC sponsored work. It will also address how various Weed Resistance Groups and HRAC may work together in the future. The HRAC supports the management of weed resistance programs at the most local level possible as management of weed resistance ultimately resides with the end user of herbicides. The knowledge and understanding available to these people is the key to final success of our efforts.

Background to Herbicide Resistance in Weeds

Herbicides, along with other crop protection products, have increased the efficiency of food and fiber production in most world areas. With the ever expanding requirements for food and fiber, coupled with the shrinking producer and production area base, such crop protection tools are critical. Except for very specialized cases, it is not practical or economically feasible to produce crops without herbicides.

As the use of herbicides has intensified the following issues have arisen. Herbicides are so effective that they are often misused versus the label, and often the same product has been used repeatedly in a continuous cropping system or across multiple crops. Different herbicide brands may be used, but they have the same mode of action. These types of situations have led to poor weed control and resistance to herbicides, as first reported by Ryan (1970).

Further driving weed control and weed resistance problems are reduced tillage practices. These programs are essential for soil conservation and economy of farming, but exacerbate weed control difficulties. Continuous cropping programs and orchard and vine crops where crop competition is not optimized and sequential use of similar herbicides occurs can enhance weed resistance issues.

The fact that herbicide resistance in weeds has occurred is not very surprising. Native or introduced plants that we classify as weeds have characteristics that favor their survival. They are usually genetically diverse, have high seed or other reproductive mechanism rates, develop rapidly to the reproductive stage and are competitive. This enables them to persist despite the use of effective weed management programs.

As our understanding of weed resistance has increased, it is evident that biotypes of weeds resistant to many herbicides are found in natural populations. The selection pressure created by the use of herbicides favors the resistant biotypes. Despite the widespread occurrence of herbicide resistant weeds, no herbicide has been lost to agriculture solely on the basis of resistance.

Despite the above rationalization regarding weed resistance to herbicides, it does occur and it must be managed. Herbicides must be preserved as valuable production tools. Developing a better understanding of weed resistance will enable the development of improved weed resistance management strategies. Many researchers around the world are engaged in this activity.

Industry Approach to Weed Resistance

The agrochemical industry, especially the basic producers of herbicides, are equally interested in understanding and managing herbicide resistance. These companies, in 1989, formed the Herbicide Resistance Action Committee (HRAC), as industry cannot continue the discovery, development, sales and stewardship of our herbicides without economic return. There is no incentive for a company to encourage the use of its products in a manner that renders them ineffective and reduces their value.

The HRAC has been an effective forum for sharing knowledge, developing resistance management tools and strategies, addressing labeling and advertising issues and creating funding to share with our colleagues in academia and research institutes. The HRAC is an industry-directed and funded group operating under the auspices of the International Group of National Association of Agrochemical Manufacturers (GIFAP). This provides an opportunity to liaise with our sister committees, the Fungicide, Insecticide and Rodenticide Resistance Action Committees.

Those individuals driving the formation and success of the HRAC were Dr. Alan Jutsum of Zeneca, Dr. Helmut Walter of BASF and Dr. Homer LeBaron of CIBA. Without their efforts, the HRAC would not be nearly as advanced and influential as it is today and may even not exist.

The HRAC has focused its effort from 1989 to now on three main areas of weed resistance. Herbicide resistant weeds problems were known to occur across three key herbicide modes of action. These were the acetolactate synthase enzyme (ALS) resistant weeds, the acetyl coenzyme A carboxylase (ACCase) resistant weeds and the triazine resistant weeds.

The HRAC directed its efforts through sub-committees that addressed specific resistance issues in each area. Each working group made proposals for research funding, assay development, documentation of the extent of the weed resistance problem and the development of management programs.

The ALS mode of action herbicides have been a major breakthrough in technology advancement in herbicide discovery and development. They are used at very low rates, are targeted to the ALS specific enzyme and show a wide range of crop selectivity. Saari *et al.* (1994) have recently described the issues surrounding weed resistance to ALS inhibitor herbicides. The primary commercial herbicides in this group are the imidazolinones, sulfonylurea and triazolopyrimidine families. The plant resistance mechanism is in almost all cases due to a difference in the target site enzyme that modifies the binding ability of the herbicides. The resistance problems with ALS herbicides have occurred primarily where they are used repeatedly in monocultural cropping systems.

The ACCase inhibiting herbicides demonstrate their greatest value in post emergence grass control. Two classes of chemistry, the aryloxyphenoxypropanoates and the cyclohexanediones, typify these effective herbicides. The Grass Herbicide Resistance Working Group (GHRWG) subcommittee has directed initiatives in this area. The resistance problems as defined by Devine and Shimabukuro (1994) have occurred primarily where there was continuous use of these herbicides, independent of crop rotations.

The Triazine Resistant Weed Group (TRWG) is the third subcommittee. Weed resistance to the triazine herbicides was the first reported, primarily due to the widespread use of these products in corn, vine and orchard crops.

The contributions that the HRAC has made to understanding weed resistance and the implementation of management strategies have been primarily in the ALS and GHRWG areas. These were new and expanding opportunities and needed immediate attention.

The HRAC work has taken many forms. We have supported academic research programs in Australia, England, Canada and the US, and these programs have been directed at understanding mechanism of action, developing diagnostic tools, distribution of resistance, multiple and cross resistance.

A further area of emphasis has been in the collection, summarizing and dissemination of information on herbicide resistance. Much is known and some is published, but in many cases information was not available to the people who implement programs. The HRAC has supported the preparation, printing and distribution of several publications and newsletters.

To further enhance information exchange the HRAC has supported technical meetings such as the International Symposium on Weed and Crop Resistance to Herbicides at Cordoba, Spain. These funds are used to support travel stipends, publications and the sessions where information is shared and discussed.

In addition to this external funding which totals over $250,000 since our founding, the HRAC has spent considerable time coordinating our own programs and publications. In addition to the direct contributions from HRAC our local industry affiliates have sponsored equal or greater work on resistance management.

HRAC: The Future

The HRAC plans to change its course and move away from a focus on the mode of action subcommittees. While effective, we do not want to duplicate efforts or miss critical areas. Last year efforts were accelerated to forge stronger ties to Country or Area working groups. The HRAC believe that there must be ways to enhance the information available to local academic and extension personnel who are recommending and implementing on farm weed resistance management programs. The HRAC want to promote cooperative work in a mutually beneficial manner.

To date this plan has focused on three world areas. These are Asia-Pacific (primarily Australia), Europe and North America. The HRAC have ongoing interactions with key individuals and working groups in these areas to seek areas of cooperative work. We see the HRAC as a group who can foster and sponsor liaison between geographic areas and issues. We are at the early stage in this transition and need inputs on how it can work for all involved. An area of concern is representation and input from other Asian countries, from Africa and South America.

The HRAC is supportive of the initiatives underway in Australia and Canada to group herbicides according to their mode of action. These initiatives covering labelling and recommendations are a major step forward in weed resistance management programs. While it may not be possible, it would be preferable to have global standardization of these groupings to facilitate their use in extension and regulatory activities.

The HRAC along with its sister RACs provided input into the way resistance management should be handled in the European Union Registration Directive, 91-414 EC.

These inputs unfortunately were not incorporated, leaving it uncertain how industry will cope with the country plans emerging from the directive. We support strong product stewardship, yet we want to invest our research money in developing information that prevents or solves problems.

The HRAC encourages the development and implementation of weed resistance monitoring programs. There have been initiatives in this area but to date no consistent program appears to be operational. There is good research documenting specific cases of resistance. But to our knowledge there is no simple yet comprehensive system available to track the occurrence and spread of resistant weeds nor to aid in their identification.

The HRAC want to be an integral part of the development of integrated weed management strategies. These programs must be straightforward and carry no economic penalties for producers. They will be a joint effort of researchers and advisors from both industry and academia.

Conclusions

In summary, the implementation of the above-mentioned programs will require many levels of education and funding. We must, as a scientific group, educate each other and align on key management strategies. We must educate advisors including vendors on these practices. Finally we must educate the final user of herbicides. All the knowledge we develop will be of little value if it is not implemented in sustainable grower programs. In addition, industry will continue to seek new herbicide products that offer solutions to existing or future weed problems, and we will apply our learning to make these not only cost effective and easy to use, but also valuable weed resistant management tools.

References

Devine, M.D. and Shimbakuro R.H. (1994) Resistance to acetyl coenzyme A carboxylase inhibiting herbicides. *In: Herbicide Resistance in Plants: Biology and Biochemistry* (S.B. Powles and J.A.M. Holtum, eds.), 141-169. Lewis Publishers, Boca Raton, FL.

Ryan, G.F. (1970) Resistance of common groundsel to simazine and atrazine. *Weed Science* **18**: 614-616.

Saari, L.L., Cotterman J.C. and Thill, D.C. (1994). Resistance to acetolactate synthase inhibiting herbicides. *In: Herbicide Resistance in Plants: Biology and Biochemistry* (S.B. Powles and J.A.M. Holtum, eds.), 83-119. Lewis Publishers, Boca Raton, FL.

CHAPTER 34
PREVENTION AND CONTROL OF HERBICIDE RESISTANT WEEDS IN AUSTRALIA

G.S. GILL
University of Adelaide, Roseworthy S.A. 5371, Australia

Abstract

Herbicide resistance has now developed in several monocot and dicot weed species in southern Australia. Of these, annual ryegrass (*Lolium rigidum*) has developed resistance to most of the selective herbicides available for its control and, therefore, has had the biggest impact on farm profitability. The seed-bank of ryegrass is relatively short-lived and can be depleted to low levels by effective management in the pasture phase. However, it is much more difficult to manage resistant populations of this weed under continuous cropping. Consequently, there has been a dramatic rise in the use of alternative herbicide groups, such as the dinitroanilines to which resistance is still uncommon. Innovative techniques such as capture of weed seeds during the harvesting operation and preventing seed-set of weeds in crops by the application of non-selective herbicides (e.g. paraquat after anthesis of ryegrass) are also being adopted by some farmers. So far, dicot weeds have only evolved resistance to the ALS inhibiting herbicides and can, therefore, still be managed effectively with several other herbicides. There is a growing appreciation in the farming community of the need to reduce reliance on selective herbicides and, instead, employ a wider range of weed control methods to delay the onset of herbicide resistance in weed populations.

Introduction

The development of selective herbicides in the seventies and thereafter, allowed farmers to increase the area under crop production. Prior to this, cropping was less frequent, because a pasture phase of 3-4 years was generally used to reduce weed density before cropping. The reliance of farmers on selective herbicides for weed control can be gauged from the fact that in Western Australia, on average, each cropped hectare is sprayed twice each year for weed control (Gill, 1995). Similar herbicide usage is not uncommon in the other parts of southern Australia. The region is also heavily infested with ryegrass (*Lolium rigidum*), a species with tremendous genetic variability. This combination of heavy reliance on herbicides and a genetically variable species has resulted in the rapid development of resistance.

The first report of herbicide resistance in a weed in Australia was in 1982, when Heap and Knight reported evidence for diclofop resistance in ryegrass. A survey carried out on the behalf of the Agricultural and Veterinary Chemicals Association in 1992 found that herbicide resistant ryegrass infested about 275,000 ha on more than 2000 farms in southern Australia. Out of this about 80% of the area was in the state of Western Australia.

Ryegrass is still the main species causing concern due to resistance. However, there is a growing number of wild oat populations (*Avena fatua* and *A. sterilis* ssp. *ludoviciana*) with resistance to one or more herbicide groups. Resistance to ALS (acetolactate synthase)-inhibiting herbicides has also evolved in four broad-leaved species: *Sonchus oleraceus, Sisymbrium orientale, Brassica tournefortii* and *Fallopia convolvulus*.

Annual Ryegrass

Populations of ryegrass with resistance to virtually all selective herbicide groups have been documented in Australia (Hall *et al.*, 1994). In the farming areas, resistance is now common to the aryloxyphenoxypropionate (APP), cyclohexanedione (CHD) and sulfonylurea (SU) herbicides, the three main groups used for selective control of ryegrass.

Some APP resistant ryegrass populations have also been shown to possess non-target site cross-resistance to herbicides with different modes of action, e.g. SUs. The mechanism of such cross-resistance has been shown to be enhanced metabolism, probably catalysed by the same Cyt P_{450} enzyme operating in wheat (Christopher *et al.*, 1991, 1992). It is not uncommon for APP resistant ryegrass populations to show some level of cross-resistance to the SU herbicides (Table 1).

Ryegrass has repeatedly shown the capacity to rapidly develop resistance to the various herbicide groups registered for its control (Gill, 1995). For the SU herbicides, resistance has usually become detectable after only 3-4 applications, with a slightly longer exposure required for the APP and CHD herbicides (Gill, 1995).

Management in the Pasture Phase

Pasture management with grazing by sheep and the use of non-selective herbicides offers an excellent opportunity to reduce the seed-bank of ryegrass. This option is finding greater acceptance in the farming community following the recent rise in the wool prices. Studies conducted by Davidson (1992) showed that grazing combined with paraquat application after the anthesis of ryegrass ('pasture-topping') reduced ryegrass density in the subsequent year by 85%. However, the effectiveness of pasture-topping can be variable, especially if grazing intensity is inadequate. Pearce and Holmes (1976) recorded a range of

29-80% reduction in ryegrass density in different field experiments with pasture-topping in Western Australia.

Table 1. Some examples of cross-resistance to SU herbicides in APP resistant ryegrass populations (from Gill, 1995)

Population	Herbicide exposure in the field (number of applications)		Shoot weight (relative to unsprayed control)	
	APP+CHD	SU	APP	SU
1 (susceptible)	0	0	0.00	0.02
2	6	0	0.83	0.23
3	6	1	1.00	0.25
l.s.d. (p=0.05)			0.03	0.11

Management in Continuous Cropping

Faced with the problem of herbicide resistance, not all farmers have embraced the pasture option. Some reasons for reluctance to adopt the pasture option are: high infrastructure cost to implement the change, unattractive economics compared with cropping under some situations, unsuitable pasture varieties for some climatic and soil types, and higher erosion risk after grazing.

Alternative Herbicide Groups

There has been a sharp upturn in the use of dinitroaniline herbicides in the southern cropping belt, particularly in Western Australia. From less than 100,000 L in 1990, the use of trifluralin in Western Australia went up to about 780,000 L in 1994. A large part of this increase has been attributed to the need to find alternatives to the APP, CHD and SU herbicides.

Field studies on APP-resistant ryegrass with non-target site cross-resistance to the SU herbicides have generally shown yield benefits from the use of SUs like chlorsulfuron or triasulfuron (Swain *et al.*, 1993). Although there is greater early survival of ryegrass in these fields, survivors usually have a retarded root system and often fail to cope with water stress at the end of the season (Table 2).

The results from this field experiment support conclusions of Hall *et al.* (1994), that the degree of resistance conferred by enhanced metabolism is much lower than that conferred by target-site cross-resistance mechanisms. The benefits to be had from using SU herbicides on such ryegrass populations, are likely to be only short-lived.

Some ryegrass populations have only been exposed to the SU herbicides and are resistant to that group alone. In these cases, therefore, farmers have been able to switch to the APP or CHD herbicides.

Table 2. The effect of triasulfuron and two dinitroaniline herbicides on the survival of a population of ryegrass seedlings, inflorescence production and the grain yield of wheat. In an earlier bioassay, a statistically significant ($p<0.05$) level of cross-resistance was detected in this APP resistant population (Gill, 1994, unpublished data)

Treatment	Ryegrass survivors (plants/m^2) 42 DAS*	Ryegrass spikes/m^2 134 DAS	Wheat grain yield (t/ha)
Trifluralin 400 g/ha	92	189	1.36
Oryzalin 250 g+ trifluralin 250 g/ha	115	118	1.63
Triasulfuron 25 g/ha	147	71	1.83
Unsprayed control	514	390	0.53
l.s.d. (p=0.05)			0.30

*DAS= days after sowing.

Paraquat for Seed-Set Control in Grain Legume Crops

In a highly fecund species such as ryegrass, it is crucial to prevent the seed-set of herbicide resistant survivors. This realisation has resulted in the evaluation of paraquat and other non-selective herbicides to prevent seed production by weeds, particularly in grain legume crops. The use of paraquat for this purpose is already a common practice in pastures.

Research carried out so far with paraquat has generally shown large reductions in seed production by ryegrass. However, the timing of application appears to be critical, not only for seed-set control in ryegrass but also for damage to lupins. In our studies, use of paraquat at the late flowering to the milky dough stage of ryegrass reduced its seed-set by about 80-90% (Table 3).

Even when paraquat is used at the correct timing for ryegrass control, lupin yield losses of about 10% are quite common. This loss could be worth a lot of money, but in some situations the farmer would not be able to grow a viable wheat crop in the following season unless seed production of ryegrass was controlled. Use of paraquat in this manner is seen solely as a salvage option to allow cropping to continue after resistance has developed to selective herbicides.

The development of herbicide resistant transgenic crops is expected to make a valuable contribution to the management of herbicide resistant ryegrass and other weeds. Genes for glufosinate (a broad-spectrum, non-selective herbicide) resistance have been successfully incorporated into lupins at the Co-operative Centre for Legumes in Mediterranean Agriculture (CLIMA), Western Australia. Release of herbicide resistant cultivars of grain legumes will widen the choice of broad-spectrum herbicides available to farmers.

Table 3. The effect of post-anthesis application of paraquat on the seed production by ryegrass and the grain yield of lupins (Gill, 1992, unpublished data)

Treatment	Ryegrass seeds/spike	Lupin yield (t/ha)
Untreated control	12.3	1.4
Paraquat 100 g a.i./ha Time 1*	1.7	1.4
Paraquat 200 g a.i./ha Time 1	1.0	1.3
Paraquat 100 g a.i./ha Time 2	9.0	1.5
Paraquat 200 g a.i./ha Time 2	4.0	1.2
Paraquat 100 g a.i./ha Time 3	11.7	1.3
Paraquat 200 g a.i./ha Time 3	8.7	1.6
l.s.d. (p= 0.05)	7.0	0.3

*Time 1= ryegrass late flowering to milky dough stage; time 2= ryegrass soft to firm dough stage; time 3= ryegrass hard dough stage.

Widespread development of resistance to selective herbicides in Australia has rejuvenated interest in various non-chemical methods of weed control and their integration with the use of selective herbicides, i.e. integrated weed management.

Capture of Weed Seeds at Harvest

Ryegrass is a prolific seed producer, with seed production estimated at times to exceed 30,000 seeds/m^2 (Rerkasem *et al.*, 1980). Seeds of ryegrass are held firmly in the spike and do not dislodge readily at maturity. Therefore, it is possible to harvest ryegrass seeds along with the crop and prevent their return to the seed-bank.

There are several different types of commercial equipment now available for capturing ryegrass during the harvest. However, this method of weed control is still relatively uncommon with only about a dozen of these machines operating on Western Australian farms.

Our research with the Redekop® system of weed seed capture in 1992 showed that 60-80% of the total ryegrass seed production could be removed from the paddock. Studies in South Australia (Matthews J.M., personal communication), with a different machinery (Rytec®), also showed a similar level (45-63%) of capture of ryegrass seeds during the harvest operation. In a later experiment, plots harvested with the Redekop® seed-capture system were found to have 73% lower ryegrass plant density in the subsequent crop. There is little doubt, that incorporation of techniques such as weed seed capture into the farming system have the potential to reduce farmers' reliance on high-risk selective herbicides for weed control.

Burning the Stubble

Burning, as a method of weed control, is probably as old as agriculture itself. In an early study, Mason *et al.* (1968) reported doubling or better of the wheat yield due to burning of ryegrass infested paddocks. Although there were visual differences between the burnt and the unburned plots, actual weed densities were not recorded.

In a more recent study, Davidson (1994) reported that more than 90% of the surface seed-bank of ryegrass was destroyed by burning. However, the effectiveness of burning declined if the paddock had been grazed during the summer, probably due to an insulating effect of the soil on the seed buried by trampling (Table 4).

Burning is one of the cheapest weed control options available to the farmer. However, on erosion prone soil types, it is an extremely risky proposition and farmers are reluctant to employ the technique under those situations.

Table 4. The effect of grazing and stubble replacement prior to burning on the emergence of ryegrass seedlings in autumn (from Davidson 1994).

Pre-burning management	Ryegrass seedlings/m^2
No grazing	278
Grazed and stubble replaced	496
Grazed without stubble replacement	697
l.s.d (p= 0.05)	214

Wild Oats

The first report of herbicide resistance in wild oats (*Avena fatua*) was published in 1990 (Piper, 1990), 8 years after the first report in ryegrass. Although there are no reliable data available, it is generally believed that the area infested with resistant wild oats is much smaller than that affected by ryegrass. In Western Australia, for example, there are only about 50 known resistant populations of *A. fatua* as compared to > 500 resistant populations of ryegrass.

Resistance to APP and CHD herbicides has now also been confirmed in *Avena sterilis* ssp. *ludoviciana* collected from South Australia (Mansooji *et al.*, 1992).

In our investigations over the last 5 years, most of the wild oat populations screened for resistance have shown resistance to only diclofop and not to the other APP herbicides (Table 5). This is in agreement with the earlier results of Piper (1990) who had reported resistance to diclofop in a population of *A. fatua* on which fluazifop was still effective.

Table 5. Resistance profile of *Avena fatua* populations from Western Australian farms, screened for resistance to the APP herbicides (Gill 1993, unpublished data).

Category	Number of populations
Total	29
Susceptible	14
Resistant to diclofop only	14
Resistant to diclofop, fenoxaprop and haloxyfop	1

Due to the common pattern of resistance found in wild oats, farmers with diclofop resistant wild oats have usually been able to switch to other ACCase inhibiting herbicides (e.g. haloxyfop, fluazifop, tralkoxydim, etc.) or to herbicides with entirely different modes of action, such as triallate or flamprop-methyl.

As in the case of ryegrass, wild oats can also be managed effectively in the pasture phase. The half-life of wild oat seeds in Australia appears to be around 6 months (Martin and Felton, 1993). Similar results had earlier been reported from New South Wales by Philpotts (1975). In our studies in Western Australia, seed-bank of wild oats in pasture declined by about 80% per year when paddocks were grazed and paraquat was sprayed after panicle emergence of the weed (Gill, unpublished data).

Seeds of *A. fatua* and *A. sterilis* are readily shed at maturity (Thurston, 1982), usually well in advance of the crop harvest. Therefore, systems that capture weed seeds during the harvesting operation are likely to be ineffective. Wind-rowing the crop could increase the retention of wild oat seeds in the panicle, thereby increasing the effectiveness of seed capture operation, however no data are available at this stage.

Broad-Leaved Weeds

In Australia, resistance to selective herbicides has now been identified in 4 different broad-leaved species. These are: *Sonchus oleraceus, Sisymbrium orientale, Brassica tournefortii* and *Fallopia convolvulus*. All of these species have developed resistance to only the ALS inhibiting herbicides such as SU, imidazolinone and triazole pyrimidines. Development of herbicide resistance in broad-leaved weed species has not caused as much concern as with ryegrass. This could be due to the availability of a much larger range of alternative herbicides for their selective control in cereals. Some of the herbicide groups that are still effective include: phenoxys, nicotinanilides, nitriles, triazines, substituted ureas and several others.

Concluding Comments

Not that long ago, development of herbicide resistance was perceived as the beginning of financial ruin by Australian farmers. However, with the improvement in the understanding of the problem brought about by research and extension, this perception has changed remarkably. Farmers now accept resistance as an inevitable consequence of herbicide usage and another management issue they have to contend with. Management of resistant weeds certainly requires greater input from the farmer, having to combine several control methods to achieve a result that was possible in the past with a single application of a selective herbicide.

It is now generally accepted that weed management systems in Australia have relied too heavily and for too long on a few herbicide groups. Widespread development of herbicide resistance in weeds has highlighted the urgent need to widen the variety of weed control methods employed on farms, which should reduce the reliance on selective herbicides and make farming systems more sustainable.

References

Christopher, J.T., Powles, S.B., Liljegren, D.R. and Holtum, J.A.M. (1991) Cross-resistance to herbicides in annual ryegrass (*Lolium rigidum*). II. Chlorsulfuron resistance involves a wheat-like detoxification system. *Plant Physiology* **95**: 1036-1043.

Christopher, J.T., Powles, S.B. and Holtum, J.A.M. (1992) Resistance to acetolactate synthase-inhibiting herbicides in annual ryegrass (*Lolium rigidum*) involves at least two mechanisms. *Plant Physio*logy **100**: 1909-1913.

Davidson, R.M. (1992) Populations dynamics. Herbicides resistance annual ryegrass *Lolium rigidum* in pasture. *Proceedings 1st International Weed Control Congress*, 141-143. Vol. 2. Melbourne, Australia.

Davidson, R.M. (1994). Biology and control of herbicide resistant *Lolium rigidum. M.Sc. Thesis*, La Trobe University, Bundoora.

Gill, G.S. (1995) Development of herbicie resitance in annual ryegrass populations (*Lolium rigidum* Gaud) in the cropping belt of Western Australia. *Australian Journal Experimental Agriculture* **35**. n.° 1. 67-72

Hall, L.M., Holtum, J.A.M. and Powles, S.B. (1994) *In: Herbicide Resistance in Plants: Biology and Biochemistry* (S.B. Powles and J.A.M. Holtum, eds.), 243-261. Lewis Publishers: Boca Raton. Florida. 353 pp.

Heap, J. and Knight, R. (1982) A population of ryegrass tolerant to the herbicide diclofop-methyl. *Australian Journal Agricultural Research* **48** :156-157.

Mansooji, A.M., Holtum, J.A.M., Boutsalis, P., Matthews, J.M. and Powles, S.B. (1992) Resistance to Aryloxyphenoxypropionate Herbicides in two Wild Oat Species (*Avena fatua and Avena sterilis* spp. ludoviciana). *Weed Science* **40**: 599-605.

Martin, R.J. and Felton, W.L. (1993) Effect of crop rotation, tillage practice and herbicides on the population dynamics of wild oats in wheat. *Australian Journal Experimental Agriculture* **33**: 159-65

Mason, M.G., Toms, W.J. and Gartrell, J.W. (1968). Stubble burning helps control Wimmera ryegrass. *Journal of Agriculture Western Australia* **9**: 582-3.

Pearce, G.A. and Holmes, J.E. (1976). The control of annual ryegrass. *Journal of Agriculture Western Australia* **17**: 77-81.

Philpotts, H. (1975) The control of wild oats in wheat by winter fallowing and summer cropping. *Weed Research* **15**: 221-225.

Piper, T.J. (1990) *Proceedings 9th Australian Weeds Conference*, 211-215. Adelaide, Australia. Rerkasem, K., Stern, W.R. and Goodchild, N.A. (1980) *Australian Journal Agricultural Research* **31**: 649-58.

Swain, J.M., Roy, W., Gill, G.S. and O'Connell, P.J. (1992) Lessons to be drawn from an herbicide trial programm in regard to annual ryegrass *Lolium rigidum* resistance. Vol. 2. *Proceedings 1st International Weed Control Congress*, 503-510. Melbourne, Australia.

Thurston, J.M. (1982) In: *Biology and Ecology of Weeds* (W. Holzner and N. Numata, eds.), 191-199. Dr W. Junk Publishers: The Hague.

CHAPTER 35
FROM RESEARCH TO PRACTISE: STAYING AHEAD OF THE PROBLEM

J. COSTA
Monsanto España, S.A., Avda. de Burgos, 17-2ª, 28030 Madrid, Spain

Abstract

The 18 years of the use of glyphosate* as a non selective herbicide in perennial and annual crops of Spain can be considered a successful case of resistance management. For two reasons: its continuously expanding use and, second, the lowering of rates down from 1,08 kg/ha to 0,18 kg/ha a.i. It should be pointed out that glyphosate is normally applied up to 1 - 2 times a year in extensive crops, and up to 4 times a year in some intensive high quality citrus orchards, directed applications repeated. In addition to the intrinsic characteristics of the herbicide, the continued satisfactory use of glyphosate has been possible thanks to:

- Close monitoring of changes in weed spectrum.
- Development of improved recommendations / application systems.
- Combinations with complementary herbicides (MCPA, oxyfluorfen, among others).
- Eventual use of tillage, which is still the most used in extensive crops.

These measures must continue in the near future, particularly considering the likely introduction of some crop varieties resistant to this herbicide, which should encourage the adoption of environmentally sound conservation tillage practices (reduced tillage with over 30 % of soil surface covered by crop residues at planting time). Tank or ready mix combinations with other active ingredients, however, may still be very useful, and, therefore, restrictions to reduce exposure should be preferred over prohibition in the case where the current use of a product does not fit contemporary environmental standards.

---

* Roundup is a registered trademark of Monsanto

Particularities of Non Selective Herbicides

Foliar applied non-residual herbicides are typically applied in preplant of registered crops or as directed sprays in established perennial crops, but to be successful they must be able to control a wide range of weed species. Lack of residual activity has been recognized as not favouring development of resistant weeds. Furthermore, the possibility of alternating different active ingredients or tillage applications offers many possibilities to control potential resistant weeds.

Non-residual herbicides can be handled practically with the same flexibility as mechanical tillage operations from a soil management standpoint. Mechanical weed control system is still the most commonly used in extensive annual and perennial crops in Spain and it is not recommended for environmental conservation. The slow development cycle of weeds, compared with other pests, and the use of a range of herbicides with different mechanisms of action, or eventually of tillage, envisage the successful management of weed resistance in the event of it happening. Weed spectrum changes can be expected with the increasing presence of the most difficult species, and are inevitable because they have also occurred when tillage was the only weed control available. However, hard-to control weeds can be successfully managed by:

- Close surveillance of weed spectrum (floristic studies or market research).
- Follow-up of user satisfaction at individual farm level, thanks to feedback of herbicide distribution.
- Development of solutions to flora inversions before they reach unmanageable proportions.

Concerning the last point, weed control professionals from public and private sectors can be proud that the new problems posed by weed resistance or by changes in the weed spectrum in Spanish intensive citrus are today of less concern to farmers than 20 years ago (Table 1).

Glyphosate Use in Spanish Citrus

When glyphosate was introduced in irrigated citrus in Spain in 1975 it was a very successful debut because previous attempts of citrus farmers to reduce tillage by applications of other herbicides such us triazine, diuron, paraquat, and others, had failed to control perennial weeds. The survey summarized in Table 1, completed in 1977, is a good description of how serious the problem of perennial weeds was considered in most citrus orchards.

The Spanish registration rules, which establish minimum and maximum rates of use. The Official Registration Document listed rates of 6 - 12 l/ha for perennial weeds

Table 1. Survey of weed problems of 591 citrus Orchards (data from Monsanto, 1977)

Main Weed Species	Presence (%)	Difficult to Control (%)
Cyperus rotundus	78	52
Convolvulus arvensis	84	44
Cynodon dactylon	69	36
Sorghum halepense	47	23
Allium spp.	58	18
Cardaria draba	43	10
Paspalum spp.	22	11

and 3 - 6 l/ha for the control of annual weeds. One irrefutable proof that the commercial use of glyphosate was satisfactory for farmers without notice of weed resistance, was the development of lower rates of herbicide application thanks to application in low volume (20 l/ha) spinning disc Controlled Drop Application systems (Valera and Costa, 1984). As this system enabled low cost application, recommended rates decreased to 1,5 - 4 l/ha vs annual weeds, and to 6 - 10 l/ha vs. perennial weeds, and the new minimum rate of 1,5 l/ha was included in 1983 Registration document. The easy use of the new system of application also increased the frequency of applications, with lower dependence on herbicides with residual control; an important number of farmers with small plots relied on 3 - 5 applications of non residual herbicides for weed control throughout the year. Weed spectrum changes were indeed happening.

However, while some perennial weeds such as *Cynodon*, *Paspalum*, and *Sorghum* declined, other annual weed species such as *Portulaca oleracea, Urtica urens, Parietaria officinalis, Equisetum arvense, Malva sylvestris, Amaranthus retroflexus, Conyza canadensis* and *Sonchus oleraceus* increased (Ruiz-Jaén, 1987). This change prompted the development of new formulations such as a ready mixture of glyphosate + MCPA (both as isopropylamine salts) which was very important and decisive in citrus infested with *Portulaca, Equisetum* or *Urtica*. On the other hand, glyphosate translocation decreases in the mixture with MCPA. However, active ingredient translocation with the mixture was reduced in comparison with straight glyphosate, but this characteristic facilitated the approval for the use of such a mixture in the control of olive suckers since it increased tree safety or oil quality (Valera *et al.*, 1993).

The main weeds in recent citrus surveys (Carretero, 1994) are shown in Table 2. However, the farmer´s concern is less than 20 years ago, when a more limited number of herbicide choices was available.

Table 2. Main weeds in three Eastern Spanish citrus in Eastern Spain (506 samplings; Carretero, 1994).

Weeds Species	% Presence	% Cover
Portulaca oleracea	39-66	6-13
Cyperus rotundus	47-61	8-11
Setaria adherens	46-67	7-11
Convolvulus arvensis	37-68	5-9
Digitaria sanguinalis	26-40	2-4
Amaranthus viridis	37-58	2-4
Amaranthus hybridus	23-37	2-5
Amaranthus graecizans (subsp. *sylvestris)*	21-34	1-5
Echinochloa colona	18-31	1-4
Amaranthus retroflexus	0-36	0-5
Amaranthus blitoide	23-50	1-4
Chenopodium album	17-54	0-3

Apparently, no weeds listed have changed in their susceptibility to glyphosate i.e., particularly the *Amaranthus*, which took most of the space left by the controlled weeds. Other herbicides or formulations developed in citrus, such as those based on glufosinate, sulfosate, or another glyphosate dry formulations, offer additional weapons to citrus farmers for the successful management of their weed infestations.

Under heavy pressure from generic competitive formulations, Monsanto Spain efforts in application technology have been redirected to develop glyphosate formulations with unmatchable environmental safety characteristics (Roundup Plus, Roundup Sec). Furthermore, and there are market research indicates an increased rates/ha of products which have become cheaper. In this context, the development of formulations combining two active ingredients is economically difficult to justify; however, tank mixes of lower rates of complementary herbicides (such as 20 g/ha of oxyfluorfen vs. *Malva*), are promising.

Conservation Tillage in Extensive Crops

In cereal crops, mechanical tillage is still the leading weed control system in Spain, partly due to the easier access to tractors, and also because some traditional thinking associates tillage with soil quality. If tillage allow 30% at least of the soil covered with crop residues it is known to improve profitability and environmental compatibility of agriculture. Therefore, Monsanto developed recommendations to implement the conservation tillage technique under Spanish conditions. By recommending application when volunteer cereals and other troublesome grasses (*Lolium, Avena, Bromus,* etc) are at their early stages, the minimum rate registered for control of annual weeds with classic Roundup formulation was reduced from 3 l/ha in 1975 to 0,75 l/ha (270 g/ha of active ingredient) in 1988, and the amount of active ingredient could still be reduced down to 180 g/l thanks to the development of a highly efficient formulation (120 g/l a.e.) (Costa *et al.*, 1989, Salto *et al.*, 1989). Furthermore, sustainable agriculture, concerned on production as well as minimum environmental impact, should be considered. Therefore, a low volume application of 180 g/ha of glyphosate (or 33 g/ha see J. Menéndez and R. De Prado, personal communication) or tribenuron at less than 1 g/ha (controlling 50% of *Sinapis* according to Green and Streibig, 1993), seem quite acceptable inputs in comparison with a mouldboard plough which requires over 20 l/ha of diesel fuel.

Many herbicide programs in this market are adapted to use the herbicide where it is most efficient, such as control of volunteers or control of seedlings emerged at planting time, but since non-selective tillage is still being practised several times a year, spontaneous development of resistant weeds is not likely. Where tillage is completely reduced, such as in direct drill systems, the risk of producing weed infestations more difficult to control is higher, with the following suggestions to resolve them:

- Crop rotation where economically feasible.
- Control with selective herbicides applied after planting.
- Use better formulations or actives, if control is marginal.
- Go back to superficial tillage, if above suggestions are not applicable.

Concerning development of crop varieties with genetic resistance to one active ingredient, no particular problems are expected other than the need to look for other active ingredients or superficial tillage in the case that crop volunteers are considered weeds for the following crop. The relatively large number of active ingredients currently commercialized is a weapon to fight resistances in future agriculture; one example is the case of atrazine resistant *Amaranthus spp.*, which can be controlled by alachlor or acetochlor at rates as low as 250 or 50 g/ha respectively (De Prado *et al.*, 1992, 1993).

References

Carretero, J.L. (1994) La flora arvense de los cítricos españoles. *Phytoma España* **63**: 19-26.

Costa, J., Valera, A. and Nieto, J. (1989) Desarrollo de una nueva formulación de glifosato. Ensayos de campo. *Proceedings 4th EWRS Mediterranean Symposium* **I**: 166-173.

García-Baudín, J.M. and Costa, J. (1978) Principales adventicias perennes en cultivos intensivos. *Proceedings Roundup Seminar,* March 9-10, 1978, Madrid. Monsanto España, Madrid.

Green, J.M. and Streibig, J.C. (1993) Herbicide mixtures. *In: Herbicide Bioassays* (J.C. Streibig and P. Kudsk eds.), 117 135. CRC Press, Boca Raton, Florida.

Gómez de Barreda, D. (1974) No cultivo en los agrios. *Cuaderno I.N.I.A.* **2**, 77 pp. INIA, Ministerio de Agricultura, Madrid.

Gómez de Barreda, D. (1994) Sistemas de manejo del suelo en citricultura. Tratamientos herbicidas. *Sèrie Divulgació Tècnica* **26**, 386 pp. Consellería d'Agricultura, Pesca i Alimentació, Generalitat Valenciana.

De Prado, R., Sánchez, M., Jorrín, J., and Domínguez, C. (1992) Negative cross - resistance to bentazone and pyridate in atrazine-resistant *Amaranthus cruentus* and *Amaranthus hybridus* biotypes. *Pesticide Science*, **35**: 131-136.

De Prado, R., Romera, E., and Jorrin, J. (1993) Effects of chloroacetamides and photosynthesis-inhibiting herbicides on growth and photosynthesis in sunflower (*Helianthus annuus* L.) an *Amaranthus hybridus* L. *Weed Research*, **33**: 369-374.

Ruíz-Jaén, A. (1987) Malas hierbas en los cítricos de la Comunidad Valenciana. *Levante Agrícola* **277-278**: 179-200.

Salto, T., Costa, J. and García-Baudín, J.M. (1989) Eficacia a corto plazo de una nueva formulación de glifosato. *Proceedings 4th EWRS Mediterranean Symposium*, **I**: 174-178.

Valera, A. and Costa, J. (1984) Advances in ultra low volume application systems for glyphosate in perennial crops. *Proceedings EWRS 3rd Symposium on Weed Problems in the Mediterranean Area:* 315-320.

Valera, A., Costa, J., and García-Torres, L. (1993) Desarrollo de las varetas del olivo y su control herbicida. *Phytoma España*, **46**: 3-6.

CONCLUSION

CHAPTER 36
WILL HERBICIDE RESISTANCE ULTIMATELY BENEFIT AGRICULTURE?

S. O. DUKE
USDA-Agricultural Research Service, Southern Weed Science Laboratory, P.O. Box 350, Stoneville, Mississippi 38776, U.S.A.

Introduction

From 60 to 70% of the pesticides used in production agriculture in developed countries are herbicides. Herbicides have helped farmers to increase yields while reducing labor. Without herbicides, labor would be a major cost of crop production in developed countries. The relatively low food prices in these countries can be partly attributed to the effectiveness of herbicides. However, the potential environmental and toxicological costs of herbicides have raised questions about our agricultural dependence on these magic bullets. In developed countries, herbicide resistance may ultimately change the economic and technological equations to favor an agriculture less dependent on chemical inputs. Compared to insecticides, resistance to herbicides has evolved very slowly, principally because of the longer life cycle of weeds compared to insects. However, after a relatively long lag phase, the number of cases of herbicide resistance in developed countries continues to grow logarithmically, threatening the use of several herbicide classes over wide areas. Growing multiple resistances exacerbate the situation. Farmers and the herbicide industry have been slow to confront the situation, and, in many places, are not effectively implementing prevention and/or amelioration strategies. Compared to past years, few new herbicides with new molecular target sites are on the horizon to substitute for herbicides or herbicide classes to which economically important weeds have evolved resistance. At the present growth rate, the problem of evolved herbicide resistance will become a major concern of most farmers in developed countries in the near future. Thus, herbicide resistance of weeds may eventually be the biggest impetus for developing methods of weed management that rely less on herbicide inputs. Herbicide resistance will also result in use of methods of using herbicides that will minimize the likelihood of the evolution of resistance. A scenario that I think is likely in some parts of North America is a shift to reduced or no tillage agriculture, in which weeds are controlled only as needed with herbicides with a very short-lived selection pressure, or to which resistance is not likely to evolve rapidly. Ideally, this type of agriculture would take advantage of environmentally benign cultural practices such as cover crops (allelopathic and otherwise) and crop rotations, of improved herbicide application technology to kill only the weeds that are competing with the crop, of models to predict economic weed thresholds, of highly potent herbicides with short-lived selection pressure, and of herbicides to which weeds apparently evolve resistance only very slowly. The latter approach may involve the use of crops genetically engineered to be resistant to these herbicides. Biocontrol of weeds may eventually play a significant role in this more integrated approach to weed management.

How herbicide resistance will ultimately influence agriculture will depend on many factors and will vary with crop and geographic location. To better focus on this speculative subject, this chapter will deal only with major agronomic crops in developed countries. Even with these restrictions on the subject, the uncertainty in making such predictions should be recognized. Thus, this chapter is written more as food for thought, rather than as firm predictions. Polish it as I might with the best available information, I am still through a glass gazing dimly. A first step in prognostication is a realistic evaluation of the past and current situations as they relate to what one is predicting.

The Current Situation

The weeds

Nature abhors a vacuum, and, thus, any successful effort to greatly reduce weed pressure on crops and weed interference with other human activities will be temporary. Weeds in agriculture have always been and will continue to be problems. Despite costly efforts to minimize the effects of weeds on crop quality and yield, they cost farmers billions of US dollars annually in the United States alone (Bridges, 1992). Nevertheless, the level of funding for weed science research would suggest that there is relatively little concern with this cost (Duke, 1992).

The most prevalent trends in the current weed situation are the evolution of herbicide-resistant weeds and weed species shifts. Weed species shifts can be due either to introduction of exotic weeds or creation of new ecological niches in agroecosystems by substantially reducing populations of present weed species.

The evolution of herbicide-resistant weeds is occurring at an alarming rate that is well documented in many of the other chapters in this volume and elsewhere (*e.g.*, LeBaron, 1991). There are already cases of significant economic impact of this trend. For example, in cotton alone, dinitroaniline herbicide-resistant *Amaranthus palmeri* costs farmers in one small state of the United States, South Carolina, approximately US$ 12 million annually (Murdock, in press). Similar biotypes of this species are being found in other states. Evolution of multiple resistance complicates and intensifies the difficulty of the problem by creating unpredictable resistance profiles for which it is difficult to devise management strategies (Matthews, 1994). Multiple resistance is a significant problem in Australia, and its incidence is growing in North America. Because of under-reporting of its incidence, the present cost of herbicide resistance is probably greater than can be documented.

At the current rate of increase in the evolution of herbicide resistance, most farmers will incur significant economic losses due to this problem within the next decade.

In the U.S., most of the weed species that influence agriculture are not native. New, non-native weed species continue to change the spectrum of weeds with which farmers must cope. For example, *Solanum viarum* has become a significant problem in the southern U.S. since it was introduced in the 1980s (Mullahey *et al.*, 1993). Furthermore, native weeds that have been relatively unimportant in the past have become problematic as they have filled the agroecosystem niches vacated by the weeds for which herbicides have been designed to control. Clearly, the biosphere is sufficiently adaptive to provide constant employment for weed scientists.

Weed Management

Herbicides are the dominant method of weed management in agronomic crops, with at least one annual application on virtually every cultivated hectare in developed countries. The evolution from reliance on cultural and mechanical weed management to chemical methods has reduced significantly the cost and labor required for weed management. In most developed countries, herbicides represent from 60 to 70% of pesticide sales and use (Figure 1). There is little likelihood of economical alternatives to herbicides during the next twenty years (Duke *et al.*, 1993). Compared to efforts in entomology and plant pathology, relatively little public sector research effort is focused on finding alternatives for herbicides (Duke, 1996), despite the fact that herbicides comprise the overwhelming majority of pesticides used.

Now biotechnology offers a new option in how we might manage weeds with herbicides by imparting herbicide resistance to normally herbicide-susceptible crops. There are two methods to do this with biotechnology. Mutant genes that confer resistance can be selected for in cell or tissue culture, or foreign genes that confer resistance can be transferred to the crop of interest by transgenic methods (Dyer *et al.*, 1993). Herbicide-resistant crops generated by the first method are already available, and a variety of those generated by the second method will become available before the end of the twentieth century.

Generation of herbicide-resistant crops has been the most intensively exploited area of plant biotechnology in the U.S. (Figure 2). This work has been the focus of the ire of various environmental and other advocacy groups (*e.g.*, Goldburg *et al.*, 1990). Both environmentalists and biotechnologists have claimed sweeping effects of herbicide-resistant crops on agriculture if allowed to come to fruition. Others have questioned the potential effects of these crops on agriculture (Duke *et al.*, 1991, Duke, 1996). This technology offers to expand the market niche for herbicides already approved for other uses. If successful, these crops will probably reduce and delay efforts to find non-chemical alternatives to weed management (Goldburg *et al.*, 1990).

Relatively little herbicide used ever reaches a weed (Hall, 1985). This inefficiency costs the farmer money in several ways: most of the herbicide is wasted; crop

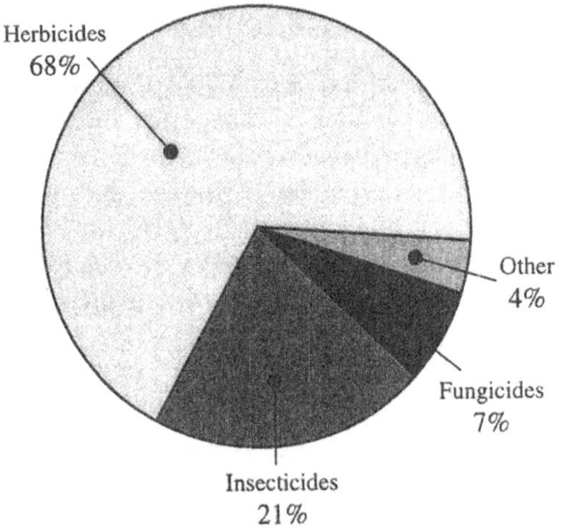

Figure 1. Pesticide sales in the U.S. in 1993 (data from Kirschner, 1994).

damage can result from herbicide contacting the crop; and law suits are often filed because of herbicide drift effects on crops and other non-target organisms off the using farmers' property. Furthermore, a proportion of the herbicide that is not intercepted by weeds can move to ground and surface water or the atmosphere before it is degraded. Large amounts of soil-applied, residual herbicides are used as prophylactic measures, before the extent of a weed problem is known.

For the last half of the twentieth century, other weed management tools have been of considerably less importance than herbicides in developed countries (Duke *et al.*, 1993). Commercial materials for biocontrol of weeds in agronomic crops are essentially unavailable. The few products that were once available and those that might have been introduced (Boyette and Abbas, 1994) have disappeared or failed to be commercialized because of the relatively high cost and reduced utility of biocontrol products compared to herbicides at this time. Although there is a growing list of patented biocontrol agents (Watson, 1993), considerably more research will have to be done to make biocontrol agents competitive with herbicides (Zorner *et al.*, 1993).

Cultural methods, such as planting higher crop densities to shade weeds more rapidly, can be used to improve weed control and/or to reduce herbicide requirements. Other methods such as use of cover crops to suppress weeds are being used by a few farmers (*e.g.*, Smeda, 1993). Utilization of the allelopathic potential of crops and cover crops has been studied as another method to suppress weeds (Chou, 1994). Planting higher crop densities can reduce weed pressure by increasing competition by the crop for environmental resources.

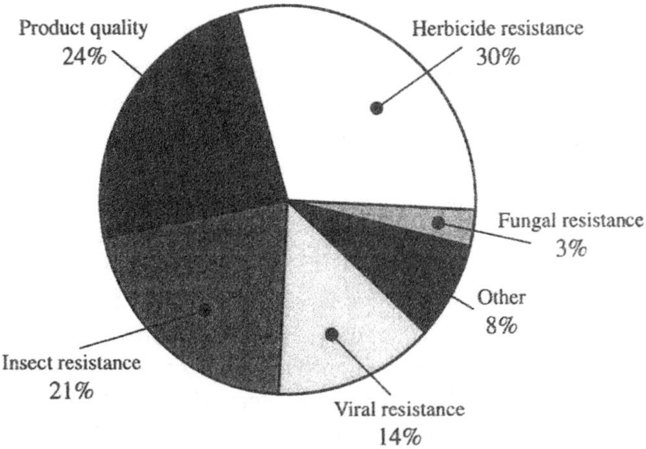

Figure 2. Permits for transgenic crops in the U.S. from 1987 to 1994 (data from *Science* (1994) 266: 1472).

Tillage is the major weed management supplement to the herbicides currently used. This centuries-old weed management technology has done more damage to agricultural lands by causing loss of soil than any other aspect of agriculture. This hidden cost of weed control is seldom factored into the costs of weeds or of agriculture. Manual methods of weed control, such as hoeing and hand-pulling, are seldom used in agronomic crops in developed countries because of labor costs.

Driving Forces for Change Other Than Herbicide Resistance

There are several forces associated with the pressure to make agriculture in the developed world more sustainable that will strongly impact how we manage weeds. These forces will have profound influences on our responses to the evolution of herbicide-resistant weeds.

Pesticide Use Reduction

There is strong public pressure to reduce pesticide inputs in agriculture. This has resulted in use reduction mandates in some countries of northern Europe (Bellinder *et al.*, 1994). The use rate components of these mandates have largely been met by switching to low use rate herbicides. In other countries there is more effort to reduce the environmental and toxicological risks associated with herbicides, but certainly reducing the volume of a particular herbicide used will reduce the risks associated with that herbicide.

The discovery of economical low use rate herbicides, such as many of the acetolactate synthase or protoporphyrinogen oxidase inhibitors, to substitute for all of the high use rate herbicides now being used will provide the potential for greatly reducing herbicide use rates. Most herbicide manufacturers that are engaged in discovery have low use rate as a requirement for new herbicides.

Herbicides are often over applied because the recommendations of the manufacturer are tailored for the worst case scenario. With better information about the weeds and their growth stages in a field and environmental factors, effective weed management can often be accomplished with reduced rates of herbicides (DeFelice *et al.*, 1993). Computer-aided decisions can help the farmer to maximize weed management effectiveness, while minimizing herbicide use (*e.g.*, Rydahl and Thonke, 1993).

The overwhelming majority of herbicides are applied in much the same way that they have been for the past half century. However, improved methods of applying herbicides that more precisely target weeds have been recently introduced or are under development (*e.g.*, Houtsma, 1994). This technology promises to substantially reduce herbicide use rates. Ultimately, devices will be developed that will identify a weed species and direct the proper herbicide at the proper dose to kill that individual weed.

After a herbicide reaches the target plant, relatively little of it usually penetrates the cuticle or cell wall to reach its actual site of action. For a herbicide to kill a plant, it must enter the plasma membrane or cytoplasm of the plant's cells. Better formulation and herbicide design can increase the amount of herbicide that penetrates the plant. For example, ultra-low volume formulations of herbicides in oils can, in some cases, reduce the amount of herbicide needed for effective weed management by more than 50% (McWhorter *et al.*, 1992). Further research to maximize the uptake and translocation of herbicides has the potential to significantly reduce herbicide use rates.

Reduction of Tillage

One of the principal objectives of tillage is weed management. As mentioned above, loss of soil due to tillage is the most pronounced environmental destruction caused by any agricultural practice. Greatly reducing or completely eliminating tillage can conserve soil and often increases soil fertility because of better crop residue management. In developed countries, large amounts of fossil fuel are expended in tillage operations. The weed seed soil bank is more short-lived in no-till agriculture. An additional environmental benefit of eliminating tillage is that pesticide mobility is decreased because of increased retention to organic residues (*e.g.*, Locke and Harper, 1991; Locke, 1992). Impeding pesticide mobility can be beneficial by reduction of movement to surface and ground water (Hall *et al.*, 1984). These facts are being recognized, and there is strong and growing pressure in the U.S. to greatly reduce tillage as a tool in crop and weed management. Cultural methods, such as higher crop densities and cover or smother crops are commonly used in no-till agriculture to give the crop a competitive edge over weeds. However, with current weed management methods, decreasing tillage usually increases total herbicide usage in order to maintain acceptable weed control.

Integrated Pest Management

Integrated pest management (IPM) requires that weed management be considered in the context of other pest and crop management methods. For example, non-tillage agriculture will result in greater crop and, perhaps, weed residues that might increase problems with insects and/or plant pathogens. However there are cases in which IPM considerations might reduce overall pesticide use in unpredictable ways. For example, some herbicides are also effective fungicides, suggesting that the herbicide could be used for dual purposes, provided the crop is resistant to the herbicide. Glufosinate is such a herbicide, and glufosinate can apparently provide some protection against fungal pathogens in crops engineered to be resistant to the herbicide (Uchimiya *et al.*, 1993).

There are many other known and potential interactions between weed management methods and insect and plant pathogen status. These factors are seldom considered in designing a weed management strategy. Eventually, good decision aid software will allow taking these sometimes hidden factors into account to provide recommendations that optimize the outcome that the farmer and society has chosen.

Herbicide Resistance's Influences

With this background, what will agriculture's response be to rapidly evolving herbicide resistance? I think that it will result in a more rapid departure from the prevailing approach to weed management in developed countries over the past half century; continue use of the same herbicides in non-rotated crops until better and/or less expensive herbicides are introduced. Herbicide resistance will eventually, but increasingly force farmers to adopt greater diversity in every aspect of weed management.

Crop rotation will be more advantageous because farmers can rotate herbicides and cultural methods of weed management. Whether crop rotation is used or not, farmers will have to rotate herbicides with different mechanisms of action and/or apply mixtures of different herbicides. There will be greater reliance on cultural methods such as higher crop densities and cover crops for weed suppression.

Much of the herbicide discovery effort over the past few decades has been spent in finding loopholes in the patent protection of competitors in order to produce similar herbicides with the same mode of action. Within the same company, much effort is often spent on extending the market niche of an existing herbicide family. As a result of weeds evolving resistance to existing herbicide classes, the herbicide industry is likely to spend more effort on finding herbicides with new molecular sites of action, thus providing greater diversity of herbicide choices. New herbicides are very likely to be low use rate herbicides for the reasons mentioned above.

Models predict that even the best strategies for delaying evolution of herbicide resistance cannot prevent evolution of resistance, as long as the herbicide is used. Furthermore, no fitness penalties appear to be associated with many of the resistances that are evolving. Thus, when widespread, some of the resistances may eventually greatly attenuate or eliminate the use of some herbicides. If so, this will create great demand for alternatives from the public sector, especially if the private sector cannot deliver an economic alternative. Pressure for alternatives and shifts in the weed management economic equation could result in sufficient research funding to develop weed biocontrol sufficiently to contribute significantly to weed management.

Ultimately, systems approaches to weed management, executed by use of management software that incorporates principles of IPM, will allow farmers to make decisions based on complex strategies that encompass multiyear considerations. Development of the knowledge base for effective, site-specific software will be a daunting task for scientists.

Prognosis

Evolution of herbicide resistance in weed populations will play a strong role, along with the other driving forces mentioned above, in affecting weed management strategies in the future. For the most part, I think that these changes, including the ones caused by herbicide-resistant weeds will be positive. Farmers will have to use more diverse methods with more concern of how their methods will impact weed management in future years. There will be less dependence on herbicides and even more pressure to find non-chemical alternatives. Short-term economic loss due to herbicide-resistant weeds will eventually be supplanted by a more sustainable and stable agriculture that will lead to more long-term savings for farmers and the public.

References

Bellinder, R.R., Gummesson, G. and Karlson, C. (1994) Percentage-driven government mandates for pesticide reduction: The Swedish model. *Weed Technology* **8**: 350-359.

Boyette, C.D. and Abbas, H.K. (1994) Weed control with mycoherbicides and phytotoxins. A nontraditional application of allelopathy. *American Chemical Society Symposium Series* **582**: 280-299.

Bridges, D.C., Ed. (1992) *Crop Losses Due to Weeds in the United States - 1992.* Weed Science Society of America, Champaign, IL.

Chou, C.-H. (1994) Allelopathy and sustainable agriculture. *American Chemical Society Symposium Series* **582**: 211-223.

DeFelice, M.S., Brown, W.B., Aldrich, R.J., Sims, B.D., Judy, D.T. and Guethle, D.R. (1989) Weed control in soybeans (*Glycine max*) with reduced rates of postemergence herbicides. *Weed Science* **37**: 365-74.

Duke, S.O. (1992) Weed science - the need and the reality. *Phytoparasitica* **20**: 183-186.

Duke, S.O. (1996) Herbicide-resistant crops- background and perspectives. *In*: *Herbicide-Resistant Crops - Agricultural, Environmental, Economic, Regulatory, and Technical Aspects* (S.O. Duke, ed.), 1-10. CRC Press, Boca Raton, FL.

Duke, S.O., Holt, J.S., Hess, F.D. and Christy, A.L. (1991) Herbicide-resistant crops. *Comments from CAST,* no. 1991-1. Council for Agricultural Science and Technology, Ames, IA, 24pp.

Duke, S.O., Menn, J.J. and Plimmer, J.R. (1993) Challenges of pest control with enhanced toxicological and environmental safety. *American Chemical Society Symposium Series* **524**: 1-13.

Dyer, W.E., Hess, F.D., Holt, J.S. and Duke, S.O. (1993) Potential benefits and risks of herbicide-resistant crops produced by biotechnology. *Horticultural Reviews* **15**: 367-408.

Goldburg, R., Rissler, J., Shand, H. and Hassebrook, C. (1990) *Biotechnology's Bitter Harvest.* Environmental Defense Fund, Biotechnology Working Group, City of New York, 73pp.

Hall, F.R., Ed.(1985) *Improving Agrochemical and Fertilizer Application Technology*, 146 pp. Agricultural Research Institute, Bethesda, MD.

Hall, J.K., Hartwig, N.L. and Hoffman, L.D. (1984) Cyanazine losses in runoff from no-tillage corn in 'living' and dead mulches vs. unmulched, conventional tillage. *Journal Environmental Quality* **13**: 105-110.

Houtsma, J. (1994) Visionary sprayers outsmart weeds. *Farm Industry News* **27**(4): 4.

Kirschner, E.M. (1994) Agricultural chemical producers rebound from floods of 1993. *Chemical Engineering News* **72**(40), 13-15.

LeBaron, H M. (1991) Distribution and seriousness of herbicide-resistant weed infestations worldwide. *In: Herbicide Resistance in Weeds and Crops* (J.C. Caseley, G.W. Cussans and R.K. Atkin, eds.), 27-43. Butterworth-Heinemann Ltd., Oxford.

Locke, M.A. (1992) Sorption-desorption kinetics of alachlor in surface soil from two soybean tillage systems. *Journal Environmental Quality* **21**: 558-556.

Locke, M.A. and Harper, S.S. (1991) Metribuzin degradation in soil: II-Effects of tillage. *Pesticide Science* **31**, 239-247.

Matthews, J.M. (1994) Management of herbicide resistant weed populations. *In: Herbicide Resistance in Plants: Biology and Biochemistry* (S.B. Powles and J.A.M. Holtum, eds.), 317-335. Lewis Publishers, Boca Raton, FL.

McWhorter, C.G., Barrentine, W.L. and Hanks, J.E. (1992) Postemergence grass control with herbicides applied at ULV in paraffinic oil. *Weed Technology* **6**: 262-268.

Mullahey, J.J., Cornell, J.A. and Colvin, D.L. (1993) Tropical soda apple (*Solanum viarum*): a weed threat in subtropical regions. *Weed Technology* **7**: 723-727.

Murdock, E. C. (1995) Herbicide resistance: historical perspective and current situation. *Proceedings Weed Science Society North Carolina.* **13:** 3.

Rydahl, P. and Thonke, K.E. (1993) PC-plant protection: optimizing chemical weed control. *Bulletin OEPP/EPPO Bulletin* **23**: 589-594.

Smeda, R.J. (1993) Potential of spring-seeded sorghum-sudangrass for weed management in soybean. *Proceedings Southern Weed Science Society* (USA) **46**: 57.

Uchimiya, H., Iwata, M., Nojiri, C., Samarajeewa, P.K., Takamatsu, S., Ooba, S., Anazi, N., Christensen, A.H., Quail, P.H. and Toki, S. (1993) Bialaphos treatment of transgenic rice plants expressing a *bar* gene prevents infection by the sheath blight pathogen (*Rhizoctonia solani*). *Bio/Technology* **11**: 835-836.

Watson, A.K., Ed. (1993) *Biological Control of Weeds Handbook.* Weed Science Society of America, Champaign, IL, 202pp.

Zorner, P.S., Evans, S.L. and Savage, S.D. (1993) Perspectives on providing a realistic technical foundation for the commercialization of bioherbicides. *American Chemical Society Symposium Series* **524**: 79-86.

INDEX

O

P

R

S

T

GPSR Compliance

The European Union's (EU) General Product Safety Regulation (GPSR) is a set of rules that requires consumer products to be safe and our obligations to ensure this.

If you have any concerns about our products, you can contact us on ProductSafety@springernature.com

In case Publisher is established outside the EU, the EU authorized representative is:

Springer Nature Customer Service Center GmbH
Europaplatz 3
69115 Heidelberg, Germany

Batch number: 10755284

Printed by Printforce, the Netherlands